Designing apparel for consumers

The Textile Institute and Woodhead Publishing

The Textile Institute is a unique organisation in textiles, clothing and footwear. Incorporated in England by a Royal Charter granted in 1925, the Institute has individual and corporate members in over 90 countries. The aim of the Institute is to facilitate learning, recognise achievement, reward excellence and disseminate information within the global textiles, clothing and footwear industries.

Historically, The Textile Institute has published books of interest to its members and the textile industry. To maintain this policy, the Institute has entered into partnership with Woodhead Publishing Limited to ensure that Institute members and the textile industry continue to have access to high calibre titles on textile science and technology.

Most Woodhead titles on textiles are now published in collaboration with The Textile Institute. Through this arrangement, the Institute provides an Editorial Board which advises Woodhead on appropriate titles for future publication and suggests possible editors and authors for these books. Each book published under this arrangement carries the Institute's logo.

Woodhead books published in collaboration with The Textile Institute are offered to Textile Institute members at a substantial discount. These books, together with those published by The Textile Institute that are still in print, are offered on the Woodhead web site at: www.woodheadpublishing.com. Textile Institute books still in print are also available directly from the Institute's website at: www.textileinstitutebooks.com.

A list of Woodhead books on textile science and technology, most of which have been published in collaboration with The Textile Institute, can be found towards the end of the contents pages.

We are always happy to receive suggestions for new books from potential editors. To enquire about contributing to our Textiles series, please send your name, contact address and details of the topic/s you are interested in to sarah.lynch@woodheadpublishing.com. We look forward to hearing from you.

The team responsible for publishing this book:

Commissioning Editor: Francis Dodds
Publications Coordinator: Lucy Beg
Project Editor: Elizabeth Moss
Editorial and Production Manager: Mary Campbell
Production Editor: Adam Hooper
Project Manager: Newgen Knowledge Works Pvt Ltd
Copyeditor: Newgen Knowledge Works Pvt Ltd
Proofreader: Newgen Knowledge Works Pvt Ltd
Cover Designer: Terry Callanan

Woodhead Publishing Series in Textiles: Number 151

Designing apparel for consumers

The impact of body shape and size

Edited by
Marie-Eve Faust and
Serge Carrier

The Textile Institute

WP
WOODHEAD
PUBLISHING

Oxford Cambridge Philadelphia New Delhi

Published by Woodhead Publishing Limited in association with The Textile Institute
Woodhead Publishing Limited, 80 High Street, Sawston, Cambridge CB22 3HJ, UK
www.woodheadpublishing.com
www.woodheadpublishingonline.com

Woodhead Publishing, 1518 Walnut Street, Suite 1100, Philadelphia,
PA 19102–3406, USA

Woodhead Publishing India Private Limited, 303 Vardaan House, 7/28 Ansari Road,
Daryaganj, New Delhi – 110002, India
www.woodheadpublishingindia.com

First published 2014, Woodhead Publishing Limited

British Library Cataloguing in Publication Data
A catalogue record for this book is available from the British Library.

Library of Congress Control Number: 2013950019

ISBN 978-1-78242-210-5 (print)
ISBN 978-1-78242-215-0 (online)
ISSN 2042-0803 Woodhead Publishing Series in Textiles (print)
ISSN 2042-0811 Woodhead Publishing Series in Textiles (online)

The publisher's policy is to use permanent paper from mills that operate a sustainable forestry policy, and which has been manufactured from pulp which is processed using acid-free and elemental chlorine-free practices. Furthermore, the publisher ensures that the text paper and cover board used have met acceptable environmental accreditation standards.

Typeset by Newgen Knowledge Works Pvt Ltd, India

Contents

Contributor contact details *xi*
Woodhead Publishing Series in Textiles *xv*

Part I Identifying shape, size, body volume and psychological aspects of designing apparel 1

1 Body shape and its influence on apparel size and consumer choices 3
E. A. Gribbin, Alvanon Inc., USA

1.1 Introduction 3
1.2 Background to sizing 4
1.3 Body shape research 7
1.4 Applying the data 10
1.5 The reality of retail 12
1.6 The consumer experience 13
1.7 Future trends 14
1.8 Conclusion 16

2 Creation of ready-made clothing: the development and future of sizing systems 17
S. P. Ashdown, Cornell University, USA

2.1 Introduction 17
2.2 The relationship of apparel products to the body 18
2.3 The development of sizing systems 20
2.4 Apparel industry methods of creating ready-made garments 22
2.5 Why standard sizing is complicated by body shape variation 24
2.6 An emerging mass customization model for clothing sizing 27
2.7 Conclusion 30
2.8 Future trends 30

2.9	Sources of further information and advice	31
2.10	References	33
3	**National sizing surveys: techniques, data analysis and apparel product development** K. M. King, [TC][2] (Textile/Clothing Technology Corporation), USA	**35**
3.1	Introduction	35
3.2	Survey design	38
3.3	Data collection	43
3.4	Data analysis	45
3.5	Applications for apparel	49
3.6	Other applications	55
3.7	Conclusion	56
3.8	References	57
4	**Body shape and weight distribution: the Body Volume Index (BVI) and the Body Mass Index (BMI)** R. Barnes, Select Research Ltd, England	**58**
4.1	Introduction	58
4.2	The Body Mass Index (BMI) and the development of the Body Volume Index (BVI)	60
4.3	Research on the use of the BVI	68
4.4	Assessing the value of the BVI	70
4.5	Conclusion	72
4.6	Sources of further information and advice	76
4.7	References	76
5	**Psychological and sociological factors influencing consumers' choice of apparel** M. Julien, Université du Québec à Montréal, Canada	**78**
5.1	Introduction	78
5.2	The need to look young	79
5.3	The need to look marginal and rebellious	80
5.4	The need to be authentic	81
5.5	The need to constantly change appearances	82
5.6	The need to expose the body	83
5.7	The need to look attractive	84
5.8	The need to attract attention	86

5.9 The need to look sporty 87
5.10 The need to be ethical 88
5.11 Conclusion 89
5.12 References 90

Part II Understanding sizing and shapes requirements and choices of particular customer groups 93

6 Infants and children: understanding sizing, body shapes and apparel requirements for infants and children 95
N. Zakaria, Universiti Teknologi Mara (UiTM), Malaysia

6.1 Introduction 95
6.2 Understanding different body shapes of infants and children 97
6.3 Sizing requirements for particular populations of children 99
6.4 Key dimensions identifications 104
6.5 Clustering children's body shapes and sizes 108
6.6 Classification of children's body shapes 113
6.7 Sizing system development and designation 118
6.8 Future trends 123
6.9 Sources of further information and advice 124
6.10 References 126

7 Older consumers: understanding sizing, body shapes and requirements of apparel for this niche 132
J. McCann, University of South Wales, Wales, UK

7.1 Introduction 132
7.2 Understanding different body shapes of older consumers 134
7.3 Sizing requirements for the older consumer: the design for ageing well project 138
7.4 Examples of specific styles for particular body sizes and shapes amongst older consumers: aesthetic needs 144
7.5 Examples of specific styles for particular body sizes and shapes amongst older consumers: needs of the body 147
7.6 Future trends 149
7.7 Sources of further information and advice 150
7.8 References 151

8 Overweight and obese consumers: shape and sizing to design apparel that fits this specific market 153
L. M. Boorady, SUNY – Buffalo State, USA

8.1 Introduction 153
8.2 Understanding different body shapes of obese consumers 154
8.3 Sizing requirements for obese consumers 159
8.4 Specific styles for the obese consumer 164
8.5 Future trends 166
8.6 Sources of further information and advice 167
8.7 References 167

9 Asian and Caucasian: designing apparel for these two different ethnic groups 169
Y. A. Lee, Iowa State University, USA

9.1 Introduction 169
9.2 Understanding different body shapes: the case of North American college students from Asian and Caucasian ethnic groups 175
9.3 Sizing requirements for different ethnic populations 178
9.4 Examples of specific styles for different ethnic populations 183
9.5 Future trends 183
9.6 Sources of further information and advice 186
9.7 References 187

10 Males: understanding sizing requirements for male apparel 189
J. Bellemare, Université du Québec à Montréal, Canada

10.1 Introduction 189
10.2 Key issues for comparison 190
10.3 Understanding different male body shapes 200
10.4 Sizing requirements for male apparel 207
10.5 Examples of sizing systems for male apparel 211
10.6 Future trends 214
10.7 Sources of further information and advice 218
10.8 References 219

11	**Male and female consumers: segmenting consumers in the apparel market by body shape and other factors** P. Robinet, IFTH, France and S. Carrier, Université du Québec à Montréal, Canada	**221**
11.1	Introduction	221
11.2	Techniques to cluster body shapes	221
11.3	Body shapes amongst the female French population	224
11.4	Body shapes amongst the male French population	228
11.5	Conclusion	233
11.6	References	234
12	**Pregnant women: understanding pregnant women's shape, sizing and apparel style preferences** M.-E. Faust, Philadelphia University, USA	**235**
12.1	Introduction	235
12.2	Identifying target markets	236
12.3	Key stages in pregnancy	238
12.4	Pregnancy and apparel choice	243
12.5	Maternity wear: a brief history	244
12.6	Meeting the needs of a target market: body measurement and fit satisfaction	245
12.7	Case study: pregnant women in Hong Kong	246
12.8	Conclusion	253
12.9	References	253
13	**Plus size Black and Latino women: the implications of body shape and size for apparel design** C. J. Patterson, Northwestern University, USA	**256**
13.1	Introduction	256
13.2	Plus size models in the fashion world	257
13.3	Causality and reality: body size among Black and Latino women	261
13.4	Plus size clothing designers who hire Black and Latina models	266
13.5	Conclusion and future trends	269
13.6	References	270

14	**Intimate apparel: designing intimate apparel to fit different body shapes** K. Shin, The Hong Kong Polytechnic University, China	**273**
14.1	Introduction	273
14.2	Measuring breast volume	277
14.3	Designing intimate apparel to fit different body shapes	282
14.4	Smart materials in intimate apparel	284
14.5	Technologies used in intimate apparel	286
14.6	Conclusion	288
14.7	References	288
15	**Headwear: designing headwear to fit the size and shape of Western and Asian populations** R. Ball, The Hong Kong Polytechnic University, China	**292**
15.1	Introduction	292
15.2	Types of headwear	294
15.3	Head physiology	296
15.4	Sizing systems for headgear	298
15.5	Head anthropometrics and design	300
15.6	Head models	303
15.7	Future trends	305
15.8	Sources of further information and advice	306
15.9	References	306
	Index	*309*

Contributor contact details

(* = main contact)

Editors

Marie-Eve Faust
Philadelphia University, School of Business Administration
Fashion Merchandising Management
School House Lane & Henry Ave.
Philadelphia, PA 19144, USA

E-mail: faustm@philau.edu

and

Université du Québec à Montréal (UQÀM)
École supérieure de mode de Montréal
École des Sciences de la Gestion (ESG)
Case postale 8888, succursale Centre-ville
Montréal (Québec) H3C 3P8, Canada

E-mail: faust.marie-eve@uqam.ca

Serge Carrier
École supérieure de mode de Montréal
École des Sciences de la Gestion (ESG)
Université du Québec à Montréal (UQÀM)
Case postale 8888, succursale Centre-ville
Montréal, (Québec) H3C 3P8, Canada

E-mail: carrier.serge@uqam.ca

Chapter 1

Edward A. Gribbin
Alvanon Inc.
145 West 30th Street
New York, NY 10001, USA

E-mail: ed.gribbin@alvanon.com

Chapter 2

Susan P. Ashdown
Department of Fiber Science & Apparel Design
Cornell University
133 Human Ecology Building
Ithaca
New York, NY 14853, USA

E-mail: spa4@cornell.edu

Chapter 3

Kerry Maguire King
[TC]² Textile/Clothing Technology Corporation
Cary, NC, USA

E-mail: kking@tc2.com

Chapter 4

Richard Barnes
Select Research Ltd
42 Calthorpe Road
Edgbaston
Birmingham
B15 1TS, UK

E-mail: richard@select-research.com

Chapter 5

Mariette Julien
Département de Marketing
École supérieure de mode de Montréal
École des Sciences de la Gestion (ESG)
Université du Québec à Montréal (UQÀM)
Case postale 8888, succursale Centre-ville
Montréal (Québec) H3C 3P8, Canada

E-mail: julien.mariette@uqam.ca

Chapter 6

Norsaadah Zakaria
Department of Textile Technology
Universiti Teknologi Mara (UiTM)
Shah Alam, Selangor, Malaysia 40450

E-mail: saadahzakaria@gmail.com

Chapter 7

Jane McCann
Visiting Professor Smart Clothes
University of South Wales, Wales, UK

E-mail: jane.mccann47@gmail.com

Chapter 8

Lynn M. Boorady
Fashion and Textile Technology Department
SUNY – Buffalo State
1300 Elmwood Avenue
Buffalo
NY 14222, USA

E-mail: booradlm@buffalostate.edu

Chapter 9

Young-A Lee
Department of Apparel, Events, and Hospitality Management
Iowa State University
1073 LeBaron Hall
Ames
IA 50011–1121, USA

E-mail: ylee@iastate.edu

Chapter 10

Jocelyn Bellemare
Département de Marketing
École supérieure de mode de Montréal
École des Sciences de la Gestion (ESG)
Université du Québec à Montréal (UQÀM)
Case postale 8888, succursale Centre-ville
Montréal (Québec) H3C 3P8, Canada

E-mail: bellemare.jocelyn@uqam.ca

Chapter 11

Patrick Robinet*
Institut Français du Textile et de Habillement

1 Rue de la Louisiane
49300 Cholet, France

E-mail: probinet@ifth.org

Serge Carrier
École supérieure de mode Montréal
École des Sciences de la Gestion (ESG)
Université du Québec à Montréal (UQÀM)
Case postale 8888, succursale Centre-ville
Montréal (Québec) H3C 3P8, Canada

E-mail: carrier.serge@uqam.ca

Chapter 12

Marie-Eve Faust
Philadelphia University, School of Business Administration
Fashion Merchandising Management
School House Lane & Henry Ave.
Philadelphia, PA 19144, USA

E-mail: faustm@philau.edu

and

École supérieure de mode de Montréal,
École des Sciences de la Gestion (ESG)
Université du Québec à Montréal (UQÀM)
Case postale 8888, succursale Centre-ville
Montréal (Québec) H3C 3P8, Canada

E-mail: faust.marie-eve@uqam.ca

Chapter 13

Courtney J. Patterson, M.A.
Department of African American Studies
Northwestern University
Crowe 5–128
1860 Campus Drive
Evanston
IL 60208, USA

E-mail: courtnej@u.northwestern.edu

Chapter 14

Kristina Shin
Institute of Textiles and Clothing
The Hong Kong Polytechnic University
Hung Hom, Kowloon
Hong Kong SAR, China

E-mail: tckris@polyu.edu.hk

Chapter 15

Roger Ball
Asian Ergonomic Lab
School of Design
The Hong Kong Polytechnic University
Innovation Tower, Floor 7
Hung Hom, Kowloon
Hong Kong SAR, China

E-mail: roger@sizechina.com

Woodhead Publishing Series in Textiles

1 **Watson's textile design and colour Seventh edition**
Edited by Z. Grosicki
2 **Watson's advanced textile design**
Edited by Z. Grosicki
3 **Weaving Second edition**
P. R. Lord and M. H. Mohamed
4 **Handbook of textile fibres Volume 1: Natural fibres**
J. Gordon Cook
5 **Handbook of textile fibres Volume 2: Man-made fibres**
J. Gordon Cook
6 **Recycling textile and plastic waste**
Edited by A. R. Horrocks
7 **New fibers Second edition**
T. Hongu and G. O. Phillips
8 **Atlas of fibre fracture and damage to textiles Second edition**
J. W. S. Hearle, B. Lomas and W. D. Cooke
9 **Ecotextile '98**
Edited by A. R. Horrocks
10 **Physical testing of textiles**
B. P. Saville
11 **Geometric symmetry in patterns and tilings**
C. E. Horne
12 **Handbook of technical textiles**
Edited by A. R. Horrocks and S. C. Anand
13 **Textiles in automotive engineering**
W. Fung and J. M. Hardcastle
14 **Handbook of textile design**
J. Wilson
15 **High-performance fibres**
Edited by J. W. S. Hearle
16 **Knitting technology Third edition**
D. J. Spencer
17 **Medical textiles**
Edited by S. C. Anand
18 **Regenerated cellulose fibres**
Edited by C. Woodings
19 **Silk, mohair, cashmere and other luxury fibres**
Edited by R. R. Franck
20 **Smart fibres, fabrics and clothing**
Edited by X. M. Tao
21 **Yarn texturing technology**
J. W. S. Hearle, L. Hollick and D. K. Wilson

22 **Encyclopedia of textile finishing**
H-K. Rouette
23 **Coated and laminated textiles**
W. Fung
24 **Fancy yarns**
R. H. Gong and R. M. Wright
25 **Wool: Science and technology**
Edited by W. S. Simpson and G. Crawshaw
26 **Dictionary of textile finishing**
H-K. Rouette
27 **Environmental impact of textiles**
K. Slater
28 **Handbook of yarn production**
P. R. Lord
29 **Textile processing with enzymes**
Edited by A. Cavaco-Paulo and G. Gübitz
30 **The China and Hong Kong denim industry**
Y. Li, L. Yao and K. W. Yeung
31 **The World Trade Organization and international denim trading**
Y. Li, Y. Shen, L. Yao and E. Newton
32 **Chemical finishing of textiles**
W. D. Schindler and P. J. Hauser
33 **Clothing appearance and fit**
J. Fan, W. Yu and L. Hunter
34 **Handbook of fibre rope technology**
H. A. McKenna, J. W. S. Hearle and N. O'Hear
35 **Structure and mechanics of woven fabrics**
J. Hu
36 **Synthetic fibres: nylon, polyester, acrylic, polyolefin**
Edited by J. E. McIntyre
37 **Woollen and worsted woven fabric design**
E. G. Gilligan
38 **Analytical electrochemistry in textiles**
P. Westbroek, G. Priniotakis and P. Kiekens
39 **Bast and other plant fibres**
R. R. Franck
40 **Chemical testing of textiles**
Edited by Q. Fan
41 **Design and manufacture of textile composites**
Edited by A. C. Long
42 **Effect of mechanical and physical properties on fabric hand**
Edited by H. M. Behery
43 **New millennium fibers**
T. Hongu, M. Takigami and G. O. Phillips
44 **Textiles for protection**
Edited by R. A. Scott
45 **Textiles in sport**
Edited by R. Shishoo
46 **Wearable electronics and photonics**
Edited by X. M. Tao
47 **Biodegradable and sustainable fibres**
Edited by R. S. Blackburn

48 **Medical textiles and biomaterials for healthcare**
Edited by S. C. Anand, M. Miraftab, S. Rajendran and J. F. Kennedy
49 **Total colour management in textiles**
Edited by J. Xin
50 **Recycling in textiles**
Edited by Y. Wang
51 **Clothing biosensory engineering**
Y. Li and A. S. W. Wong
52 **Biomechanical engineering of textiles and clothing**
Edited by Y. Li and D. X-Q. Dai
53 **Digital printing of textiles**
Edited by H. Ujiie
54 **Intelligent textiles and clothing**
Edited by H. R. Mattila
55 **Innovation and technology of women's intimate apparel**
W. Yu, J. Fan, S. C. Harlock and S. P. Ng
56 **Thermal and moisture transport in fibrous materials**
Edited by N. Pan and P. Gibson
57 **Geosynthetics in civil engineering**
Edited by R. W. Sarsby
58 **Handbook of nonwovens**
Edited by S. Russell
59 **Cotton: Science and technology**
Edited by S. Gordon and Y-L. Hsieh
60 **Ecotextiles**
Edited by M. Miraftab and A. R. Horrocks
61 **Composite forming technologies**
Edited by A. C. Long
62 **Plasma technology for textiles**
Edited by R. Shishoo
63 **Smart textiles for medicine and healthcare**
Edited by L. Van Langenhove
64 **Sizing in clothing**
Edited by S. Ashdown
65 **Shape memory polymers and textiles**
J. Hu
66 **Environmental aspects of textile dyeing**
Edited by R. Christie
67 **Nanofibers and nanotechnology in textiles**
Edited by P. Brown and K. Stevens
68 **Physical properties of textile fibres Fourth edition**
W. E. Morton and J. W. S. Hearle
69 **Advances in apparel production**
Edited by C. Fairhurst
70 **Advances in fire retardant materials**
Edited by A. R. Horrocks and D. Price
71 **Polyesters and polyamides**
Edited by B. L. Deopura, R. Alagirusamy, M. Joshi and B. S. Gupta
72 **Advances in wool technology**
Edited by N. A. G. Johnson and I. Russell
73 **Military textiles**
Edited by E. Wilusz

74 **3D fibrous assemblies: Properties, applications and modelling of three-dimensional textile structures**
J. Hu
75 **Medical and healthcare textiles**
Edited by S. C. Anand, J. F. Kennedy, M. Miraftab and S. Rajendran
76 **Fabric testing**
Edited by J. Hu
77 **Biologically inspired textiles**
Edited by A. Abbott and M. Ellison
78 **Friction in textile materials**
Edited by B. S. Gupta
79 **Textile advances in the automotive industry**
Edited by R. Shishoo
80 **Structure and mechanics of textile fibre assemblies**
Edited by P. Schwartz
81 **Engineering textiles: Integrating the design and manufacture of textile products**
Edited by Y. E. El-Mogahzy
82 **Polyolefin fibres: Industrial and medical applications**
Edited by S. C. O. Ugbolue
83 **Smart clothes and wearable technology**
Edited by J. McCann and D. Bryson
84 **Identification of textile fibres**
Edited by M. Houck
85 **Advanced textiles for wound care**
Edited by S. Rajendran
86 **Fatigue failure of textile fibres**
Edited by M. Miraftab
87 **Advances in carpet technology**
Edited by K. Goswami
88 **Handbook of textile fibre structure Volume 1 and Volume 2**
Edited by S. J. Eichhorn, J. W. S. Hearle, M. Jaffe and T. Kikutani
89 **Advances in knitting technology**
Edited by K-F. Au
90 **Smart textile coatings and laminates**
Edited by W. C. Smith
91 **Handbook of tensile properties of textile and technical fibres**
Edited by A. R. Bunsell
92 **Interior textiles: Design and developments**
Edited by T. Rowe
93 **Textiles for cold weather apparel**
Edited by J. T. Williams
94 **Modelling and predicting textile behaviour**
Edited by X. Chen
95 **Textiles, polymers and composites for buildings**
Edited by G. Pohl
96 **Engineering apparel fabrics and garments**
J. Fan and L. Hunter
97 **Surface modification of textiles**
Edited by Q. Wei
98 **Sustainable textiles**
Edited by R. S. Blackburn
99 **Advances in yarn spinning technology**
Edited by C. A. Lawrence

100 **Handbook of medical textiles**
Edited by V. T. Bartels
101 **Technical textile yarns**
Edited by R. Alagirusamy and A. Das
102 **Applications of nonwovens in technical textiles**
Edited by R. A. Chapman
103 **Colour measurement: Principles, advances and industrial applications**
Edited by M. L. Gulrajani
104 **Fibrous and composite materials for civil engineering applications**
Edited by R. Fangueiro
105 **New product development in textiles: Innovation and production**
Edited by L.Horne
106 **Improving comfort in clothing**
Edited by G. Song
107 **Advances in textile biotechnology**
Edited by V. A. Nierstrasz and A. Cavaco-Paulo
108 **Textiles for hygiene and infection control**
Edited by B. McCarthy
109 **Nanofunctional textiles**
Edited by Y. Li
110 **Joining textiles: Principles and applications**
Edited by I. Jones and G. Stylios
111 **Soft computing in textile engineering**
Edited by A. Majumdar
112 **Textile design**
Edited by A. Briggs-Goode and K. Townsend
113 **Biotextiles as medical implants**
Edited by M. W. King, B. S. Gupta and R. Guidoin
114 **Textile thermal bioengineering**
Edited by Y. Li
115 **Woven textile structure**
B. K. Behera and P. K. Hari
116 **Handbook of textile and industrial dyeing. Volume 1: Principles, processes and types of dyes**
Edited by M. Clark
117 **Handbook of textile and industrial dyeing. Volume 2: Applications of dyes**
Edited by M. Clark
118 **Handbook of natural fibres. Volume 1: Types, properties and factors affecting breeding and cultivation**
Edited by R. Kozłowski
119 **Handbook of natural fibres. Volume 2: Processing and applications**
Edited by R. Kozłowski
120 **Functional textiles for improved performance, protection and health**
Edited by N. Pan and G. Sun
121 **Computer technology for textiles and apparel**
Edited by J. Hu
122 **Advances in military textiles and personal equipment**
Edited by E. Sparks
123 **Specialist yarn and fabric structures**
Edited by R. H. Gong
124 **Handbook of sustainable textile production**
M. I. Tobler-Rohr
125 **Woven textiles: Principles, developments and applications**
Edited by K. Gandhi

126 **Textiles and fashion: Materials design and technology**
Edited by R. Sinclair
127 **Industrial cutting of textile materials**
I. Viļumsone-Nemes
128 **Colour design: Theories and applications**
Edited by J. Best
129 **False twist textured yarns**
C. Atkinson
130 **Modelling, simulation and control of the dyeing process**
R. Shamey and X. Zhao
131 **Process control in textile manufacturing**
Edited by A. Majumdar, A. Das, R. Alagirusamy and V. K. Kothari
132 **Understanding and improving the durability of textiles**
Edited by P. A. Annis
133 **Smart textiles for protection**
Edited by R. A. Chapman
134 **Functional nanofibers and applications**
Edited by Q. Wei
135 **The global textile and clothing industry: Technological advances and future challenges**
Edited by R. Shishoo
136 **Simulation in textile technology: Theory and applications**
Edited by D. Veit
137 **Pattern cutting for clothing using CAD: How to use Lectra Modaris pattern cutting software**
M. Stott
138 **Advances in the dyeing and finishing of technical textiles**
M. L. Gulrajani
139 **Multidisciplinary know-how for smart textiles developers**
Edited by T. Kirstein
140 **Handbook of fire resistant textiles**
Edited by F. Selcen Kilinc
141 **Handbook of footwear design and manufacture**
Edited by A. Luximon
142 **Textile-led design for the active ageing population**
Edited by J. McCann and D. Bryson
143 **Optimizing decision making in the apparel supply chain using artificial intelligence (AI): From production to retail**
Edited by W. K. Wong, Z. X. Guo and S. Y. S. Leung
144 **Mechanisms of flat weaving technology**
V. V. Choogin, P. Bandara and E. V. Chepelyuk
145 **Innovative jacquard textile design using digital technologies**
F. Ng and J. Zhou
146 **Advances in shape memory polymers**
J. Hu
147 **Design of clothing manufacturing processes: A systematic approach to planning, scheduling and control**
J. Gersak
148 **Anthropometry, apparel sizing and design**
D. Gupta and N. Zakaria
149 **Silk: Processing, properties and applications**
Edited by K. Murugesh Babu
150 **Advances in filament spinning**
Edited by D. Zhang
151 **Designing apparel for consumers: The impact of body shape and size**
Edited by M.-E. Faust and S. Carrier

Part I

Identifying shape, size, body volume and psychological aspects of designing apparel

1

Body shape and its influence on apparel size and consumer choices

E. A. GRIBBIN, Alvanon, Inc., USA

DOI: 10.1533/9781782422150.1.3

Abstract: Apparel fit remains a major consumer frustration and barrier to online sales growth, not because of sizes and measurements, but because few brands and retailers properly understand and address body shape. This chapter explores the background on industry sizing and grading protocols, recent body shape research and how the industry is using it to evolve and improve clothing fit. We also examine how the increasing shift to e-commerce is impacted by fit in terms of suppressed sales and higher returns. We discuss how many companies are using new technologies and smarter marketing to better engage and inform shoppers about fit, size and shape.

Key words: body shape, sizing, e-commerce, fit technology, body scanning.

1.1 Introduction

As surprising as it may seem to consumers who are shopping for clothing, fit has little to do with sizing or measurements. Consumers, especially women, but also the relatively small percentage of men who shop for their own clothing, tend to embark upon the shopping experience with the idea, firmly planted in their brain, that they are a particular size when they shop for clothing. When consumers walk into a shop and are attracted by a particular style, their first reaction is usually to walk over, wanting to touch the garment, feel the fabric, pick it up. What's interesting to observe, however, is that most consumers do not pick up the first garment on a pile or on a rack; rather, they look through the pile or rack for 'their' size; that's the one they pick up.

Intellectually, of course, most consumers know that different brands fit differently, and, perhaps, even that different styles within a single brand will fit differently. But, that does not stop them from thinking that they are a particular size, and that they are that size regardless of whether or not they've gained a little weight. This tiny bit of irrationality in the shopping process leads to a significant amount of consumer frustration and dissatisfaction on the one hand, and a significant number of returns, and lost customers, to brands and retailers on the other hand.

Even more surprising, perhaps, is that the majority of merchandisers, product developers, technical designers, sourcing agents, and manufacturers themselves equate fit with sizing and measurements. Throughout a garment's development cycle, people check 'fit' by laying a garment flat on a table and measuring it to see if it falls within the tolerance of its manufacturing specification for key measurements. If a sample or prototype measures out properly, it is shipped to the brand headquarters where a live fit model tries it on; only then does the brand realize that it actually does not fit. Why? Because the shape or balance or depth is not right. Trying to solve the fit dilemma by using measurements and patterns, working in a two-dimensional world, simply does not work.

Fit – proper fit – is not about size or measurements; it's all about shape. The body is three-dimensional and when clothing patterns are developed that actually represent the three-dimensional shape of real human bodies, the odds that clothing will fit increase sharply. In this chapter, we will explore a brief background on apparel sizing; recent research on body shape; how the apparel industry has reacted to, or responded to, this research (or not, in many cases); and what the future might hold for consumers as we find ever more sophisticated methods for gauging body shapes, and then building apparel to match those shapes. If the apparel and retail industries get there, and that is still a very big 'if', consumer frustration with apparel fit may become a thing of the past.

1.2 Background to sizing

Sizing of clothing, for all intents and purposes, is an arbitrary protocol designed by someone in the apparel creation-supply chain to sell products to more people. We use the term 'someone' because it could be almost anybody, or in some cases, nobody. There are a significant number of fashion brands, importers and retailers who, basically, buy what they like from vendors on the open market and then re-sell that product to their customers. These 'buyers' generally have no say, or even interest, in the sizing specifications, grading or protocols of the product they are selling; and, since they buy from multiple vendors, they often get very different sizing – not a good thing for building customer loyalty perhaps, but for many of these buyers, especially those in the 'fast fashion' arena, often considered 'disposable' fashion, fit is not their value proposition. Rather, what is hot, new and just off the runway, at an unbelievably low price, is their value proposition and they generally get away with whatever fit and sizing they have, no matter how much it varies.

The majority of traditional, or established, brands and retailers do, however, pay a significant amount of attention to fit and sizing. They understand the clear connection between fit and conversion rates at retail, full-price sell-throughs (and, conversely, markdowns), returns, customer retention, and brand loyalty. Even among these established companies, however, it is not always clear who is responsible for establishing fit or sizing protocols. For some, the creative or design team is responsible; for others, it's the technical design team; for still others, this responsibility is outsourced to overseas suppliers or vendors. And, in many companies (too many companies, in our opinion), fit is a team effort, established by committee, with all of the subjectivity and inconsistency that that approach would suggest.

The industry is certainly aware, generally, of the diversity of sizes and shapes of real people out there in the consuming world. Size protocols, however they are established, supposedly allow them to reach out and connect with, and hopefully sell to, the majority of these consumers. Unfortunately, the methodology for building these size protocols hinges on two things: the number of sizes a brand is willing to (or able to) produce, based on available retail space, development costs and inventory costs; and grade rules – the mathematical formulae for establishing the difference between any two sizes. The general idea behind establishing grade rules is that a brand wants to fit as many people in their target demographic as possible, and do it with the fewest number of sizes. For simple purposes of return on investment, not to mention productivity of retail space, the fewer sizes they have, the better off they are. On the other hand, missing sales opportunities because people who are bigger than the largest size, smaller than the smallest size, or whose body shape just does not match the shape of that brand's designs, is a significantly greater long-term cost to a company than that of unproductive inventory.

Every brand tries to strike that perfect balance, having a range of sizes that covers the most people possible, with the fewest number of sizes, all the while trying to ensure that sizes are not so close together that consumers cannot tell the difference (we will discuss the impact of sewing tolerance on this shortly), and not so far apart that a high percentage of consumers find themselves 'between sizes' where one size is too small and the next size too large to provide a flattering fit. This is of particular concern with alpha, or letter-sizing; the intervals between S, M, L sizes are typically double the intervals between numeric sizes (such as 8, 10, 12). While this is perfectly normal and acceptable for certain types of styles and fabrics (knit tops or sweaters, for instance, due to the amount of stretch in the fabric and drape-intent of the product), brands risk losing sales if their grade intervals are too far apart.

These size protocols may have some commonalities within particular countries but vary widely around the world. In the US, for example, the majority of women's brands offer sizes 4 through 16, with many offering smaller sizes (2, 0, even 00) and larger sizes (18, 20, and then plus sizes). The grade intervals between sizes are typically 1" between the smallest sizes (0–10), 1 ½" between the middles sizes (12–16), and 2" between the larger sizes (18 and up). In Europe, or in China, however, the grade intervals tend to be a standard 4 cm (about 1 5/8") between all sizes, no matter how small or large. In the UK and Australia, the grade intervals tend to be 5 cm (about 2") between all sizes. As global e-commerce expands, these varying protocols will make shopping for apparel even more difficult for consumers unless the sellers figure out how to effectively communicate fit through their marketing.

There are two key factors behind why so many consumers complain about not finding clothes that fit their body shape when they buy clothing off-the-rack or online. One is that the core body shape that most brands start with represents a very small percentage of the real people who make up their target customer demographic. Some of this emanates from the industry's use of professional fit models who maintain a 'perfect shape' and represent the 'ideal' woman or man; these models are often more a muse for the brand than a reflection of real customers. They reflect what the industry calls an 'hourglass' shape for women (think Marilyn Monroe) because for many years clothing 'standards' maintained by the US Government reflected this hourglass shape. Unfortunately, those 'standards' were based on data originally collected from young men and women in their late teens and early twenties who joined the military at the onset of World War II. Most industry professionals have known for years that these so-called standards are outdated, but until about ten years ago, with the development of new body scan technologies (Size USA, completed in 2003, was the first truly representative study of the American population since 1942), there was no empirical evidence as to how the population had actually changed. Everyone knew that growing obesity rates, fast food consumption, remote controls and computer games, increased physical activity such as jogging, yoga and outdoor activities all had impacts – in both directions – on the nation's body shapes, but no one had the proper information that could quantify the change. Add to those factors the rapid influx of immigrants from Mexico, Central America and Asia especially, in the last thirty years, and it becomes pretty clear that a sizing standard from the 1940s was no longer representative of real shoppers in America.

The second factor for fit frustration is grade rules. Grade rules, the formulae for creating different sizes based on a single base or core size, were originally invented in Germany in the early 1900s by clothing production engineers. There was no data available on body shapes, so these formulae were very logical, mathematical and balanced, and made for ease of manufacturing. For the most part, these are the same grade rules that the industry uses today. The issue with this is that while grade rules are linear – assuming that key body measurements all increase or decrease by the same amount for people who need larger or smaller sizes – people are not. One of the key learning for the industry from the modern body scan studies that have been around for about ten years now is that people do not grade linearly. The body shapes of women who need a size 4, or 10, or 16 are all quite different; the industry, by using linear grade rules however, ensures that the shape of clothing is exactly the same at every size. What this usually means for consumers is that a higher percentage of people who need the core size (usually a size 8/10 or M for women or a size 40 or M for men) will find that the shape of the product tends to match the shape of their body. On the other hand, for consumers who need larger or smaller sizes, a higher percentage of people are likely to complain about sizing and fit; this is primarily because the shape of the clothing does not match well with the shape of their bodies.

1.3 Body shape research

Soon after [TC]^2, a non-profit think tank in North Carolina devoted to the US apparel and textile interests, completed the Size USA body scan survey in 2003, people in the industry started analyzing the data. It supported the assumption that Americans had gotten significantly larger over the past sixty years. The Center for Disease Control data on US obesity rates had told the industry this for years. But the industry's response was, basically, what has come to be known as 'vanity sizing' – decreasing the size label but keeping the measurements the same. According to most industry veterans, the measurements that comprised a women's size 8 in the early 1970s are essentially the same measurements that comprise a size 2 or 4 today. What did not change over those years, however, was the basic shape – the hourglass shape – that was the starting point for most clothing patterns.

In 2005, NC State University, which has one of the largest and most prolific apparel and textile programs in the US, conducted the first Body Shape Analysis, using the Size USA data, in conjunction with Alvanon, a New York-based company that had become a leading manufacturer of technical fit mannequins. Dr Cindy Istook of NC State and Janice Wang, the CEO of Alvanon, collaborated to identify the body shapes of the US population and then portray them three-dimensionally as virtual bodies (avatars) and as physical mannequins. A key finding from the study was that only about eight percent of American women actually had an hourglass body shape – the shape which the vast majority of clothing brands were using to design their products (see Fig. 1.1).

Now, at least, the industry had concrete data to support the fact that body shapes had changed and this could be a key reason why apparel return rates were high, markdowns – especially of smaller and larger sizes – were high,

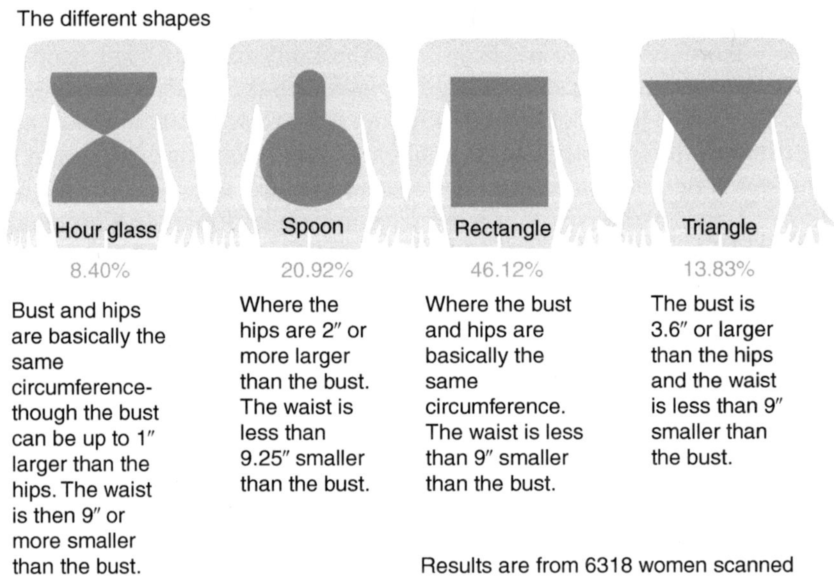

1.1 Percentage of women by key body shapes in the 2003 Size USA Project. (*Source*: Janice Wang (Alvanon, Inc.) and Cindy Istook (N.C. State University), 2005.)

conversion rates at retail were less than desirable, and customer retention was difficult, at best.

Intellifit Corporation was a Philadelphia-based technology start-up that created the first body scanner using radio waves rather than laser or white-light technology to scan shoppers at retail. The benefit of using radio waves (similar to cell phone technology but at much lower power levels) was that consumers did not have to undress to get scanned; this allowed Intellifit to put scanners in malls and other public locations and collect large amounts of data quickly and unobtrusively. About 150000 people were scanned across the US and Canada between 2004 and 2008, some for the proprietary benefit of certain retailers. Charming Shoppes, the largest specialty retailer in the US devoted to plus-size consumers, commissioned one such study, scanning over 30000 women in Lane Bryant stores across the US. David's Bridal, the largest bridal chain in the US, commissioned another, scanning over 10000 brides-to-be in their stores in various areas of the country. For these particular clothing categories – plus-sizes and bridal – fit was always a key part of their value message for customers. Plus-size women have long felt underserved by traditional retail and been frustrated by the lack of clothing to fit their body shapes. A bride wants to look absolutely perfect on that special day, and how the wedding dress fits – how it makes her look – is a critical factor. Now, both of these companies had the data on the real body shapes of women coming into their shops, and they were determined to do something about it.

Levi Strauss – unlike retailers or brands in specialty categories – is about as mainstream and iconic of an American brand as there is. Everyone wears jeans. The inventor of the jean, however, was in the same position as those other companies – according to their customers, fit was the key factor in why a person chose a particular pair of jeans. In fact, fit was so important in jeans that the NPD Group, a large consumer research firm, reported that the average women tried on 15 pairs of jeans before buying one, trying to ensure that they had the best fit. Because jeans fit so close to the body, and because denim has a somewhat rigid construction, shape – not measurements – would be critical in determining if a particular jean fits or not.

Levi's, between 2004 and 2006, put Intellifit scanners in 12 flagship stores to collect data on body shape and help shoppers find the best fitting jeans while they were there. Levi's had long recognized that body shapes varied widely, and that consumers wore jeans very differently, based on personal wearing preference.

Because the jeans category was their core business, they could afford to go deeper in terms of shapes and styles than many other retailers who were focused on multiple product categories. Walking into a Levi's store and seeing the wide range of styles, all designated by unique numbers (501, 505, 514, 519, etc), gave consumers the confidence that this brand would have a shape to fit, and flatter, them. Levi's had different styles, and fit blocks/patterns for higher rises, lower rises, slimmer fits, and more relaxed fits. As we will explore in detail shortly, in 2009, they were about to take their understanding of body shape one giant leap forward.

Starting in 2006, Alvanon, the New York-based leader in fit mannequins, launched a consulting division, today called AlvaInsight, which interprets body shape data for clients all over the world. In addition to using Intellifit data and Size USA data, Alvanon acquired additional data from the IFTH (Institut Français du textile et de l'habillement) in France, Size Germany, Size Mexico and others on its way toward building the largest database of consumer body measurements in the world. Alvanon supplemented scan data by assembling additional supporting data from organizations like the World Health Organization, the UK's National Health Service, the Centers for Disease Control in the US, and other national statistics bureaus around the world. For the purpose of understanding body shapes for makers of children's wear, Alvanon also collected numerous pediatric and academic studies from around the world and recently participated with a consortium of British retailers and Select Research (UK) to analyze the Shape GB study of children in the UK.

AlvaInsight uses the data primarily to help brands and retailers determine the best body shape, or shapes, to use as the basis for their standard of fit, and then helps those same brands develop the optimum fit mannequins, grade rules and block patterns, based on their analysis of the data relevant to a particular brand. In 2009, Alvanon launched its own AlvaScan division, which completed an independent size study of 35 000 consumers in seven key cities across China and additional studies in Germany, Australia, and the UK. Future studies are planned to analyze the prevalent body shapes in India and Brazil in 2013 and 2014, respectively.

1.4 Applying the data

As soon as the industry became aware of the vast differences between the body shapes in the population and their core apparel fit standards, questions arose over how best to respond and apply the data to improve their fit. While there are a few notable exceptions, which we will explore here, the vast majority of retailers and brands use only one core body shape as the basis for the fit of all of their apparel.

The good news for consumers is that, over the past six years, more and more brands have taken advantage of body shape analyses and modified their core body shapes from the classic hourglass shape to what we would call a 'modified hourglass' shape. While the hourglass shape has at least an 8–9" difference between the bust and waist measurements, and a 10–12" difference between the waist and hip measurements, the modified hourglass increases the waist by anywhere from 2–3" to create a more realistic shape. The percentage of women who fall into this shape jumps to about 33% of the population, from about 8% for the classic hourglass.

While the change – increasing the waist by 2–3" – may seem significant, it is actually a subtle change for many brands. The reason is that most apparel products are actually separates – tops or bottoms. The waist in tops is generally not designed to be very close to the body, so the change here is a subtle one; the waist change in bottoms is made in relation to a hip measurement which does not change. Here, the shape change actually fits a significantly higher percentage of people comfortably, and with less fit compromise (gapping at the waist, or baggy hips and thighs) than fit based on the previous hourglass shape. With these changes, as subtle as they are, more consumers are likely to find a good fit in apparel today than they would have 5–10 years ago.

As we will explore briefly in the next section, retailers are often extremely resistant to change. With fit, resistance to change actually serves a benefit. Consumers come to rely on a brand because of the fit; if the brand changes the fit, they are likely to lose at least some percentage of their most loyal customers. That is why any change in fit must be approached cautiously and with subtlety. Brands should use their base fit as the starting point for any change in core fit, and then use body shape data and analysis to determine how they can best modify their core shape and grading in a way that will not alienate their current customers. All brands and retailers want to attract new customers, but to do so by alienating current customers is usually a sure recipe for failure.

So, while most retailers still only use one core body shape as the basis for their fit standard, a handful of innovative brands have pushed the envelope to create different shapes to accommodate an even more diverse group of customers. Lane Bryant was one of the first. In 2005, Lane Bryant used their proprietary body scan data to launch a line of jeans called 'Right Fit' jeans. These jeans acknowledged and addressed the three significantly different body shapes found in the US plus-size population. Lane Bryant launched a website that guided women through the process of determining what their shape was, and which jeans – red, yellow, or blue – would fit them best. In a further innovation, new to the plus-size world (but previously used by Chico's in women's regular sizes), they sized the Right Fit jeans in sizes 1, 2, 3, 4, and 5, rather than the more conventional (at the time) 16W through 24W.

In 2009, Levi Strauss took this philosophy of serving different body shapes to the next level, using body scan data from 37000 women in 14 countries around the world. They launched 'Curve ID', which quickly became the most successful new product launch in Levi's illustrious history. In store or online, they made it easy for any woman to determine if she was a 'slight curve', 'demi-curve' or 'bold curve'. In 2012, they even launched a 'supreme curve' in the US market only, acknowledging that there were a significant percentage of even curvier women in the US who wanted a great fit for their shape. The key success factor in Curve ID is that Levi's created significant differentiation between the different shapes so that they were not cannibalizing their own business or confusing their customers. Women who fit in the demi-curve, for example, would not really fit well in any of the other curves. So, rather than just sell to the same, relatively small pool of consumers, Levi's was able to reach out and connect with four quite significant and different pools of consumers around the world.

Most recently, in August 2012, Marks & Spencer, the iconic British department store, launched their own line of jeans, taking advantage of a recent body shape analysis/study of their own unique British customer demographic. Their 'Body Shape Denim' offers three distinctly different body shapes, marketed as the 'Marilyn', 'Eva', and 'Lana', and in its original rollout was selling out of stock almost daily. It is clear that brands that acknowledge and understand body shape differences in their target customer demographics, and develop products to serve them, gain market share and brand loyalty in the process.

1.5 The reality of retail

Today's retail is, to a large degree, a real estate game. Or, to be fair, yesterday's retail was a real estate game from which, unfortunately, many retailers have yet to evolve.

The real estate analogy is that, in the bricks and mortar world, one commits some percentage of space to a particular category and each category has to be 'productive' in order for the real estate investment to pay off. If a retailer opens a shop with 5000 square feet of space, that space has to produce a certain amount of profit for the retailer to be successful. At lower price points, that means you will likely see almost every square foot covered with merchandise because moving volume will be critical to achieving profitability for them. In higher price point shops, you might see a more 'relaxed' layout or display, without bulging racks of clothing; because the product itself is more profitable, the retailer can sell lower volume and still be successful.

Why, you might think, does this have anything to do with sizing, fit, or body shape? Well, because of limited real estate and an existing track record of sales, most retailers feel that they just cannot physically accommodate extra sizes, or worse, additional body shapes, without taking something else out of the shop. The technical design teams, even the merchants themselves, may know that the market will support additional sizes or shapes, creating the ability to actually grow the core customer base – a good thing for any retailer or brand – yet the finance and retail management may not be willing to take the risk on replacing product with a known ROI (return on investment) with product that has only the potential of growing the business. The truth is that many retailers are reluctant to change the status quo or take the risk that the unknown, however logical and promising, is preferred to the known.

The growth of e-commerce has been particularly good for addressing this dilemma, however, and a significant number of retailers are taking advantage of this. Retailers as wide-ranging as Gap, Target, Wal-Mart, Levi's, Kohl's, and Nike all offer extended sizes online that they do not offer in the retail shops. Lands' End and JC Penney were pioneers of this concept and remain among the leaders in e-commerce among traditional apparel companies largely because of their ability to 'serve the under-served'. They have been able to build successful, targeted e-commerce campaigns for the shoppers who need and buy those extended sizes and command envied brand loyalty in the market as a result.

While bricks and clicks can address the need for extended sizes, bricks alone still struggle with meeting the real needs of today's diverse shoppers. Fit Logic is a Texas-based company with an innovative approach to solving the fit dilemma – three different body shapes for every size – but has struggled to gain traction with traditional retailers, largely because of their resistance to additional inventory. While it is clear that consumers have different – in some cases, very different – body shapes, traditional retail has and will continue to struggle with how to properly meet their needs and ultimately grow their businesses.

1.6 The consumer experience

The worlds of fashion design, product development, sourcing, and development are pretty opaque to the average consumer. Even with the rise of popular cable TV shows such as Project Runway and Fashion Police or behind-the-scenes features on shopping networks such as QVC and HSN, most consumers don't have a clue about how fit is established or executed in the clothes that they buy (and often return). Fit to a consumer is really as simple as how the garment looks on them (and often not how it 'feels' – more on this to follow). They know nothing about fit models, specifications, grade rules, blocks, quality control or, even worse, sewing tolerance. Sewing tolerance, in particular, has to be a frustration point because for the most part, the products that consumers buy are made to pretty exacting tolerances.

Consumers won't find noticeable differences between two of the same model iPhones, wing chairs, Toyota Camrys, or even boxes of cereal. Their expectation in shopping for apparel is that, at least within a particular brand, a 10 is a 10 or a 40 is a 40. Yet, they can walk into any retail shop, pick up two of the exact same style jeans, in the same size, and they can fit totally differently. Since typical sewing tolerances are about half the distance to the next size (0.5" tolerance when there is a 1" grade between two sizes, such as a size 8 and a size 10, or a men's 32 and 33, for example), a consumer can find one size 8 that fits like a 10, and one that fits like a 6. There are continuing technological developments in techniques such as laser seaming and automated manufacturing that are gradually reducing sewing tolerances, but since fashion is constantly pushing the envelope with new and innovative fabrics, and fabric characteristics can vary so significantly, this is not an issue that will go away any time soon.

Consumers tend to think that if a garment they like and try on (in 'their' size, of course) does not flatter them, then that brand will not fit them well. The assumption is that the brand is designing for, and catering to, 'someone else'. This is not good, of course, for establishing or maintaining brand loyalty, and when it comes to building a successful fashion brand over a long period of time, loyalty – getting customers to come back regularly over time – is the holy grail. Kurt Salmon Associates, a large global consultancy based in Atlanta, published a statistic based on a large consumer survey that many technical designers around the world can cite verbatim: 85% of consumers say that they return to a particular brand specifically because of how it fits them.

The unfortunate thing for consumers though, and for the brands trying to sell to them, is that the inherent nature of fashion is change. Every season will see new styles and designs, fabrics with more or less stretch, and silhouettes with varying degrees of intended shape. Inevitably, to a consumer, garments in 'their' size will seem to fit differently (they actually will, by intent, fit differently), and that will frustrate the shopper, and often push the shopper to try another brand. One of the keys for brands and retailers, in terms of addressing this issue, is simple marketing and communication, yet many fail to understand, appreciate or tackle this concept. Often, how a brand communicates fit intent, such as slim, classic or relaxed, for example, is the difference between a happy customer and a lost sale.

1.7 Future trends

There have been many commercial attempts to solve the size and fit dilemma for consumers whose body shape just did not match up well with most off-the-rack clothing. My Virtual Model allowed a customer to submit measurements to a website that created a customized avatar of their body for trying on clothes. Others such as My Best Fit, Get My Size, and My True Fit relied on algorithms and supporting databases to convert a person's self-reported body measurements into the best fitting size in any brand. My Shape and Fit Logic recognized that shape was more important than size or even measurements, but identifying the right shape was one thing; finding actual product at retail to match that shape was another. Some of these early attempts to solve the problem also failed to take into account personal wearing preferences or error factors in processing self-reported measurements; others addressed these issues but not to a level that engendered actual consumer confidence.

Some of the earliest attempts focused on mass customization as the answer. Made4Me.com was a pioneer in this area, attempting to collect consumer body measurements to make custom shirts and pants that would fit their body perfectly. Unfortunately, the process for collecting body measurements was flawed because, even with excellent well-designed measuring tools, consumers seem to be incapable of accurately taking and/or reporting their measurements. The other drawback was that the pattern alteration process on the back end was still very two-dimensional, making simple adjustments that did not properly capture the shape and balance characteristics necessary to create a garment that really matched the consumer's body shape. Even though that technology was premature, the concept was a hit with consumers.

Other companies, such as Archetype Solutions, had initial marketing and commercial success with mass customized garments, but for a number of key reasons most of their technology also failed to catch on. First, those who sought out customized fit were usually people who could not find clothes to fit at retail; if an off-the-rack garment was not going to fit their body shape well, a two-dimensionally altered version of what was basically the same pattern was not going to be much better. The second factor was that the consumer's expectations in ordering a custom-fit garment were quite high (almost like their dreams were finally being answered); receiving a garment after three weeks of high expectations that was not perfect disappointed many who tried the process. Third, the apparel supply chain is not truly set up to make one-off garments, preferring mass volume to single orders; finding reliable sources to make custom-fit garments at reasonable prices was difficult. And, fourth, the vast majority of consumers who buy apparel do not want to wait two, three or four weeks to get it. At the higher end of the market, custom men's suits and bridal gowns, for example, there are exceptions; but most shoppers make spontaneous purchases when they find something they like, or they are buying for a specific need or event that is (usually) imminent.

Entrepreneurs have been undeterred, however, by early failed attempts to solve the sizing and fit dilemma. Today, companies such as Metail (UK) and Upcload (Germany) are using simple photo technology to help consumers find clothes that fit them well. Me-Ality, [TC]^2, Bodymetrics, and Styku all have programs that purport to find the best fit off-the-rack for people who step into their body scanners. All of these technologies, as of this writing, are still in their early stages; retailers and brands are actively testing them for both in-store and potential online sizing and fit solutions, but the jury is still out on whether any, or all, will succeed in being adopted in a broad way. Unfortunately, any approach that tries to help a consumer find the closest fitting size in a given brand still cannot ensure that the individual consumer will agree with their recommendation. The consumers who gravitate towards these exciting new technologies in the first place tend to be people who cannot necessarily find a good fit off-the-rack to begin with. Since most brands, as discussed earlier, still only build to one body shape, these consumers are still likely to be frustrated with their 'expert' recommendation more times than not.

Meanwhile, research into customizing patterns in three dimensions, using body scan and advanced pattern-making technologies, continues. Today, companies like [TC]^2, Styku, and Human Solutions have all developed versions of body scanners that can translate a person's shape – in addition to their measurements – into a three-dimensional pattern that actually replicates the shape, balance and depth of the body that was scanned. These are also early stage developments, but it is likely that these technologies will gradually become more common, enabling at least some form of mass customization to become more common in the future where fit – to a specific person's body shape – is the key value proposition.

1.8 Conclusion

In the past ten years, the apparel and retail industries have come a long way in terms of understanding fit, its relation to body shape, and the critical role both play in the success of every retailer and brand. While companies will continue to explore new, innovative technologies trying to solve the fit frustration that many consumers face on a regular basis, there is no 100% solution. Fit, like beauty, is often in the eye of the beholder. Consumers will continue to be the only relevant judge of whether something fits them or not, no matter what the industry does. What the industry can do is to continue to work to make fit the 'science' part of their product development equation. Design and merchandising will always be, rightfully so, an 'art'; but the companies that find a way to make a science out of fit, establishing it accurately based on body shape and executing it consistently throughout their supply chains, will be the ones that consumers reward with their business and their loyalty.

2
Creation of ready-made clothing: the development and future of sizing systems

S. P. ASHDOWN, Cornell University, USA

DOI: 10.1533/9781782422150.1.17

Abstract: Appropriate sizing systems are a key component in production of the ready-made garments that comprise the majority of the clothing we wear. However, the creation of appropriate sizing systems is a complex task given the challenge of variation in body shape and size, the inadequacy of current size codes for communicating size information, and the variety of clothing styles, of clothing fits and of fit preferences. In this chapter we discuss apparel products and their special relationship to the body, the historical development of sizing systems, current methods of creating sizing systems, and the impact of developing technologies on concepts of sizing.

Key words: apparel sizing, apparel fit, apparel technology, mass customization.

2.1 Introduction

Ready-made clothing, available off-the-rack in retail stores or shipped directly from a warehouse to the customer, is the primary paradigm for clothing production used by the apparel industry in the world today. The development of sizing systems makes this model of clothing production and distribution possible. Sizing systems are designed to provide the best fit for an apparel company's customers, providing enough variation to accommodate all customers, but limiting the number of sizes for efficiency of production and distribution. Sizing systems must provide an appropriate interface; an effective method of communication and size selection for the customer so that the process of choosing and validating the correct size is efficient. The number of sizes in a system is an important factor, as more sizes will add to both manufacturing and distribution costs (and will require extra floor space in a bricks-and-mortar retail space). More sizes may give the consumer more choices, but may also introduce difficulty in locating their 'correct' size if there are a multitude of sizes and shapes from which to choose. The challenge of a sizing system is therefore to provide good fit for the greatest number of customers with the fewest sizes, especially difficult given the variation of body shape and size within the population. The strategies to address these challenges will differ by the apparel product category, based on variables such as the range of body variation in the target market, fit requirements and preferences, the styles being produced, material properties and design features.

Emerging technologies and strategies such as body scanning, virtual fit evaluation, and mass customization offer some exciting future directions for clothing sizing systems and are examined to conclude this chapter.

2.2 The relationship of apparel products to the body

The need for clothing to relate closely to the body gives it a special place in the range of consumer goods produced; other consumer products have a lesser challenge in their distance from, and relationship to, the body. All other objects which we use, from tools, to furniture, to buildings, impact the body less. Most products can be designed as 'one size fits all' (such as hand tools or computer keyboards); other products can be made adjustable for different body types (such as chairs or automobile cabins). However, we expect our clothing to follow every contour of our body, to remain comfortable as we move, and at the same time to provide attributes such as comfort, durability, and aesthetic pleasure. We demand a lot from clothing, requiring good fit to fill each requirement. As first described by Susan Watkins, our clothing forms a portable environment, close to the moving active body, with extraordinary requirements of fit in order to be physically and psychologically appropriate for our needs (Watkins, 1995).

Of course, if worldwide apparel needs are considered it becomes clear that some apparel forms have very different sizing requirements. Wrapped clothing may fit many different body sizes and shapes with just one size of rectangle; minimally shaped styles such as the traditional Japanese kimono require fewer sizes for satisfactory fit. Generally, sizing issues are greatest in the tailored, closely fitting apparel typical of Western clothing styles. Those apparel products for which looser fits are satisfactory, such as t-shirts and bathrobes, can provide fit satisfaction with fewer sizes. Certain apparel products make use of design features, such as the tab on the back of a baseball cap, that make it possible to sell them in one size only. At the other end of the continuum, highly structured garments such as blue jeans, tailored suits, and evening gowns have demanding fit requirements. These garments are difficult to make in a range of sizes appropriate for all body sizes and shapes, and are good candidates for mass customized or customized sizing systems, systems with more sizes to fit different body types. Conceived of as part of a sizing continuum, custom-fitted clothing can also be thought of as

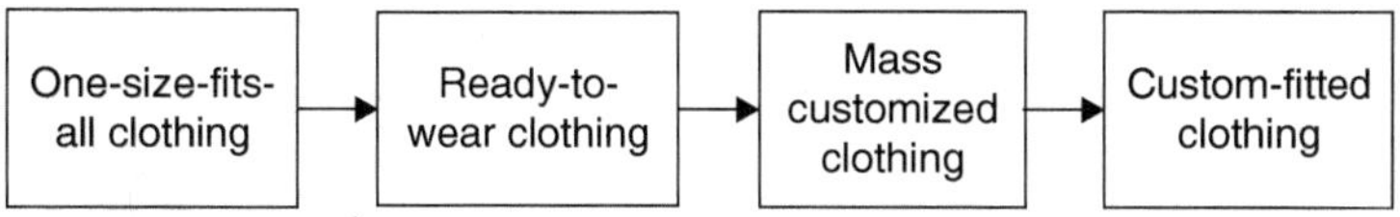

2.1 Clothing is provided using many different sizing strategies; these categories of sizing systems are arranged in order from the fewest to the greatest number of sizes provided.

a 'sizing system' that requires the greatest number of sizes, a unique size for each individual in the population (see Fig. 2.1).

Optimizing the design of an apparel sizing system is not possible without a good understanding of apparel fit, the measure of the body/clothing relationship. Therefore, the first step in the process of creating or improving a sizing system is to clearly identify the fit required for the particular style and target market, and to determine how fit will be measured in order to test the success of the sizing system. Both qualitative and quantitative methods have been used for fit assessment (Fan *et al.*, 2004). However, quantitative methods can require time consuming and expensive data collection and can be difficult to validate. Qualitative methods that address both expert fit assessment and the assessment of the wearer can be productive data collection methods (McConville, 1986; Petrova and Ashdown, 2012).

Functional clothing provides special challenges to clothing fit requirements. The user group for the clothing may be quite extensive, with a single clothing item needed for both men and women. As women increasingly take part in activities traditionally closed to them, such as active combat in the armed forces, or firefighting as a civilian occupation, sizing systems must be changed or augmented to provide protective clothing that fits them well (Robinette *et al.*, 1991). Clothing designed for protection, that also needs to perform functional tasks effectively, can require specific fit characteristics (Coca *et al.*, 2008). Clothing fit for elite athletes can also be quite demanding in garments designed to extend the athlete's performance (Tucker and Dugas, 2008).

Status, group membership and belonging can also be a concern in fit preferences, as these factors can be signaled by clothing's relationship to the body. For example, hip-hop styles are predicated on both style and extreme fit characteristics of garments (Koppel, 2007). Other groups, often categorized as special needs individuals, have non-normative bodies that require a very different relationship to the body (for example wheelchair bound, or osteoporotic women) and some special categories of people in the population are not well served by current providers (for example older people, little people, or obese people). Yet, these groups desire normative clothing styles and clothing fits that relate to their bodies in ways similar to clothing fits of the population as a whole.

2.3 The development of sizing systems

Before the middle of the eighteenth century, most clothing was custom-made either within the family, or by a tailor or dressmaker (Aldrich, 2008). In the pre-industrial world, ready-made clothing was limited to few articles including sailors clothes, called 'slops', sold to sailors who were only in port briefly, and leather breeches for men, a utilitarian item that was difficult to sew at home (Kidwell and Christman, 1974). In times of war, ready-made clothing 'suitable for officers or soldiers' was available. However, with the industrial revolution, the invention of new textile production and garment construction machinery, the introduction of the concept of gathering workers under one roof for a reliable workforce, and the development of mass production methods provided conditions conducive to the creation of ready-made clothing (Kidwell and Christman, 1974).

The transition of the primary method of providing clothing from a custom-made process to a ready-made system, by which most clothing is produced for an unknown customer, was a slow process which began with tailors' patternmaking methods ultimately leading to a proportional grading system. By the 1820s, clothing for men and boys was widely available as ready-made, and by the 1830s most tailors kept a large stock of ready-made garments for men (Aldrich, 2008). The development of department stores, and of wholesale firms supplying smaller retail operations, in the 1840s and 1850s provided new distribution outlets that encouraged development of further ready-made styles and varieties of garments, including women's cloaks and corsets, items that were difficult to sew in the home. In the 1870s and 1880s, the inception of the mail order catalog provided yet another distribution channel for ready-made clothing, driven by US and Canadian growth as the population began to inhabit vast rural areas of North America (Santink, 1990; Emmet *et al.*, 1950). Between 1860 and 1920 many categories of clothing, from business suits to outerwear to sportswear to work clothes to nightclothes to underwear, were available as ready-made items for men and boys. Ready-made clothing for women and girls was developed more slowly, but by 1910 women's styles were also available for most categories of clothing (Kidwell and Christman, 1974).

Early sizing systems relied heavily on the use of alterations to customize fit. There are indications that in major department stores, and in some catalog operations, garments were sold as a general template that was then altered to fit the individual, with the cost of alterations included in the cost of the garment (Santink, 1990; Jeacle, 2003). In 1899 the trade journal The Cutter Up acknowledged that alterations should be made to ready-made garments to perfect fit (Kidwell and Christman, 1974). Ready-made garments were advertised as only being appropriate for the 'proportional' figure – that is, for one body shape. Altman's fall and winter 1888–1889 catalog showed ready-made dresses in proportionate sizes, stipulating that they 'can only accept bust, waist, and skirt front length measurements'. A woman had to be the exact proportions of the sizes offered in order to fit into one of these ready-made dresses without alterations (Kidwell, 1979).

Sizing information from catalogs and advertisements show variation in the range of body sizes and shapes included in early sizing systems. Original sizing systems varied greatly as each manufacturer set sizes by trial and error. Women's sizes were listed by bust measurement, and could range from 32" to 46", with Stout sizing available up to a 56" bust. A 1910 listing

2.2 This advertisement in the November 1915 McCall's Magazine shows a ready-made coat that was sold in Misses sizes 14, 16, and 18 years, and in women's sizes 34 to 44 bust measure (*McCall's Magazine*, November 1915, p. 91).

in Sears catalog showed women's suits in sizes ranging from 32 to 44, in 3 proportions (regular, stout, and small) for a total of 39 sizes. Men's suits were available ready-made for regular, stout, or tall men, with sizes indicated by chest, waist, and inseam measurements (Kidwell and Christman, 1974). Misses sizes in catalogs in the early part of the twentieth century were listed as '12 to 20 years old', so these early size labels indicated age, not body measurements. It was very common for items to be sold in sizes for both Misses and Women (see Fig. 2.2). Misses sizes were still listed as ages in the 1934 Sears catalog, but the Misses sizing in the 1941 Sears catalog used a size code, not an age (Cherry, 2008). By the 1950s, women's briefs were being sold as small, medium, or large. The 1966 Sears catalog still provided a wide range of size categories, showing dresses in four categories: Misses, Misses Petite, Tall Misses, and shorter Women's sizes (half sizes from 14 1/2 to 22 1/2), a total of 20 sizes. JC Penney, just a year later, showed dresses in Junior/ Misses sizes ranging from size 5/6 to 17/18 for a range of 7 sizes only (Cherry, 2008).

By the 1920s and 1930s, department stores were beginning to consider the cost of alterations to be out of control, and started to put procedures in place to hold manufacturers and buyers accountable for poor choices that resulted in high alteration rates. Department stores began to pay attention to the development of sizing systems that would fit the majority of their customers with few alterations (Jeacle, 2003). In the 1950s and 1960s, alteration services were still available in most large department stores, but some alteration costs were not included in the cost of the garments. However, the availability of in-store alteration services had almost entirely disappeared by the 1980s. The increased speed of fashion cycles, the consumer's changing expectations of fit, and the increasing competition in the apparel industry led to lower cost construction methods that made alterations more difficult, and contributed to a gradual decrease in the use of alterations to perfect the fit of clothing. Today, the alteration of ready-made clothing tends to be the exception instead of the rule.

2.4 Apparel industry methods of creating ready-made garments

Basic industry practices for creating sizing systems, developing and fit-testing patterns, grading, developing production patterns, and controlling variation in ready-made clothing are well established. Most companies do not conduct an independent study of their customer to establish a sizing system, as this has historically been a very expensive process. Often, either size standards developed by testing and standards agencies are used (LaBat, 2008), or sizes are developed by knocking off the sizing system used by another company. The process of creating a knock-off sizing system is fairly simple, and can be based on either a body measurement size chart or on measurements from a full size set of garments.

Once the size system is established the next step is to create a set of grade rules and garment specifications. Grade rules define the incremental changes between sizes, and are typically expressed as a set of X–Y coordinate numbers that describe the movement of the cardinal points on a garment pattern (Moore *et al.*, 2001). Skilled graders can apply these rules to a variety of garment styles, can adjust the points between the cardinal points, and can modify the grade rules to accommodate variations that occur in the transition from the 2D pattern to the 3D garment for a variety of styles.

Measurement specifications are developed for each style, and are used to maintain quality in the manufacturing process. These specifications include detailed instructions on how to measure the garment flat on the table (as small differences in the measuring process can result in different values), and will list each measurement for each size in the range. Tolerances for each measurement are also included. Tolerances are an important feature of specifications because of the inherent variation that will occur in the manufacturing process, variation that is unavoidable given the flexibility of clothing materials and the lack of affordable automation in the apparel manufacturing process. Once the grade rules and sizing specifications for the garments are developed, there is little need in the design and manufacturing process for a body measurement chart, and many companies will not maintain such a chart unless it is needed for marketing (this is an important tool for direct marketers such as online and catalog retailers). The one set of body measurements that is important for all companies is the set that defines the company's fit model. One size, the base size, is chosen from the size range for development of prototype garments, and a fit model who has these measurements is hired (Workman and Lentz, 2000). One of the smaller sizes is generally chosen as the base size, as the prototype garment can then also be used for marketing purposes.

Once the size system, grading rules, and general garment specifications are developed, individual styles are created by constructing a prototype garment from the proposed fabric, and perfecting the fit of this prototype on the fit model. This perfected pattern is then used as the base size that is graded up and down to create patterns for all of the sizes in the range. These patterns may be modified further for ease of manufacturing if the style is adopted (Keiser and Garner, 2012). As the great majority of grading systems used are strictly proportional, the sizes in the range will only fit customers who are different from the fit model in overall body dimensions, but are identical in body proportion. Many of the fit problems encountered by apparel companies can be traced to this dependence on sizes that are proportionally developed based on one fit model.

A common confusion in the study of sizing systems is the confounding of size labeling (the size number) with the actual size (the dimensions of the clothing). As the size standards published by various countries are not mandatory, the number on the size tag can be chosen by each company to apply to their own dimensions. Size labels in some cases are not directly linked to body measurements (i.e., the 2, 4, 6, 8, 10, etc, range from US Misses sizes), and if a company does not follow a standard then any number can be used to indicate any set of garment dimensions. This has led to a widespread practice known as vanity sizing, in which apparel companies deliberately label garments of a larger than usual size with a smaller number, thus appealing to the customer's desire to fit into a smaller garment. The lack of consistent sizing across apparel companies can be confusing for the customer, who must learn what garment dimensions are associated with the size labels for each company, and which is their size for that particular company. Vanity sizing can also be extended to size labels that are associated with body measurements. For example, a men's pair of pants labeled with a waist size of 34 may actually fit a man with a body waist dimension of 36 or larger.

However, the labeling used within a sizing system to match a customer to their size is a separate issue from the problem of creating sizing systems that provide sizes to fit the range of body size and shape variations in the target market. Studies to improve or establish sizing systems must carefully distinguish between these two issues – first creating a sizing system with appropriate clothing dimensions, then communicating with effective size labeling so that the customer can easily find their size in the system.

2.5 Why standard sizing is complicated by body shape variation

In recent years, apparel companies have lacked reliable data on actual body sizes and shapes in the population needed to develop and assess data based sizing systems, due to the expense of conducting valid anthropometric studies. Data that were available were often outdated, or were collected on a different population (such as the military) from a company's target market. Although some apparel companies have carried out their own studies to collect body measurements of their customers, this is not common practice. Once data are collected, knowing how to translate these complex numbers into useful information on sizing and fit is also a factor.

Since the year 2000, many countries and some individual apparel firms have conducted anthropometric studies using 3D bodyscan technology, collecting demographic data as well, so that relevant subsets of the database can be identified, including body shape groups (CAESAR, n.d.; SizeGermany, 2007; SizeUK, 2004; [TC]^2, 2009). Several methods of categorizing body shape have been developed (Connell *et al.*, 2006;

Simmons *et al.*, 2004). Body shape categorization systems based on body circumferences or body silhouettes have the advantage of being simple systems with relatively easy categorization. However, they cannot distinguish between body shapes with different arc measurements (for example fuller or flatter buttocks) nor can they identify the postural variations that can affect fit. The complexity of the problem precludes easy solutions to questions of sizing and fit. The extreme variation of body sizes, body shapes, and body postures in the population, the difficulty of clearly identifying the target market for the company and of finding valid data on the body size and shape of the target market, the complexity of categorizing body proportions, and the complicating factor of variations in fit preference by customers are all issues that need further study. Academic research has only begun to unravel the issues facing an apparel company that wishes to develop or improve their sizing system based on anthropometric data.

Sizing systems developed for multiple body proportion categories must address a full range of issues related to sizing. These issues can be categorized into four main areas; body dimension data and conversion of these data into a sizing system; fit issues including fit preferences; design, manufacturing, and distribution issues; and size selection and communication issues (see Fig. 2.3). The graphic representation of the issues around sizing in Fig. 2.3 was developed as a guide for research into apparel sizing. Population measures are categorized in three ways (tailor's measures, traditional anthropometry, and 3D scan studies) as the databases developed using different methods and measurement categories provide different types of data that must be analyzed and applied appropriately to the problem of developing or perfecting sizing systems. It is expected that 3D scan databases will be increasingly available due to the reduced cost of this method, and the fact that scans can be retained, and new measurements can be taken as their importance is discovered, makes this method of data collection very attractive.

Design, production, and distribution issues are important as they encompass basic concepts as varied as whether a company makes tailored or less fitted clothing, whether their materials are primarily structured wovens or more flexible knits, how much control they have over manufacturing and quality issues, how garments are made, warehoused, and delivered, and whether function or aesthetics are more important to their customer.

Fit issues are important in order to quantify the goal of the sizing system (to satisfy both the company's fit aesthetic and the fit preference of their customer), and also to create a standard to assess the effectiveness of the sizing system. Both the initial identification of fit (on the fit model) and final quantification of fit (in consumer fit studies) are desirable. Communication of fit and sizing are critical to the development of an effective sizing system. A labeling system that successfully matches the best size in the range that fits a customer without repeated visits to the dressing room to try out different

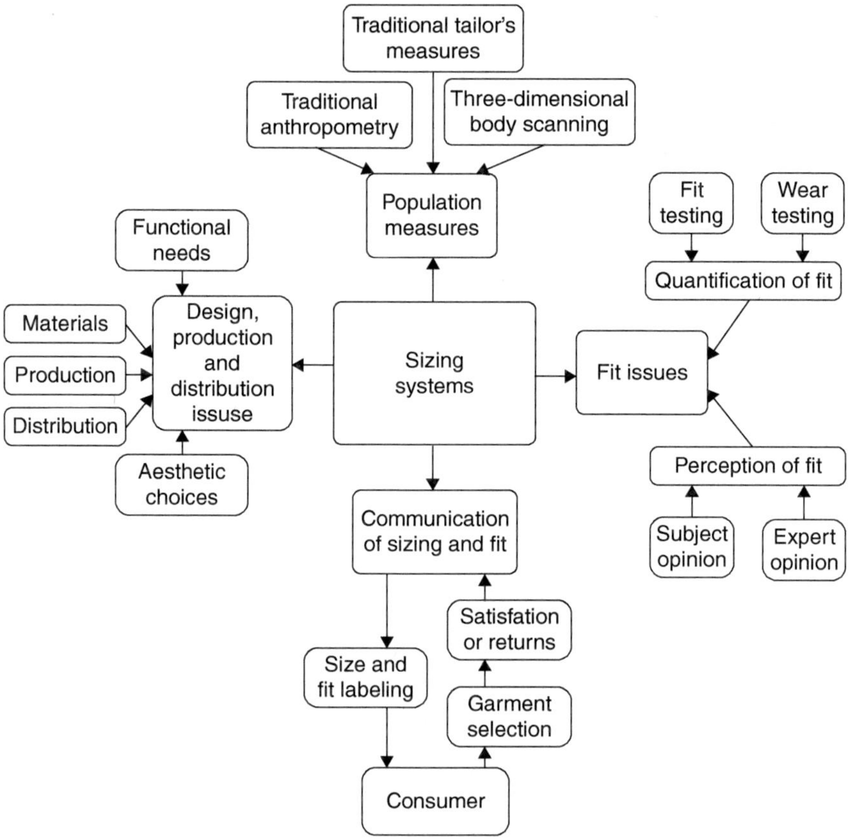

2.3 A conceptual framework for research on sizing and fit (Ashdown, 2007).

sizes is the first step. However, return communication from the customer is equally important to understand why garments are not purchased or are returned, and to collect fit data to improve the sizing system.

A final issue with apparel sizing is the difficulty of changing an existing sizing system. Any change that results in misfit for already satisfied customers in order to try to attract new customers will be counterproductive. Indeed, an apparel company has another important decision to make – whether they will change their sizing to satisfy their loyal customers as they age, with concurrent changes in body size and shape, or whether to stay with their original demographic. These are difficult decisions that often require the development of a new brand or specially defined 'fits' of selected styles.

Given all of the difficulties of creating sizing systems for ready-made clothing, it seems that it would be impossible to provide good fit for everyone in the population, and indeed the difficulties increase as more people are included in the target market for a firm. Therefore, most apparel companies create their sizing systems by defining their target market in a way that gives them a loyal following of customers, who appreciate the styles, fit characteristics, level of quality, and cost provided by the company. Different companies have fit models of different proportions; therefore each company provides good fit for those customers whose body shape most resembles their fit model. Companies often look for underserved markets – those people with identifiable body proportions who cannot find good fit – and will manufacture and market to these niche markets (von Mueffling, 1999; Scott, 2009). Companies are increasingly using body measurement information from national data sets to develop sizing systems for particular target markets, creating significant advances in the effectiveness of clothing sizing. The next section will describe a variety of emerging technologies and mass customization strategies that we expect to revolutionize the clothing industry and provide better fit for more consumers.

2.6 An emerging mass customization model for clothing sizing

Though careful analysis of target market body size and shape can be used to improve sizing systems, unless the target market exhibits little variation, the central issues of providing good fit for everyone within the sizing system remain. New technologies and increasing use of mass customization strategies in other product categories make mass customization models for clothing production and distribution both possible and acceptable to consumers.

2.6.1 Mass customization and sizing

Mass customization is a business strategy that provides individual and customized choices within a mass production framework while keeping costs under control. The goal is for the consumer to get the exact product that they want, when they want it, and at a reasonable price. Many products, including clothing, are now being designed, produced, and distributed for consumers and businesses based on mass customization strategies, such as Brooks Brothers' shirts and suits, Nike shoes, various brands of office furniture and modules, and specialized manufacturing equipment. For clothing, style, color, and fit options as well as individualized clothing can be part of a mass customization process.

We modified the conceptual model for sizing and fit to address this advance toward mass-customized target market sizing resulting from

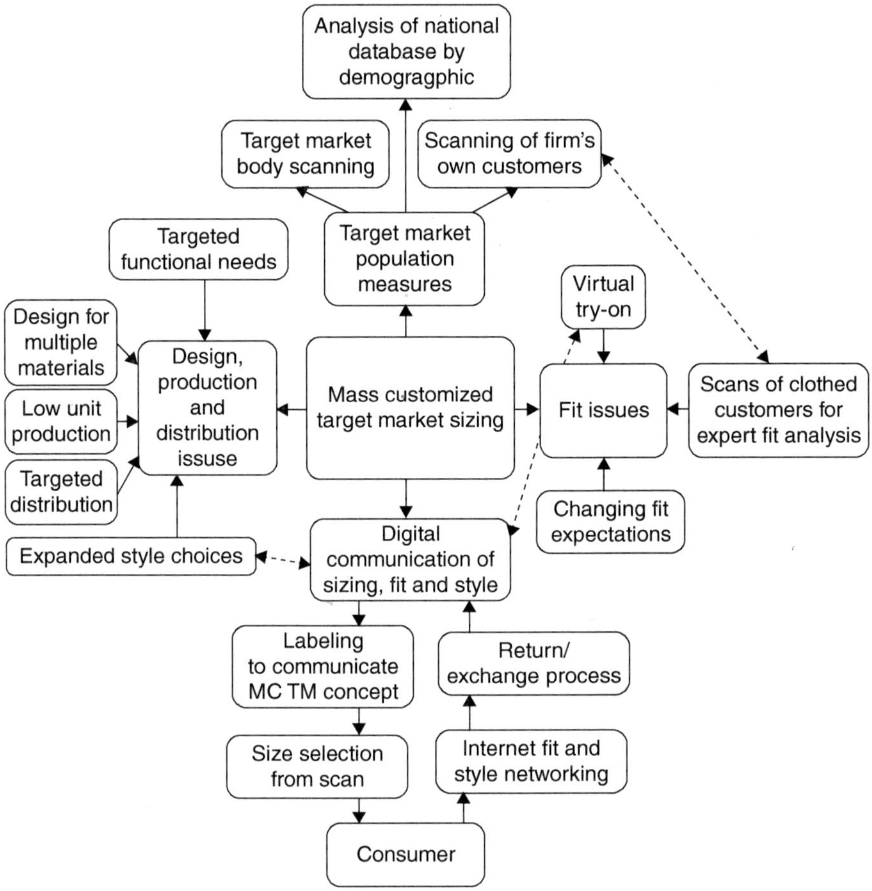

2.4 Conceptual model adjusted to address issues of a mass-customized system (Ashdown and Loker, 2010).

emerging technologies and business strategies (see Fig. 2.4). When we consider the sizing issue in particular, mass customization offers many more combinations of body measurements (or SKUs) than ready-to-wear can profitably offer, but can make use of production and distribution methods that are less expensive than custom sizing. For example, offering short, regular and tall options for every size in the range increases the options for the consumer with a minimum of production adjustments. Lands' End swimsuits offered in different torso lengths, and Levi's early mass-customized jeans are examples (Rifkin, 1994). With advanced technologies specific to the clothing industry, more complex and individualized sizing options have begun to emerge in the design, production, distribution, and communication segments of the industry.

2.6.2 Mass customized sizing and fit

Body scanning, virtual try-on, computer assisted design (CAD) and manufacturing (CAM) technologies have become more compatible and can now accommodate mass-customized clothing specific to individuals and target markets. National databases of body measurements offer companies data sets particular to their target markets. These focused, refined data sets can be used along with CAD and CAM technologies to enable quicker and more accurate development of patterns, virtual try-on with individual consumers' body scans for fit assessment, and more combinations of size dimensions to provide more choice of fits within a mass-customized production system. The closer connection that consumers have with the sizing process through access to their body scan and virtual try-on technologies, as well as choices of style and color, etc, brings them additional knowledge about sizing and fit.

2.6.3 Mass-customized production

Production processes are still catching up with the mass customization model. It is very expensive to retool an industry, particularly an industry that is dependent on outsourcing to developing countries with their low skill, low wage basis. Those apparel firms that have committed to mass customization, usually both large and entrepreneurial firms with clothing products where custom sizing and styling is a competitive advantage, have either adopted technology that supports an entire system of mass-customized products or separated the mass-customized and mass production lines, a more incremental approach. To accommodate one-off sizing and styling requirements, equipment that adjusts to low unit production and multiple materials is needed, requiring a very different factory set up. Bar codes at the item level can be used to send the items to the right operator using an overhead transport system, or a stand-up, hand-off modular manufacturing system can be implemented. These systems are also more time efficient, delivering a finished garment in hours rather than days or weeks and thus matching the goals of mass customization. As the final product is customized to the customer's dimensions, the delivery and return policies are also greatly modified, requiring some adjustment in communication to consumers and to their expectations.

2.6.4 Communication of sizing and fit with mass-customized clothing

Both the clothing industry and consumers will have to modify their communication systems in a mass-customized sizing system. The consumers will have more customized choices that will require more information that can only be provided by the industry. For example, consumers will need to become familiar with the technologies that deliver these new choices such as body scans used in virtual try-on, virtual and collaborative fit evaluations with feedback from both experts and friends, and product configurators that offer styling, color, and sizing options virtually. New approaches to expectations for mass-customized products will be necessary so that consumers understand both the exciting choices and the limitations of mass customization. Delivery time and return policies need to be re-crafted to be acceptable for both customer and business needs.

2.7 Conclusion

The development of new methods of identifying and categorizing body shapes, and new paradigms for sizing systems, can create changes in the methods of production and distribution of clothing that rival the shift from custom-made clothing to ready-made clothing in the early twentieth century. New technologies such as 3D body scanning, virtual fit, computer-aided design of patterns, the ability to create new grading systems, and new manufacturing systems can all contribute to development of new and effective ways to provide good fit for everyone, of any body shape or size.

2.8 Future trends

Currently, apparel companies have increased availability of anthropometric data from various populations in recent years, allowing them to acquire data on the range of body shapes in their target markets relatively easily. Body scan technology to gather data and new research in how to use anthropometrics to create effective sizing systems will result in effective methods for fitting more people successfully. The Levi Strauss jeans fitting system called Curve ID is an example of effective data collection, successful development of a new sizing system, and, perhaps most importantly, development of a method for customers to easily use the system, and a strong multinational marketing campaign to introduce the new sizing concept (Tschorn, 2010).

Scan-at-home and related developments are bringing 3D image technology away from the stand-alone scan technology. Using these systems, a customer may be able to create a 3D image that faithfully represents their size and shape at home, either by using Kinect ® technology or by taking a series of silhouette and profile photographs using a simple digital camera, which are then transformed into a 3D image (Kim, 2012). These images can then be used as an avatar for virtual fit initiatives.

New methods of measuring and analyzing body shape, using data from 3D scans and advanced statistical analysis methods, will create much more reliable results that reflect the nuances of complex proportional and postural variation in the population. For example, recent work in sorting women into body shape categories based on multiple body measurements and unique analysis methods can provide a highly nuanced version of body shape that quantifies both profile and silhouette shape (Song and Ashdown, 2011) (see Fig. 2.5). Although this complex analysis process precludes simple visual assignment of a customer to a shape category, a computerized system for which a customer is scanned, automatically measured and assigned to a shape group, and given the information on their group designation, is a viable system. These more complex shape classifications are important to the balanced fit of garments on the body, as was demonstrated in a study of pants fit using this shape analysis method (Song and Ashdown, 2012). However, more work is needed to identify all of the issues involved in creating an effective sizing system for these complex body shape definitions.

Virtual fit can provide an important link for size selection, fit analysis, and style consultation. As these new technologies are improved they will contribute greatly to the important characteristic of size and fit communication, a vital consideration for the success of any sizing system, but particularly critical for educating the consumer to new sizing paradigms.

2.9 Sources of further information and advice

Books on how to create sizing systems, and sizing standards for different countries, are generally available. Various resources are listed in the chapter references in Sizing in Clothing; Developing Effective Sizing Systems for Ready-To-Wear Clothing from Woodhead Publishing. Petrova (2008) describes methodologies for creating sizing systems from anthropometric data.

Anthropometric data from 3D body scan studies are also available in many countries. Textile Clothing Technology Corporation [TC]^2, www.tc2.com, conducted the survey SizeUSA in which 11 000 individuals in 12 locations across the US were scanned in 2002–2003. Data can be purchased from this study based on demographics and shopping preferences. Many similar studies have been conducted around the world using [TC]^2 scanners, so current information on what surveys have been completed and who to contact in other countries may be available from this source. Both [TC]^2 and Alvanon (www.alvanon.com) will provide consultant services in apparel sizing and fit worldwide. Many other consultants in apparel technology can be found by contacting local educational institutions with apparel programs, or conducting web searches for technical apparel consulting.

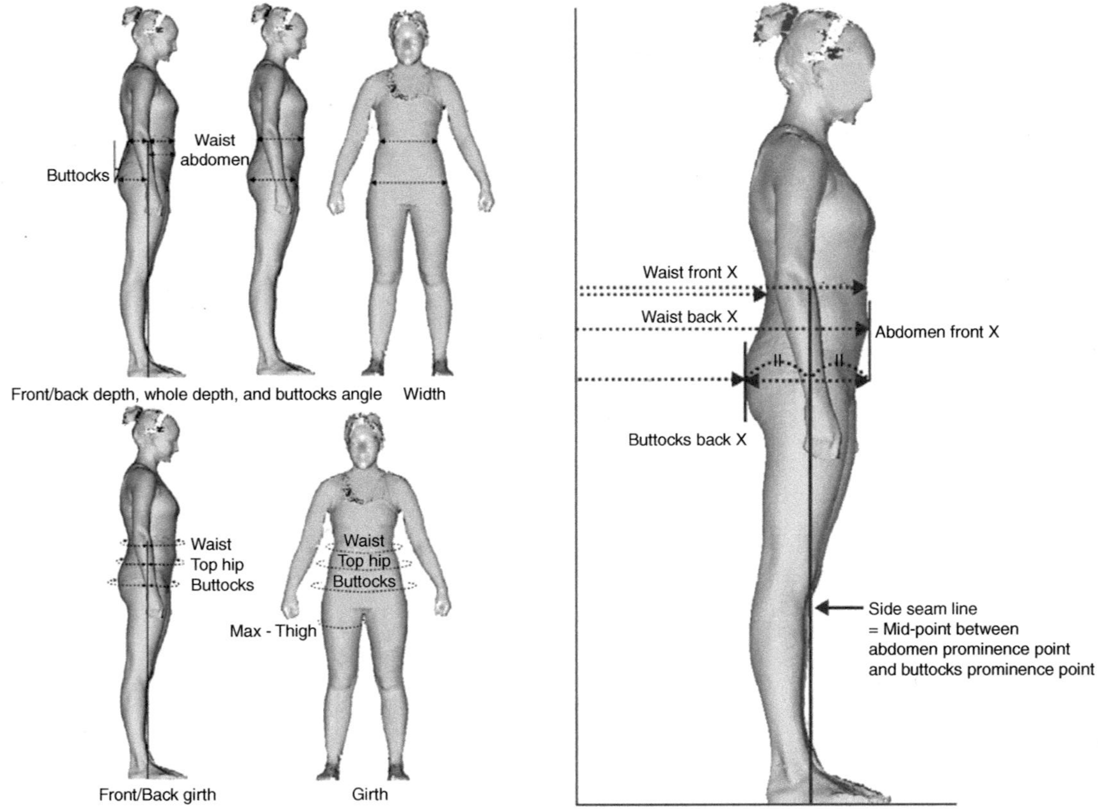

2.5 Detailed arc measurements, depths, and widths are needed for effective body shape categorization. This set of measurements for the lower body was automatically derived from body scan data and used to categorize lower body shape (Song, 2011).

2.10 References

Aldrich W (2008), 'History of sizing systems and ready-to-wear garments', in S P Ashdown *Sizing in Clothing; Developing Effective Sizing Systems for Ready-to-Wear Clothing*, Cambridge, Woodhead Publishing Limited, 1–56.

Ashdown, S P and Loker, S (2010) 'Mass customized target market sizing: Extending the sizing paradigm for improved apparel fit', *Design Practice*, **2**, 2, 147–73.

Ashdown S P, ed. (2007), *Sizing in Clothing: Developing Effective Sizing Systems for Ready-to-Wear Clothing*, Cambridge, Woodhead Publishing Limited.

CAESAR (nd), CAESAR: The most comprehensive source for body measurement data, Available from: http://store.sae.org/caesar/#3dna [accessed 24 March, 2013].

Cherry R (2008), *Catalog: An Illustrated History of Mail-order Shopping*, New York, Princeton Architectural Press.

Coca, A, Roberge, R, Shepard, A, Powell, J B, Stull, J O, and Williams, W J (2008), 'Ergonomic comparison of a chem/bio prototype firefighter ensemble and a standard ensemble', *European Journal of Applied Physiology*, **104**, 351–9.

Connell, L J, Ulrich, P V, Brannon, E L, Alexander, M, and Presley, A B (2006), 'Body shape assessment scale: Instrument development for analyzing female figures', *Clothing and Textiles Research Journal*, **24**, 2, 80–95, doi: 10 1177/0887302x0602400203.

Emmet, B, Jeuck, J E, and Rosenwald, E G (1950), *Catalogues and Counters: A History of Sears*, Roebuck and Company, Chicago, University of Chicago Press.

Fan J, Yu W, and Hunter, L (2004), *Clothing Appearance and Fit: Science and Technology*, Cambridge, Woodhead Publishing Limited.

Jeacle, I (2003), 'Accounting and the construction of the standard body', *Accounting, Organizations and Society*, **28**, 357–77.

Keiser, S J and Garner, M B (2012) *Beyond Design: The Synergy of Apparel Product Development*, New York, Fairchild Publications.

Kidwell, C B and Christman, M C (1974) *Suiting Everyone: The Democratization of Clothing in America*, Washington DC, Smithsonian Institution Press.

Kidwell, C B, (1979), *Cutting a Fashionable Fit*, Washington DC, Smithsonian Institution Press.

Koppel, N (2007) 'Are your jeans sagging? Go directly to jail', *New York Times*, 30 August, 2007, Available from: http://www.nytimes.com/2007/08/30/fashion/30baggy.html?_r=0 [accessed 20 March 2013].

Kim E (2012), 'Can't zip those jeans? Bodymetrics can help', Innovation Nation, @ CNNMoneyTech, 25 July 2012, Available from: http://money cnn com/2012/07/24/technology/bodymetrics/index htm [Accessed 22 March 2012].

LaBat K (2008) 'Sizing standardisation', in S P Ashdown, ed., *Sizing in Clothing; Developing Effective Sizing Systems for Ready-to-Wear Clothing*, Cambridge, Woodhead Publishing Limited.

McCall's Magazine, November (1915), Boston Store Chicago, 91.

McConville, J T (1986), 'Anthropometric fit testing', In R L Barker and G C Coletta, eds, *Performance of Protective Clothing (ASTM STP 900)*, Philadelphia, American Society for Testing and Materials.

Moore, C L, Mullet, K K, and Young, M P (2001), *Concepts of Pattern Grading: Techniques for Manual and Computer Grading*, New York, Fairchild Publications.

Petrova, A and Ashdown, S P (2012), 'Comparison of garment sizing systems', *Clothing and Textiles Research Journal*, **30**, 4, 315–29.

Petrova A (2008), 'Creating sizing systems', in S P Ashdown, ed., *Sizing in Clothing; Developing Effective Sizing Systems for Ready-to-Wear Clothing*, Cambridge, Woodhead Publishing Limited.

Rifkin, G (1994), 'Digital blue jeans pour data and legs into customized fit', *The New York Times*, Business Section, 8 November 1994.

Robinette, K M, Mellian, S A, and Ervin, C A (1991), 'Development of sizing systems for navy women's uniforms', TR 183, Armstrong Laboratory, Wright-Patterson Air Force Base, Ohio 45433, Navy Clothing and Textile Research Facility.

Santink, J L (1990), *Timothy Eaton and the Rise of His Department Store*, Toronto, University of Toronto Press.

Scott A (2009), 'Rebecca & Drew makes custom shirts for women' Luxist, 11 February 2009, Available from: http://www.luxist.com/2009/02/11/rebecca-and-drew-makes-custom-shirts-for-women/ [accessed on 22 March 2013].

Simmons, K P, Istook, C L, and Devarajan, P (2004), 'Female figure identification technique (FFIT) for apparel, Part I: Describing female shapes', *Journal of Textile and Apparel, Technology and Management*, **4**, 1, Available from: http://www.tx.ncsu.edu/jtatm/volume4issue1/articles/Istook/Istook_full_105_04.pdf [accessed on 24 March 2013].

SizeGERMANY (2007), Zum Size GERMANY-Portal, Available from: http://www.sizegermany.de/ [accessed on 12 January 2010].

SizeUK (2004), 'UK National Sizing Survey Information Document', Available from: http://www size org/SizeUKInformationV8 pdf [accessed on 12 January 2010].

Song, H K and Ashdown, S P (2011), 'Categorization of lower body shapes based on multiple view analysis', *Textile Research Journal*, **81**, 9, 914–31.

Song, H K and Ashdown, S P (2012), 'Development of automated custom-made pants driven by body shape', *Clothing and Textiles Research Journal*, **30**, 4, 315–29.

[TC]^2 (2009), 'SizeUSA', Available from: http://www.tc2.com/sizeusa.html [accessed on 13 January 2010].

Tschorn A (2010), 'Levi's debuting Curve ID jeans', *Los Angeles Times*, 21 August 2010, Available from: http://articles.latimes.com/2010/aug/21/image/la-ig-curves-20100821 [accessed on 24 March 2013].

Tucker, R and Dugas, J (2008), 'The Speedo LZR racer: feedback on a developing debate, and the value of technique', *The Science of Sport*, Available from: http://www.sportsscientists.com/2008/03/speedos-lzr-swimsuit.html, [accessed on 29 July 2009].

von Mueffling, D (1999), 'Bloomingdale's or bust! The rise of Shoshanna Lonstein (Jerry's ex)', *New York Observer*, 18 January 1999, Available from: http://observer.com/1999/01/bloomingdales-or-bust-the-rise-of-shoshanna-lonstein-jerrys-ex/ [accessed on 21 March 2013].

Watkins, S M (1995), *Clothing: The Portable Environment*, 2nd ed. Ames, Iowa, Iowa State University Press.

Workman, J E and Lentz, E S (2000), 'Measurement specifications for manufacturers' prototype bodies', *Clothing and Textiles Research Journal*, **18**, 4, 251–9.

3

National sizing surveys: techniques, data analysis and apparel product development

K. M. KING, [TC][2] (Textile/Clothing Technology Corporation), USA

DOI: 10.1533/9781782422150.1.35

Abstract: This chapter provides a summary of technology features that support the use of 3D body scanning for sizing surveys, including measurement extraction and body visualization capabilities. Using the SizeUSA study as a point of reference, this chapter addresses aspects of survey design and data collection for national sizing surveys and describes the product development opportunities associated with analysis of data collected. This chapter also touches on the value of data from national surveys for applications outside the apparel sector.

Key words: apparel/garment/clothing, apparel/garment/clothing fit, apparel/garment/clothing size, size analysis, sizing study, anthropometry, anthropometric study, SizeUK, SizeUSA, SizeMexico.

3.1 Introduction

A quick conversation with the average consumer on the topic of shopping for clothing is likely to reveal anecdotal evidence of a level of consumer frustration associated with garment fit and sizing. This issue is well known to brands, retailers and manufacturers, and has been confirmed by consumer surveys undertaken by Kurt Salmon Associates, which report (1999) that almost two-thirds of the consumers surveyed in 1998 indicated '...difficulty finding clothes that fit well' (p. 11). More recently, product developers have included consistency of sizing and fit among their list of concerns in the product development area, as brands and retailers have increasingly adopted global, multi-vendor sourcing strategies and have, in many cases, outsourced the pattern development process. In response to such issues, the apparel industry has historically relied on sales and product return data and research strategies, including wear tests and focus groups to obtain consumer insight, in support of development that results in apparel product that better serves the customer's needs. These research strategies offer merit in that they may uncover selling patterns and help to determine consumer likes and dislikes. However, established methods rarely uncover the reason for lost sales. In effect, product developers don't know what they don't know. As a result, development teams are left to ponder best paths with limited support for their conclusions.

Thus, large scale sizing surveys have been viewed by the apparel industry as a key avenue for obtaining more in-depth information about consumers in reference to body dimension, shape and size. Though opportunities for conducting large scale surveys via manual methods for body measurement have been limited, the introduction of 3D body scanning technology has made large scale surveys feasible. This chapter will highlight the value of information obtained from national surveys and will describe a variety of applications for data analysis. The chapter will also provide insight into strategies for conducting national surveys, including study design and data collection.

3.1.1 Value of sizing surveys

Large scale sizing surveys are a significant undertaking. However, if well executed, the data collected offers tremendous value to brand owners, retailers and manufacturers for optimization of garment fit and sizing strategies. Body measurement data coupled with demographic information such as gender, age, ethnicity, income level and geographic location can be analyzed in reference to target customer populations with the aim of confirming current or proposed sizing strategies and improving garment fit for a given demographic. Analysis can also be undertaken to support strategic decisions regarding size and product assortments in reference to a target customer by brand, store and/or region. Furthermore, body measurement data that is statistically significant in reference to a population may also be used to inform industry discussions regarding sizing guidelines or standards where there is support for such initiatives.

Although the consumer is perceived to be the primary benefactor, all members of the supply chain may experience reward when access to accurate consumer information leads to greater efficiency. In the context of growing emphasis on reduced product development and manufacturing cycle times, reduced consumption of materials and energy, and elimination or reduction of waste at all levels, information that enables more informed decision making with respect to consumer wants and needs is a tremendous asset to all members of the chain.

3.1.2 National sizing survey initiatives

In the US, the earliest sizing surveys were conducted by government agencies, and were generally used to support development efforts for apparel and equipment in reference to military populations. As previously noted, the introduction of 3D body scanning technology has been the primary enabler for surveying large populations. In the mid-1990s, the CAESAR survey, which refers to the Civilian American and European Surface Anthropometry Resource, was conducted jointly by the Society of Automotive Engineers and the US Air Force using 3D body scanning technology developed by Cyberware. The sampling strategy included data gathering in three NATO countries: the Netherlands, Italy and thc US. The final report (Robinette *et al.*, 2002) indicates that the US portion of the study resulted in the collection of data from 2375 subjects.

Though the CAESAR survey was a significant accomplishment, later efforts have been more comprehensive in nature, offering greater relevance for apparel applications. In the late 1990s, SizeUK was initiated in the UK through a collaboration that included participation from the UK government, 17 UK retailers and a collection of academic institutions (SizeUK, n.d.). The study, which was coordinated by the University College London and the London College of Fashion (Sizemic, 2013), used 3D Body Scanning technology from Textile/Clothing Technology Corporation (known as $[TC]^2$) to scan approximately 11 000 adult subjects at various locations around the country. Data collection commenced in 2001 (Bougourd and Treleavan, 2010) and according to Sizemic (2013) results of the study were made available for purchase in 2004.

More recently, a national children's survey was completed in the UK by Select Research, who were involved in SizeUK and who managed a private series of sizing surveys for Marks & Spencer in the late 1990s. Shape GB was a collaborative project between six major retailers (George at ASDA, M&S, Monsoon, Next, Shop Direct and Tesco), five universities (Aston University, University of Hull, Loughborough University, Manchester Metropolitan University and University of Hertfordshire) and a number of companies and organizations that specialize in retail sizing surveys. The first phase of this survey began in 2008, and for this part 3D body scanning by $[TC]^2$ was used to measure over 2800 participants that included boys and girls aged from four to seventeen. Phase II of Shape GB was launched in the spring of 2012, and involved manual measurement of babies and toddlers (R. Barnes, personal communication, 20 January 2013).

Like the SizeUK study, SizeUSA was a comprehensive initiative that involved collaboration between the apparel industry, government and academic institutions. This study was coordinated and executed by $[TC]^2$, which is a US not-for-profit firm specializing in technology development and supply chain improvement. The study began in 2003 and involved scanning almost 11 000 adult participants, including over 6000 women, at 13 locations across the US. The final SizeUSA report was issued in 2004, and body measurement information as well as 3D body scan data from the study continues to be available for purchase via $[TC]^2$.

Surveys have also been conducted in Mexico and Thailand using [TC]^2's 3D body scanning technology. The Mexican initiative was launched by Cámara Nacional de la Industriadel Vestido, the National Chamber for the Clothing Industry in Mexico known as CANAIVE, with support from [TC]^2, INEGI, the Instituto Nácional de Estadistica y Geografía (the government office in charge of statistics), and the Economic Ministry of Mexico (Textile Clothing Technology Corporation, 1 July 2010). The study, which culminated in 2011, included collection of data in 14 cities for over 16000 participants (Perez and Rodriguez, 31 August 2011).

Sizing surveys have also been conducted with 3D body scanning technology from Human Solutions, GmbH, for regions including France, Germany, Italy, Spain and Finland (M. van Genabith, personal communication, 14 January 2013). The survey in France included 3D body scanning of over 11500 participants, while SizeGERMANY encompassed collection of data for more than 13000 participants. This survey, which took place between 2007 and 2009, traveled to over thirty locations around the country and included data collection from men, women and children (Seidle *et al.*, 2009). The authors indicate that the SizeGERMANY initiative involved cooperation among various parties including Human Solutions, the Hohenstein Institutes and a list of apparel and automotive firms. In addition to the surveys described, it should be noted that Cyberware technology was used in support of a sizing survey conducted in Korea between 2003 and 2004, though Cyberware ceased operations in 2011.

3.2 Survey design

As is the case with most population research, the value of the data collected is at least partially dependent on the design of the survey. A number of factors have been considered with respect to the coordination of large scale sizing surveys. Some of the key areas have included identification of goals for the survey and potential uses for the data, population sampling strategy in reference to the identified goals, identification of a best fit technology solution/s for supporting data collection, storage and analysis, and development of a strategy for ensuring data privacy with respect to the participants surveyed. These considerations are outlined in greater detail in the following sections.

3.2.1 Survey goals

Although sizing surveys may be initiated for a variety of reasons, the overriding purpose of a national survey includes the acquisition of a pool of body measurement data that accurately represents the population of the country in question. A series of motivations or secondary goals may underpin the primary reason for conducting a national survey. These underlying factors should be considered as part of the survey design to maximize the value of the data collected given the cost and effort associated with a large scale initiative.

As demonstrated by previous surveys, national sizing study initiatives often involve partnership between industry, government and academia. The various stakeholders are likely to enter into the arrangement with the hope that the data can be analyzed to address specific population questions or product issues. For example, among partners of the European Union, consistency of size labeling supports effective trade between countries. Thus, there has been regional motivation for using national sizing data in the development of common sizing standards or guidelines. In contrast, although US brands and retailers have a general interest in sizing guidelines, they often view apparel sizing and fit as part of their competitive advantage within the marketplace. Therefore, among this group, the motivation for collecting body measurement data for the US population has primarily related to optimization of sizing strategies and garment fit in reference to target customers. The aim here is increasing consumer satisfaction and brand loyalty, thereby improving sales and reducing product returns. Additionally, supporting academics and/or government bodies may share some of the stated motivations, and stakeholders from outside the realm of apparel may be interested in using the data to inform development of standards or strategies for more industrial applications such as design of products with ergonomic and/or safety considerations in mind.

From a practical standpoint, the various motivations have helped to determine the range of participants included in a national survey, the specific body data and associated measurement strategies. While most surveys have targeted the adult population within a country, some efforts have not included collection of data among infants, children, pre-teens or teenagers. With respect to the range of body data collected, height and/or weight values may have been measured manually, and whole body scanning to obtain upper and lower body data has been standard. However, specialized strategies may have been required for obtaining data for head, hands and feet and, therefore, limited or no data may be available for various surveys in that regard. Also, although analysis of alternate pose scan data may be useful for athletic and functional apparel and for transportation, industrial and ergonomic applications, using 3D body scanners to obtain data for participants in seated or specialized body poses was not possible or feasible in the context of all national surveys completed to date. Thus, it is important to understand the limitations of each survey.

The motivations described may also have an impact on the demographic information requested of each participant during the survey process.

Demographics such as age and income level may have been collected based on categories or groupings as compared to the designation of a specific value (e.g. age group 18 to 25 rather than age 23). Stakeholders may have been interested in specific factors such as ethnicity, physical fitness/activity level, education level, employment and/or shopping preferences. Though a wealth of information is available, demographic information collected for participants of national surveys does not address all possible interests for stakeholders. In the end, questionnaires have been designed to optimize information collected without being overly intrusive and with the aim of minimizing the time it takes for each participant to complete.

Beyond development of the demographic portion of the survey, it has been important for organizers to consider the type of measurement information extracted at the end of the data collection process. For example, analysis of data in support of standards development and for comparison to existing guidelines (e.g. ASTM International, formerly known as the American Society for Testing and Materials) has required extraction of body measurements with aligning definitions. Purchasers of SizeUSA data have also requested extraction of custom measurements (i.e. measurements with a definition that differs from that found in the standard extraction). For this reason, participants in the SizeUSA database have been coded to support matching new measurement extractions with the original data set.

The method for sharing the data and survey results with stakeholders has also been important. Bougourd and Treleaven (2010) indicate that data from the SizeUK survey has been made available to purchasers and partners via on-line software tools. Seidl *et al.*, (2009) report a similar scenario in regard to results for SizeGERMANY. With respect to data format, while some stakeholders in the apparel sector may be familiar with the use of database software for conducting statistical analysis and running queries (e.g. Microsoft Access), others may be more familiar with typical spreadsheet based tools (e.g. Microsoft Excel). Thus, the ability to exchange measurement data and associated demographics between software solutions has been advantageous. Moreover, the advancement of 3D technology for apparel product development and virtual fashion suggests that going forward there may be growing interest in the analysis and use of 3D body models or avatars in addition to analysis of body measurements. In this context, the availability of 3D body data in a format that supports exchange with 3D computer assisted design (CAD) and visualization systems is valuable.

3.2.2 Population sampling strategy

In order to obtain data sets that are representative and comprehensive in reference to a given national population and survey goals, sampling strategy has been a key factor for consideration. While total number of participants surveyed is important, the demographic makeup of the sampled group is also essential to the strategy. In determining the demographic characteristics for statistical significance, organizers of the SizeUSA survey referred to the population sampling strategy for the National Health and Nutrition Examination Survey (NHANES) which was conducted in the US by the Center for Disease Control and Prevention (CDC) (Textile Clothing Technology Corporation, 2004). This strategy directed organizers to target participation based on two genders, six age groups (including ages 18 to 66+), and four ethnic categories (Non-Hispanic Black, Non-Hispanic White, Hispanic, and Other). The SizeUSA report notes for the NHANES III study:

> ... height and weight by age and ethnicity were recorded. The study published height and weight in 7 age groups from 20–80+, 2 genders, and the same 4 ethnic groups. The height and weight were published in 9 percentiles (5th – 9th). SizeUSA sampled each cell in 3 percentile groups (below 15% and 85%, above 85%) for height and weight extrapolated from the NHANES III study (p. 6).

Note that while the SizeUSA study developed a sampling strategy that targeted only four ethnic categories for statistical significance, demographic information was also collected for a broader range of ethnic groups that included the category Asian. In order to meet the sampling targets, data was collected at 13 locations that encompassed the southwest to northwest coast, the central and south central regions, and the southeast to northeast regions of the country. While the strategy described was believed to be valid for sampling the population in the US, a different strategy may be needed for sampling populations in other regions of the world. For example, where ethnic diversity is a less significant feature of a national population, factors such as number of subjects by age or by region may become paramount to the sampling strategy.

3.2.3 Technology

As described in the introductory section of the chapter, the feasibility of acquiring body measurement data from a large number of subjects is a relatively recent occurrence and is generally agreed to be a direct result of the development of 3D body scanning technology. Selection of a suitable technology solution for scanning involves a number of considerations. The ability to acquire accurate 3D data is paramount. To date, laser based scanners (e.g. Human Solutions and Cyberware) and white light scanning technologies (e.g. $[TC]^2$) have demonstrated the greatest levels of data accuracy among the variety of technologies available for whole body 3D scanning. It has also been advantageous to work with systems that are relatively portable and that can be moved to various locations with a minimum of time, effort and training for setup and takedown. Whole body 3D scanners typically have multiple scanning sensors that must be aligned in order to ensure high quality scanning results. In the context of a survey that moves to various locations, the alignment procedure must be fairly simple or, alternatively, technical staff must be available to support this activity during setup.

System cost and flexibility have been important considerations for national surveys. While laser based technologies often demonstrate high levels of accuracy, these scanners have been more costly to purchase to date. Depending on the specific arrangement with a technology vendor, this may or may not be an important issue with respect to overall cost for a project. In terms of flexibility, scanners with a smaller footprint have enhanced the ability to secure survey locations in retail settings, office buildings and universities that may attract walk-in participation.

System flexibility also pertains to supporting software. Measurement extraction is critical to the ability to obtain required information from the body scan data. Where measurement extraction is not part of the body scanner technology system or where the capability of the supplied software is limited, the ability to exchange the scan data for use in a more robust solution becomes a feature for consideration. As a point of reference, over 100 body measurements were defined as part of the standard extractions for the SizeUK and SizeUSA studies and, as previously mentioned, in the case of SizeUSA it is possible to obtain custom extractions for measurements not included in the standard list as well.

Looking forward, emerging technology solutions that integrate depth sensors (like the Microsoft Kinect device or equivalent) are lowering the cost of whole body 3D scanning. In addition to lowering cost structure, depth sensor based systems may also offer greater portability, including ease of setup and takedown. As these systems are new to the market, data accuracy will need to be demonstrated in order for them to be considered suitable for sizing survey applications. Ultimately, any accuracy or data quality issues that arise will likely be resolved through refinements to system design as the technology becomes more mature.

3.2.4 Data privacy

Development of a strategy for ensuring data privacy has been a requirement and, at a minimum, standard protocols have been followed in reference to research conducted using human participants and anonymity has been maintained in regard to saving and storage of participant data. As an example, the SizeUSA database contains scan data, body measurements and associated demographics coded by number instead of by participant name. Also, during the collection stage, participants were made aware of potential uses for the data and the coding strategy for ensuring privacy through the life of the data.

Given the personal nature of body data and the 3D body scanning process, additional effort has been undertaken by survey organizers to ensure the psychological comfort of participants. As part of a scanning procedure, participants may be asked to remove clothing and jewelry and/or change into form fitting undergarments or exercise apparel that is sometimes referred to as scanwear. While technology requirements and measurement accuracy were partial drivers for scanwear selection in regard to SizeUK, SizeUSA and SizeMexico, personal comfort was also taken into account for these surveys. As a point of comparison, participants in the SizeUK study were scanned in their own undergarments where the garments were compatible with the scanning process. Organizers of SizeUSA took a slightly different approach and provided all participants with scanwear that for women included a non-compressive, form fitting short and a soft cup bra. In the case of the SizeUSA survey, private space was supplied for dressing, with same gender staff available to help facilitate the scan as needed. Note that since additional planning and accommodations are typically required for ensuring privacy and safety of infant, children or teen populations, the SizeUSA survey focused on collecting data from adults only.

3.3 Data collection

A number of considerations have already been highlighted with respect to the data collection phase for national surveys. Additionally, organizers have developed some important strategies for achieving high levels of participation from target populations to ensure the resulting database offers value to end users. Effective management of the survey from a participant engagement, experience and demographic perspective has also been important to the success of surveys.

3.3.1 Registration and engagement of participants

Given the fact that national surveys require participation from a significant number of subjects, obtaining a statistically significant sample of the population is an achievement. In the context of the SizeUSA survey, some participants were happy, if not eager, to contribute to a comprehensive survey that lends support to improvements in garment fit and sizing among brands and retailers. Others were found to be motivated by less altruistic reasons and, with that in mind, small cash incentives and gift cards supplied by sponsors of the survey were used to attract participation.

For some of the studies mentioned, pre-registration was enabled to fill as many appointments as possible in advance of the survey's arrival to a given site. In the case of the SizeUSA study, this strategy helped to minimize the amount of time required for each location, and cities were selected, in part, to help ensure that sampling targets were achieved with respect to ethnic diversity. In addition to advance registration, participation in the SizeUSA survey was also driven by having 3D body scanning sites at retail sites, universities, offices, and manufacturing facilities. These are examples of locations where larger numbers of people are available, thereby increasing the likelihood of on-site registration of participants. Naturally, university locations enabled scanning of a younger demographic, and word spread quickly with respect to incentives, whereas retail, office and manufacturing facilities enabled collection of data from a variety of demographic groups. Where retail settings where used, sales associates, who were aware of the survey, understood the general purpose of the initiative and knew about the incentives provided, helped to encourage consumer participation. In the case of SizeMexico, the survey locations were supplied by Walmart and Suburbia Department Stores (part of the Walmart Mexico group) (L. Perez, personal communication, 16 January 2013) and the sampling strategy was supported by moving to store locations for the various regions targeted.

3.3.2 Participant experience

In addition to survey locations and financial incentives, positive participant experience has also been fundamental to driving registration for large scale surveys. Ensuring participant privacy is a vital part of this effort. The importance of scanwear selection and the provision of changing rooms for the participants have been emphasized as a key part of the privacy strategy in the case of the SizeUSA study. Ensuring a positive experience also included providing potential participants with an overview of the process as they registered and entered the survey area. Management of the flow of participants was also paramount, and SizeUSA organizers staged areas for reception and on-site registration, collection of demographic information, manual measurement of height and weight, and the 3D body scanning process. $[TC]^2$ used the company's Cary, North Carolina office as a trial site to ensure that timing and staff levels worked well prior to going live across the nation. Likewise, organizers of SizeUK conducted a proofing survey with participation from 70 men and 70 women to verify procedures (Bougourd and Treleaven, 2010).

3.3.3 Collection of demographic information

As previously emphasized, the collection of demographic information from survey participants is a feature of primary importance with respect to the value of the data arising from a national survey, as it enables the stakeholders to analyze the data in reference to target customer groups. Thus, organizers have given careful consideration to the kind of information collected, as well as the method of collection. In the context of the SizeUSA survey, demographic information was collected in paper form coded by participant ID, and analysis of data in reference to gender, age, and ethnicity have been the primary points of interest for stakeholders. Income, education level, marital status, activity level, and shopping preferences have been of interest to a lesser degree. Activity level may be a feature of growing importance for purchasers of the data in response to the rise in product development for athletic and performance apparel applications. In the case of the SizeMexico survey, information was collected via electronic means and included demographic features such as age, region, household items, and types of garments purchased. The SizeMexico initiative also included collection of demographic information related to health issues such as diabetes, hypertension, and eating disorders in reference to analysis interests from government stakeholders in the health area.

3.4 Data analysis

Ultimately, the formation of a database that allows for analysis of body measurement and shape information in reference to demographic description has been critical to the perceived value of national sizing surveys and, upon completion of the data collection procedure, a number of processes have been undertaken to prepare the database. In the case of SizeUK, SizeUSA and SizeMexico, preparatory steps centered on preliminary cleaning of the 3D data set. Specifically, more than one scan was captured for each participant of the surveys with the aim of optimizing the quality of the master data set. Thus, the 3D body scans were reviewed with the aim of removing 3D body scans of insufficient quality (i.e. scans with missing 3D data) and duplicate scans were removed by determining the best quality scan by participant identification number. This effort then enabled survey organizers to move forward with extraction of body measurements.

3.4.1 Body measurement extraction

The extraction of relevant body measurements is paramount to the ability to conduct analysis that provides value to the end users of national survey data. In the case of apparel industry stakeholders, it has been important to correlate body measurement definitions for automated extraction from 3D body scan data to that of manual methods employed by product developers. As previously mentioned, it's also been important to align extracted measurements with measurement methods specified via current industry standards and/or guidelines (e.g. ASTM or ISO). Given these requirements, flexibility with respect to measurement definitions is valuable.

In the context of SizeUK, SizeUSA and SizeMexico, software developed by [TC]^2 for automated measurement extraction was used to define and extract body measurements in reference to stakeholder needs. For each survey, one or more measurement extraction profiles (.mep files) were created in reference to the requirements, and measurements were extracted via a batch processing strategy. The extracted measurements were then sorted to identify obvious errors in primary measurements (e.g. unlikely values such as a 10" waist circumference) as an additional strategy for optimizing the quality of the data set.

In regard to flexibility, the [TC]^2 software allows for user defined measurements. This measurement extraction software references body landmarks, and definitions can be customized by editing the extraction parameters for a given definition (see Fig. 3.1). For example, when defining the position for extraction of waist circumference, the logic includes finding the smallest circumference within a given range. The software user may select from several options, one of which references the small of back. The software then allows the user to enable the waist position to float up and/or down at center front, sides and/or center back according to user defined limits. The

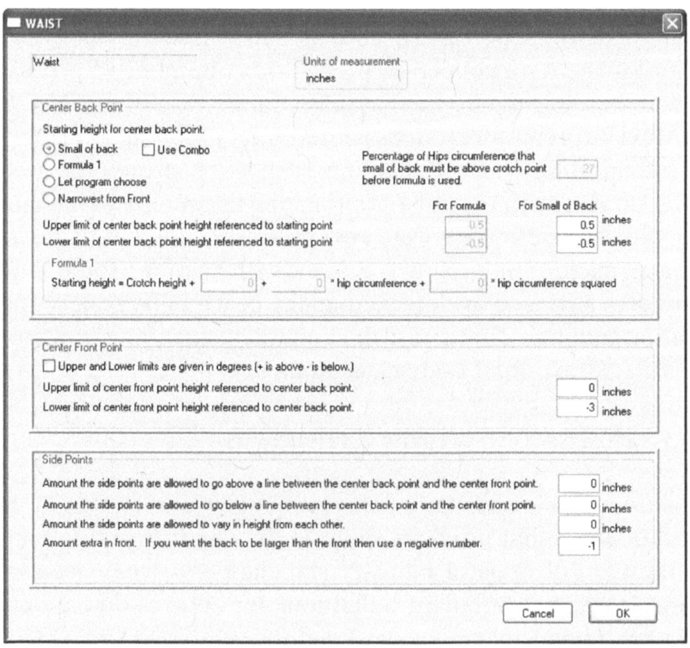

3.1 Measurement extraction parameters. (*Source*: Image provided courtesy of [TC]^2.)

logic for the definition varies by measurement and, for example, extraction of hip circumference involves finding the largest circumference between an upper limit (e.g. waist height) and lower limit (e.g. crotch level).

Users can also define new measurements based on those previously defined in the measurement extraction profile (.mep file). For example, the position for extraction of high hip circumference can be defined in reference to the waist height (e.g. 3" below waist). Measurements can also be defined through combinations or calculations. For example, a second waist measurement could be defined to incorporate negative ease of a given value – perhaps half an inch. Or, the vertical distance between two circumference measurements could be calculated by subtracting the height of the lower measurement from that of the higher measurement (e.g. vertical distance between crotch level and waist may be defined as equal to front waist height minus crotch height). The point here is that in order for the extracted measurements to be useful, the definitions for the measurements must be made available to the data's customers so that they can compare body measurements associated with sizing charts and fit models.

3.4.2 Analysis of body dimension and shape characteristics

Upon completion of the measurement extraction procedure, the resulting information can be married with the demographics for the survey to support analysis of data in reference to target populations. For the SizeUSA study, an initial round of analysis was conducted to provide general findings for the survey that included average and median values for the measurements extracted in reference to demographic groups (e.g. age groups for men and women). Standard deviations and percentile values were also calculated, and a summary of findings were included in the SizeUSA report supplied to sponsors of the survey and purchasers of the measurement data. A similar round of analysis was also conducted in reference to the SizeMexico survey upon completion of the measurement extraction process.

Beyond general analysis, the measurement data arising from national surveys can be analyzed for demographic groups in reference to size ranges. In order to conduct this type of analysis, researchers require access to body measurement specifications for the size range in question. The survey data can then be sorted or queried using the body measurement values to define fit tolerances for each size in the range. Body measurement values can be examined on an individual basis (e.g. waist only) and on a combined basis (e.g. bust/waist, waist/hip, or bust/waist/hip) to gain insight into typical body dimensions, proportions and shapes. Depending on the measurement definitions extracted, the analysis effort may include investigation of height, width, depth, and circumference values. For measurements such as waist, it may be possible to review the total circumference value, as well as the circumference separated into back and front lengths, to enhance understanding of typical body proportions. Similarly, comparison of width and depth values for waist circumference can help provide a sense of body shape at waist level. In other words, do the values for body width and depth at waist level suggest a wide, flat shape or round shape for the individual? Furthermore, visual review of 3D body scan data from the survey can serve as an additional strategy to obtaining insights into body shape characteristics (see Fig. 3.2).

Some of the analysis areas described will be discussed in greater detail in the context of applications for the data. Before launching into greater detail with respect to applications, keep in mind that it may be valuable to begin a research effort by focusing on a broad demographic (e.g. examining data for a target gender and broad age range). Once analysis has been completed for the larger population, product developers may wish to compare results to

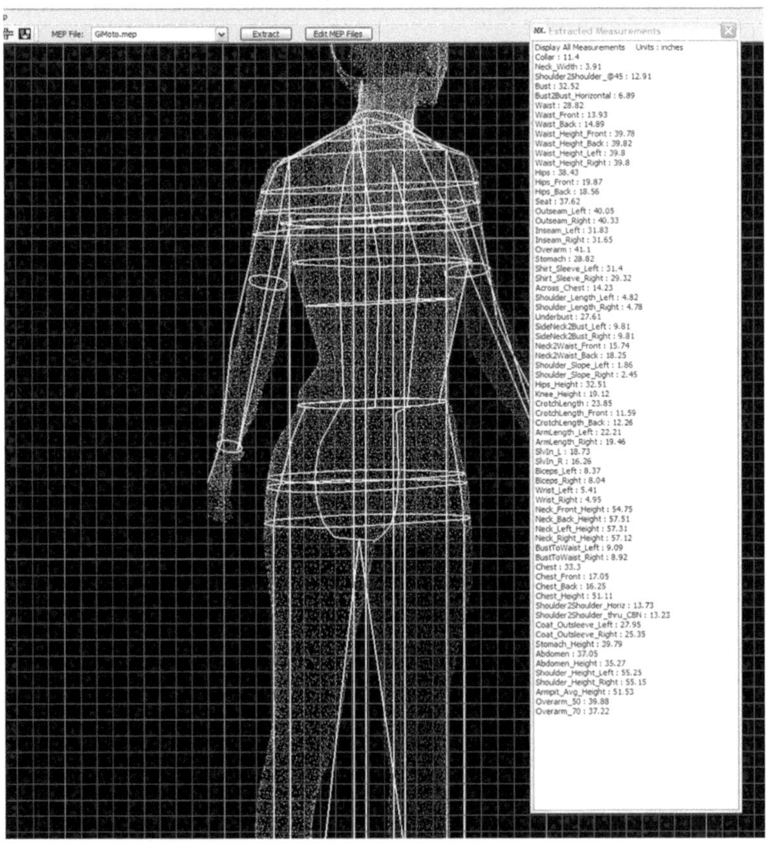

3.2 3D body scan data. (*Source*: Image provided courtesy of [TC]².)

those obtained for more specific demographic descriptions such as narrow age ranges and/or ethnic groups. The broad to narrow approach can be used to help validate sizing strategies with the aim of covering a varied population. This strategy for apparel sizing is sometimes described as democratic (i.e. a size specification that works well for a variety of demographic groups and body types). At the same time, the broad to narrow approach can also be used to identify opportunities to better serve specific demographic groups through the development of specialized garment fit and sizing.

3.5 Applications for apparel

In regard to national surveys, analysis for apparel applications has been a primary area for study among academic researchers, industry groups and supply chain participants. Although it is possible to extract a wealth of information from national survey data, applications for apparel typically center on a few primary areas. Analysis of the data for optimization of garment sizing strategies has been a focal point, and data analysis has been undertaken in regard to sizing standards, existing product lines, and new product development for apparel. As a result of advancements in software for 3D visualization and design, 3D body data from surveys is providing a basis for analysis in addition to the extracted measurement data. Specific strategies for data analysis are driven to a large degree by the application of the results. The following highlights some of the primary queries and areas for investigation.

3.5.1 Sizing standards and guidelines

Where the goals for a national survey include analysis of data in support of updates to standards or guidelines for apparel sizing, insights may be obtained by examining measurement trends for the population in reference to demographic factors such as gender and age group. Body measurement values for current standards can be compared to populations in the database to determine the percentage of the population covered by the standard. In the context of standards development, researchers may look for results that suggest the population is larger or smaller than the current specification or whether the body proportions outlined by primary measurement ratios (e.g. bust/waist/hip) align with the current standard. Researchers might also examine the data to determine the need, or inform standards development, for specific demographic groups or body types (e.g. plus size, big and tall, curvy, tall/narrow, etc). As an example of this, in January 2013 organizers of the Shape GB children's survey announced that Select Research and the sponsoring retailers collaborated with Alvanon to analyze data and produce mannequins for boys and girls, ages 4–17. Barnes notes that the analysis constitutes a major update on the 1990 published British Standard, which was based on data from 1978. He emphasizes that this consensus of opinion among major UK retailers offers a possible breakthrough in standardization of children's sizing through harmonization of the supplier base across retailers long-term (personal communication, 23 January 2013). Additionally, where data from multiple nations is available to researchers, and measurements extracted from the data align, survey results may be combined and/or compared in support of standards development for a broader geographic region (e.g. European Union surveys).

3.5.2 Development and refinement of garment sizing strategies

Outside the realm of sizing standards and guidelines, stakeholders are using survey data to inform development and optimization of sizing strategies for retail selling. In the context of the SizeUSA survey, custom analysis has been conducted by brands, retailers and product developers in reference to existing size ranges to determine correlation of specifications for key body measurements with populations in the SizeUSA database. Specifically, analysis can be conducted to determine how many survey participants for a given demographic meet the body measurement criteria for the size range. This total can be compared to the total number of participants surveyed for the same demographic, thus providing insight into missed sales opportunities. In some cases, the missed opportunity may be addressed through expansion of the size range at the top or bottom end.

Analysis can be conducted to determine correlation of measurements to a given demographic in the database on an individual measurement basis. In addition to finding the total number of participants that fit the body measurement criteria for the size range based on a single body measurement (e.g. waist or hip circumference), analysis may include how the population is distributed across the size range for the measurement examined (see Fig. 3.3). Distribution of participants across the size range lends support to discussions regarding most popular sizes for size assortments. To this end, brands and retailers may conduct size range analysis in reference to sales and return data by product type and size label to lend support to strategic decisions regarding size assortments and size offering.

In addition to analysis of body measurements on an individual basis, it is also valuable to analyze measurements on a combined basis to determine correlation to the target population. In other words, analysis can be conducted to determine how many participants fit the body measurement criteria for the size range when two or more measurements are combined. In

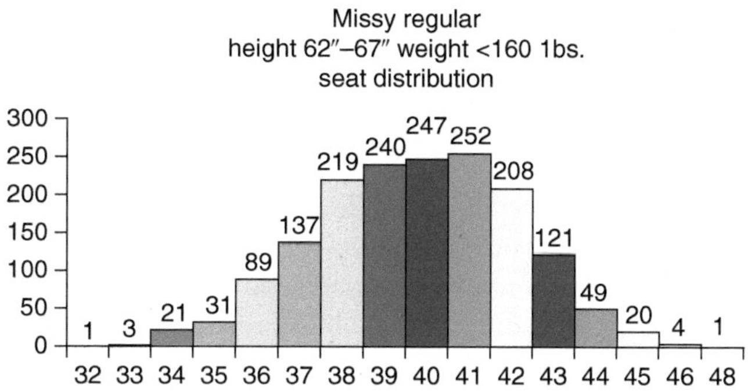

3.3 Population distribution by seat for SizeUSA women, defined as Missy Regular (Munro and White, 2007). (*Source*: Image provided courtesy of [TC]².)

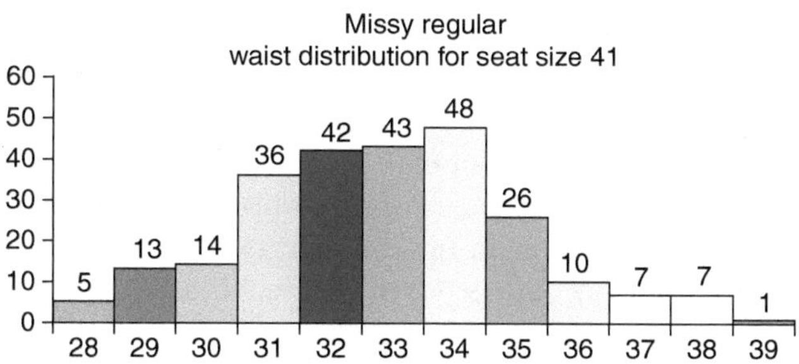

3.4 Population distribution by waist for SizeUSA women, defined as seat size 41 within the Missy Regular category (Munro and White, 2007). (*Source*: Image provided courtesy of [TC]².)

the case of ladies bottoms, analysis might be conducted in reference to combinations such as the waist and hip dimension, the waist and seat dimension (see Fig. 3.4), or the waist, hip and thigh dimension. Analysis of survey data based on measurement combinations provides insight into common body proportions by demographic group. Results of the analysis may suggest refinement of body measurement specifications to better serve the target market. Comparison of results for various demographic groups is a common point of interest with respect to body proportion and shape, and analysis may be used to identify opportunities to better serve the customer base through development of additional fits (e.g. curvy vs. straight) or product lines (mature or plus size).

In the context of fit development, the data can be analyzed with an eye toward non-linear grading strategies. This analysis might include an examination of best position for size breaks in addition to identification of optimal grading values between breaks for a given body measurement. Where specifications for garment measurements are available for key styles, the measurements can be compared to body measurement tolerances to determine range of ease for each size in the range. For example, if size 10 fits bust dimension 34.1" to 36", and the garment measures 38" at bust level, the range of ease for the garment is 2" to 3.9" at bust level. Keep in mind that this type of analysis does not provide the complete story in regard to comfort, since fabric properties and factors such as construction features also contribute. Additionally, analysis can be used to inform optimization of specifications for body measurements such as crotch height or inseam that do not directly correlate to an increase in body size (i.e. larger members of the population do not necessarily have longer legs or longer arms). In the case of inseam, the goal may be to optimize coverage of the population while minimizing material costs at the production level.

As previously stated, the availability of survey data for multiple nations is a resource for researchers and groups working on the development of sizing standards such as the international size label. This information is also an asset for brands and retailers. Businesses developing product for sale in multiple nations may examine the relevant data to determine how well their existing specification covers the population for the region in question. In this context, the research aim includes optimization of sizing specifications for a region or group of regions in order to increase product sales for the area. Where national survey data is not available, brands and retailers may select to conduct private surveys for a region to ensure that the product developed appeals to the target population in terms of garment fit and size.

In some cases, members of a product development team note recurring fit issues as part of the development process and may have thoughts or opinions regarding sizing methods and missed sales opportunities. Analysis of national survey data can be used to lend support to or validate new directions and updates to sizing in response. Where modifications to a sizing strategy are proposed as a result of an initial round of analysis, subsequent analysis can be undertaken to verify that the proposed change is likely to result in improved coverage for the target population as compared to the original specification. Product developers can take steps to further validate a change through development of test products using the revised specifications and through product evaluation via focus groups and similar channels. Clearly, the validation stage is important in that it contributes to a level of confidence that the proposed change will be successful at retail.

3.5.3 New product development

For established brands, retailers and product developers, analysis of national survey data may be undertaken to identify body dimension and shape characteristics of the population in reference to a new product or line. In this context, the target consumer may be the same as or similar to the existing customer base. In other instances, the business is targeting a new population. For both scenarios, body measurement specifications from an established product line can be used as an initial point for comparison. When developing new products, brands and retailers are sometimes interested in measurements for specific regions of the body. For example, expansion of a business into the foundations or lingerie area will drive an interest in bust measurements, whereas development of a line of athletic performance wear may drive examination of measurements that provide insight into musculature in the chest, shoulder, neck and arm region.

In some cases, analysis for new product development may be undertaken on behalf of an entrepreneur. It's common for entrepreneurs to start with product ideas that serve an identified need or niche market. However, they may not have a clear understanding of who that customer is from a demographic perspective (e.g. What is the typical age range and income level for plus size women?) or in terms of body shape and dimension (e.g. What does the population distribution look like for bust measurements among women 18 to 35?). In this context, the goal often relates to development of a sizing scheme that best fits the target population, and industry sizing guidelines (e.g. ASTM) and product lines currently in the market often serve as a valuable point of reference for conducting analysis.

3.5.4 3D visualization

Additional investigation of body shape can be conducted by reviewing the 3D body scan data for a survey participant or a group of participants that meets body measurement criteria (e.g. bust/waist/hip definition for a given size) and/or demographic description (e.g. Hispanic women, ages 18 to 25). Within the context of [TC]^2's software for 3D body scanning and measurement extraction, the body data can be opened and rotated for viewing from multiple angles. It is also possible to create slices of the data at specified heights to obtain an unobstructed view of body shape at a given level (see Fig. 3.5). Furthermore, slices can be obtained for export as .dxf files to support additional analysis of shape and dimension using standard and industry specific CAD packages.

The [TC]^2 software also supports rapid generation of avatars in reference to the 3D body scan to create a digital representation of the participant that is accurate in terms of body dimension and shape. These avatars can be exported in a standard 3D format (i.e. .obj file) for further review in 3D software solutions, including various apparel CAD applications for 3D product

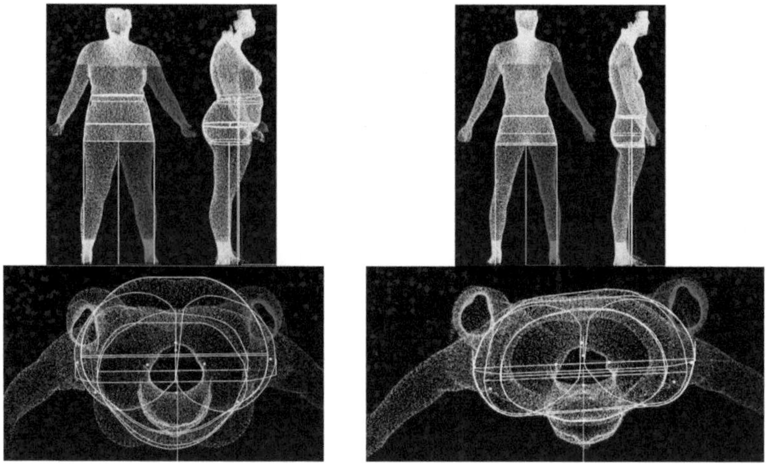

3.5 Example of a cross-section extracted from 3D body data. (*Source*: Image provided courtesy of [TC]^2.)

development. Human Solutions technology also supports generation of 3D models, called 'Scanatars', that are described as accurate representations of individuals in terms of body shape and dimension (M. van Genabith, personal communication, 2 February 2013). In the context of research, analysis of 3D data can be undertaken with the aim of understanding variations in body shape among and between demographic groups. Moving into product development, scan data from national surveys can be used to help validate fit for a range of body types and dimensions. In essence, the data provides a large pool of virtual fit models and, when viewing bodies of similar size, it is often possible to see how body shape varies despite similarities in key measurement values (see Fig. 3.6).

Looking forward, 3D data from national surveys may be used as a basis for 3D to 2D pattern development. Pattern unwrapping software has been used for more industrial applications, including development of upholstered furniture, automotive seating and aerospace (King, 19 October 2012). In the context of apparel, 3D software is moving beyond the ability to virtually stitch garment patterns and drape on virtual fit models. Vendors are now beginning to enable greater interaction between the 3D and 2D window. In some cases, it is even possible to outline pattern shapes on the 3D form and unwrap the pattern for continued development in 2D (King, 26 October 2012). As this technology area advances, 3D body data from surveys can serve as a valuable resource for 3D garment pattern development and grading. Looking ahead, 3D body data from surveys may also serve as a source for manufacturing in emerging technology areas such as 3D knitting and 3D printing.

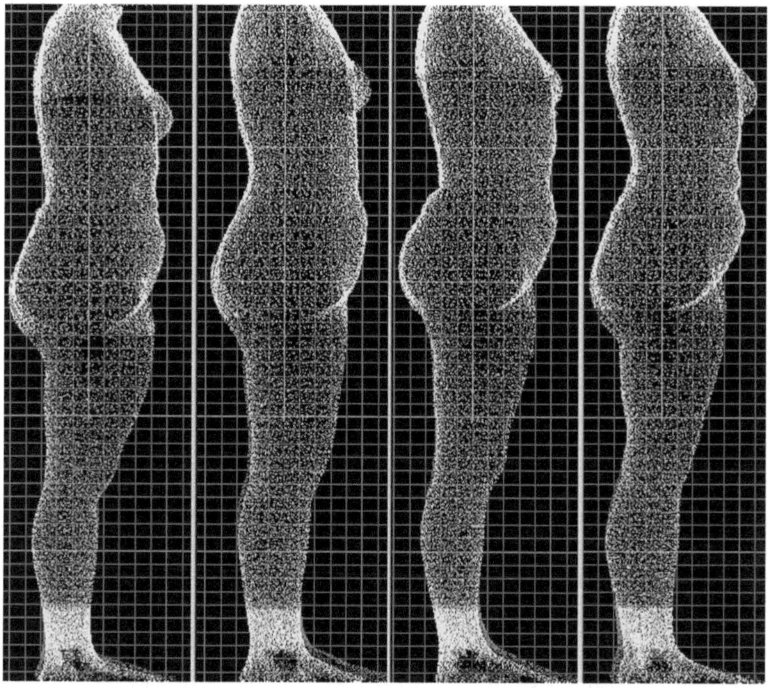

3.6 3D body scans illustrating difference in body shape despite similarity in key body measurements. (*Source*: Image provided courtesy of [TC]^2.)

3.6 Other applications

Though acquisition and use of national survey data for apparel applications is the focus of this chapter, the data has relevance to other fields as well. As for apparel use, data can be analyzed according to demographic descriptions to gain insight into body dimension and shape characteristics with respect to medical, health and fitness. In the case of obesity, it may be valuable to understand how bodies change in shape and dimension in reference to height and weight values. Furthermore, [TC]^2's measurement extraction software also enables extraction of body volumes and, in combination with key measurements, calculates a body fat percentage based on a strategy developed by the US Department of Defense. The body volume extraction includes estimates for the head, hands and feet, since these are areas for which incomplete data was obtained.

For the SizeUSA survey, participants were asked to describe their activity level as part of the demographic portion of the survey. As a result, it may also be possible to glean information regarding self perception of fitness as compared to factors such as body shape, dimension, weight and height.

Additionally, it may be possible to review the data with the aim of identifying body shape and measurement characteristics that are potential indicators of health and fitness. To that end, visual review of 3D body data may also provide insight into body shape with respect to distribution of weight and musculature.

In the areas of industrial or functional design, data from 3D body scans can be reviewed for development of products with improved ergonomics and/or functionality. Functionality can be applied to apparel applications such as uniforms or work wear (e.g. military, hotel or service industries), clothing for safety and protection (e.g. fire or ballistics), and medical use (e.g. pressure garments), and performance athletic apparel (e.g. swimwear, running, and training garments). Functionality also applies to non-apparel areas too, and data from national surveys is a potentially valuable resource for development of products that interact with the body in some way. For example, a company that develops outdoor sporting and camping equipment might select to investigate specific measurements such as waist height value and torso length value for target demographic groups (e.g. women vs. men between the ages of 18 and 45). This information could be used to optimize the fit tolerance and engineering of backpacks or sleeping pads. Similarly, developers of medical products may be in need of body measurement data to support development of devices that fit to the body (e.g. splints, canes and walkers).

Data analysis for industrial and functional reasons often requires examination of a small number of key measurements. In some cases, the required measurements for these applications do not correspond to those measurements extracted for apparel. With that in mind, the ability to obtain custom extractions post survey has been of particular value to non-apparel and functional areas. The availability of 3D body data has also been important for this area given the use of 3D CAD solutions for product design. Furthermore, national surveys that scanned participants in both standing and seated poses offer particular value with respect to development of seating for transportation and work environments.

3.7 Conclusion

The availability of body measurement and 3D data from national surveys provides an almost endless opportunity for analysis. Thus, it is important for stakeholders to gain an understanding of the wealth of data available from various surveys. Most importantly, careful consideration must be given to the research question/s in reference to consumer and market needs to ensure that the right kind of information is examined, and that the strategy for analysis will shed light on the research question/s. Furthermore, the advancement of software solutions for 3D visualization and design is a critical technology area for monitoring with respect to apparel stakeholders specifically. These systems will play a critical role in the ability to make greater use of the 3D data arising from national surveys in support of improvement to garment fit and sizing.

3.8 References

Bougourd, J. and Treleaven, P. (2010). UK National Sizing Survey – SizeUK. International Conference on 3D Body Scanning Technologies, Lugano, Switzerland.

Consumer Outlook^{SM} '99, KSA's Annual Consumer Outlook^{SM} Survey Results. (1999). Atlanta: Kurt Salmon Associates.

King, K. (19 October 2012). 3D Pattern Flattening. Retrieved from http://www.cottonsrevolutions.org/applications/blog/Technology/2012-10-19/3D-Pattern-Flattening.

King, K. (26 October 2012). 3D Pattern Flattening Part II: 3D to 2D for Apparel. Retrieved from http://www.cottonsrevolutions.org/applications/blog/Technology/2012-10-26/Pattern-Flattening-Part-II-3D-to-2D-for-Apparel.

Munro, K. D. and White, E. M. (May 22, 2007). Lower Body Measurements for Women's Apparel. Cary, NC: [TC]^2

Perez, L. and Rodriguez, A., (31 August 2011). [TC]^2 Activity – Intermoda Report. [TC]^2 Technology Communicator. Retrieved from http://www.tc2.com/newsletter/2011/083111.html.

Robinette, K. M., Blackwell, S., Daanen, H., Boehmer, M., Fleming, S., Brill, T., Hoeferlin, D. and Burnsides, D. (2002). Civilian American and European Surface Anthropometry Resource (CAESAR) Final Report, Volume I: Summary. Retrieved from http://www.humanics-es.com/CAESARvol1.pdf.

Seidl, A., Trieb, R. and Wirsching, H. (2009). SizeGERMANY – the new German Anthropometric Survey: Conceptual Design, Implementation and Results, IEA 2009, 17th World Congress on Ergonomics, 9–14 August 2009, Beijing, China.

Sizemic (2013). General Format. Retrieved from http://www.sizemic.eu/sizeuk.html.

Size UK (n.d.). UK National Sizing Survey Information Document. Retrieved from http://www.size.org/SizeUKInformationV8.pdf.

Shape GB (2013). Shape GB – Measuring The Nation. Retrieved from http://www.shapegb.org/b/.

Textile Clothing Technology Corporation (1 July 2010). Qué Talla es México? A Sizing Study of the Population of Mexico. Retrieved from http://www.tc2.com/pdf/sizemx.pdf.

Textile Clothing Technology Corporation (February 2004). The National Sizing Survey, Body Measurement and Data Analysis Reports on the US Population. Cary, NC: [TC]^2

4

Body shape and weight distribution: the Body Volume Index (BVI) and the Body Mass Index (BMI)

R. BARNES, Select Research Ltd, England

DOI: 10.1533/9781782422150.1.58

Abstract: The Body Volume Index (BVI) is a new obesity measurement, specifically designed to use 3D technology and create a viable and modern day application to measure and predict obesity or the health risk of an individual. With its ability to measure and produce replication of body shape, 3D technology allows volume and weight distribution of a patient to be assessed, thereby creating a measurement more attributable to that patient's body shape and their inherent health risks as a result of their individual weight distribution.

Key words: Body Mass Index (BMI), Body Volume Index (BVI), obesity, body shape, obesity prevention, obesity measurement, body fat, body volume and body composition.

4.1 Introduction

The number of people in the world who aren't worried about their weight at some point in their lives are a very small minority and, for those that are, a change in weight acts as a barometer for self-esteem, self-perception and whether their future health is potentially at risk. To assess when extra weight has become an issue, most people notice a difference either when they look at themselves in the mirror, when they compare themselves to old photographs or when they stand on the weighing scales. Useful as these tools are, in healthcare the use of self-image in relation to weight gain or loss is not as yet captured or encapsulated in a form that can make sense of the paradox where muscle weighs more than fat, extra weight around the stomach area is accepted as creating the greatest risk, but areas of the body where extra muscle weight is achieved through exercise are not measured. The Body Mass Index (BMI), currently used as the de facto risk indicator for a person's health, uses height and weight only, with weight centred on total weight as opposed to distributional weight. Therein is a paradox where extra weight around the abdominal area, with its emphasis on additional fat content, is not measured by BMI, and indeed by no other measurement, until now.

The Body Volume Index (BVI) is a new obesity measurement using 3D technology, developed by Select Research, the leading company in the UK for the management of large-scale 3D sizing surveys in the retail field. Select have recruited and scanned over 28000 men, women and children for major UK retailers since 1997, which has included direct involvement in management of both national sizing surveys in the UK, the first of their kind in the world.

Select were commissioned by UK retailers in 2001 to manage recruitment and sampling for SizeUK – the National Adult Survey – as we had recruited and managed five sizing surveys for M&S and Next before that. As a result of our experience, we were again commissioned by the retailers in 2008 to undertake the UK National Childrenswear Survey. This project is called Shape GB6 – in recognition of the importance of body shape, and to represent the geographical area of the representative sample.

However, in parallel to our extensive work in retail, and in a totally different field, since the start of the millennium Select have been pioneering the use of 3D scanning technology in healthcare, and specifically for the measurement of obesity with a new measurement; the BVI. The idea for BVI was originally conceived in January 2000, but development was delayed until we had completed our commitments to M&S on their Lingerie Survey, and to the seventeen retailers in SizeUK who commissioned us to help with that project in early 2001.

At that time, we recognised that use of imaging such as MRI (magnetic resonance imaging) and DEXA (dual-energy X-ray absorptiometry) had permeated healthcare and clinical decision-making and provided unparalleled data and insight into the condition of a patient, using leading edge technology to assist in the management of healthcare and patient care.

However, in contrast, risk assessment and measurement of obesity relied almost entirely on a manual measurement, the BMI, devised between 1830 and 1850 as a broad based public health measurement tool. Measuring risk based only on height and weight, with no allowance for differences in gender, age or ethnicity, the use of manual techniques and the accuracy of BMI as a risk indicator has since been called into question.[1,2] Now, in the technological age, doubts have been voiced as to whether we should still be so reliant on manual measurement for designation and evaluation of health risk.

Technology in healthcare is increasingly being used in measurement of an individual. However, in the initial stages of risk assessment for certain conditions such as obesity, healthcare professionals continue to remain reliant almost exclusively upon manual measurements. This is particularly true for obesity, where manual measurements of waist circumference, waist to hip ratio and BMI are currently used for assessment and diagnosis. However, recently published reports[3,4] have instigated a steady change of opinion in the medical profession on the usefulness of BMI and prompted debate that a long-term replacement of BMI for the diagnosis of obesity and its associated health risk is necessary.

All those who work or have worked in the apparel industry within recent years will be aware of 3D scanning and the effects this has had on data collection, production of mannequins, processing of grading and garment design and feedback to customers, including management of returns. Although there is awareness and acceptance that the usefulness of BMI is somewhat limited, technological advances in healthcare take longer to become acknowledged and manual measurement is used almost out of habit or by default. There are plans to expand and provide more widespread availability of BVI by 2014, so an extensive use of 3D scanning in healthcare can only benefit the apparel industry long term, as patients and physicians become used to 3D scanning and will, through their own choice, be able to use and apply their 3D scan to shop for bespoke clothes in stores, online, or through mobile technology. Body Volume Index is a specific application in healthcare, but is also designed to be a subsequent patient oriented apparel application allowing patients to use their own scans, subject to the usual consent and regulatory requirements. This would prove to be especially beneficial to those of a distinct body shape, who may have difficulty finding a perfect fit among garments that are generally constructed to a certain generic shape. The route to achieving that benefit does not necessarily have to start from a clothing perspective, and as BVI becomes more widely adopted, apparel manufacturers will be able to target a certain body shape or use BVI to mass customise garments to better fit individuals.

4.2 The Body Mass Index (BMI) and the development of the Body Volume Index (BVI)

The instigation for BVI came from a realisation that BMI had been developed in 1830 and that 3D scanners, available since the early 1990s and used since 1997, offered new insights into measurement that just were not possible in 1830. From an idea conceived in February 2000, progress on development of BVI started in earnest in September 2000.

4.2.1 Body Mass Index (BMI) as a measure of body shape: strengths and weaknesses

The inherent strength of BMI as a measurement tool lies in its simplicity. Using just height and weight as component parts of the measurement requires only a height measure and a set of weighing scales in order to collect the data required to provide the ratio. Although international interest in classifying subject health status according to adiposity is increasing, no accepted published ranges of percentage body fat currently exist, and attempts have been made to re-classify BMI with newly defined body fat percentage references. A study published in 2000 examined a potential new approach for developing percentage body fat ranges, by linking healthy BMI guidelines established by the National Institutes of Health and the World Health Organisation with predicted percentage body fat.[14]

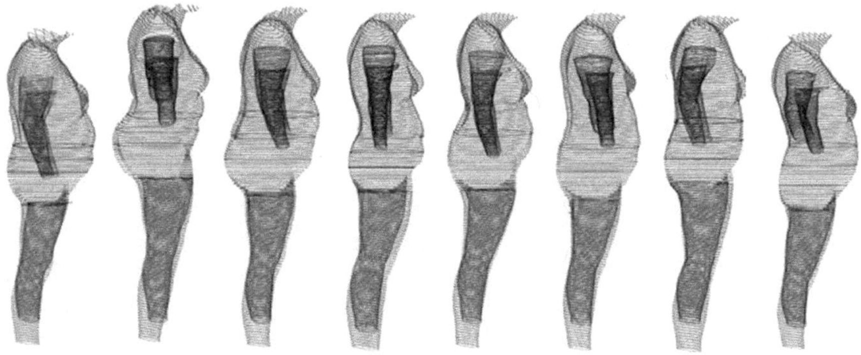

4.1 Eight women with a BMI of 30, the threshold for obesity, highlighting different body shape.

Although valid in both approach and integrity, this study using BMI as the baseline for body fat percentage calculations inherently implies that percentage body fat calculated for the whole body creates better indication of risk, but this cannot by its nature fully reflect body shape and weight distribution, which would determine where the fat is located on the body. That is the premise of BVI: to be reflective of body shape and weight distribution irrespective of a BMI rating. Differences in body shape, weight and fat distribution can be considerable even between patients of the same gender and same BMI rating (Fig. 4.1).

4.2.2 The origins of the BVI

In 1997 we were commissioned by Next, a leading UK retailer, to undertake a Womenswear Sizing Survey using 2D scanners, and this led in 1999 to completion of four separate sizing surveys (Menswear, Womenswear, Lingerie and Childrenswear) for Marks & Spencer, the UK's largest clothing retailer. Select successfully completed two sizing surveys for M&S – Womenswear and Menswear – before they commissioned us to undertake the Childrenswear Sizing Survey. The Childrenswear Survey consisted of the recruitment and measurement of 3500 children aged 0–16 in the UK, the first major collection of measurements for children's clothes since 1978.

The M&S Childrenswear Sizing Survey was set up within eight weeks and involved measurement of children primarily in the school environment, with some measurements of 0–4 year olds undertaken using interviewers employed by Select. The three previous adult surveys for M&S had used 2D and 3D scanners supplied by Telmat and Wicks & Wilson, but, after much deliberation, it was decided that given the sensitivity of using scanners whereby children would have to undress, manual measurement using trained nurses would be a more pragmatic and sensible approach for this survey.

Data collection for the survey began in late 1999, and during the initial phases of the project, it became apparent from media coverage of the sizing surveys we had undertaken for M&S that 'obesity' and 'childhood obesity' were public health issues, albeit not as prevalent and high profile as they are now. As a specialist in the 3D field, I started to contemplate the options of using 3D scanners to measure obesity in healthcare.

The non-use of 3D scanners on the childrenswear survey for M&S, and the importance and direct use of them on the previous three adult surveys, highlighted the potential of using 3D scanners for measurement of obesity and health risk in both children and adults.

As a result of this, after data collection had finished on the childrenswear survey I authorised an internal research project to investigate the potential use of 3D scanning technology for healthcare. We decided to research the concept with patients, general practitioners and hospital doctors in focus groups and individual interviews at the beginning of the millennium.

4.2.3 Research and development of the BVI concept

The aforementioned research project commenced in mid-January 2000, and started with a series of in-depth telephone interviews with patients and general practitioners. The interviews were set up by our research team and conducted by me across the country to ascertain:

- the awareness of 3D scanners given the high profile of the M&S sizing surveys;
- the use and purposes of measurement within healthcare settings;
- the potential uses of 3D scanning technology within healthcare settings;
- the barriers and restrictions to usage given the prominence of the state funded National Health Service (NHS) in the UK;
- the existing uses and applications of the BMI as a risk indicator in healthcare.

Following these initial interviews, focus groups were set up in late January 2000 in Birmingham, UK, amongst nurses, general practitioners (GPs) and orthopaedic consultants from a selection of local hospitals. The focus groups lasted 1.5 h and were open ended in terms of content to explore the potential benefits and impacts of using 3D scanners in healthcare. I moderated the focus groups at a viewing facility to allow for both filming and audio recording of the research.

From the research, it became apparent that there was a scope for 3D scanning technology in healthcare for specific applications such as prosthetics, osteoporosis, and physiotherapy. Moreover, it emerged that respondents on the whole acknowledged that at that time there was no cohesive and identifiable measurement process used within healthcare that focussed on the individual health risks of a patient. The management and use of manual techniques such as the BMI, waist circumference and the recently publicised

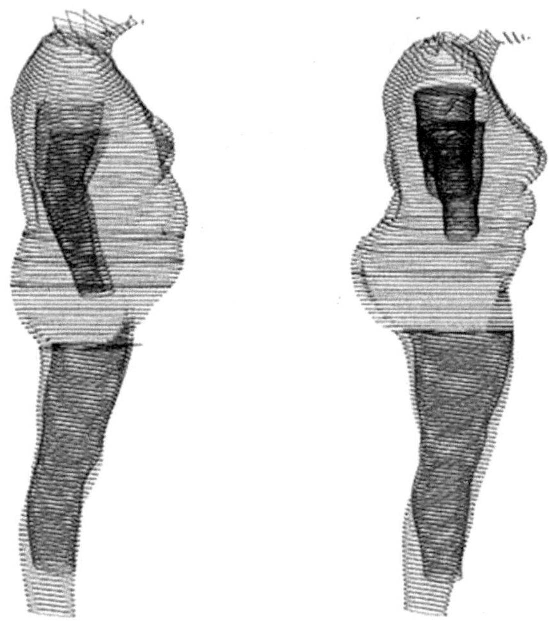

4.2 BMI does not allow differentiation between two individuals with exactly the same BMI (>30) but different weight distribution.

waist to hip ratio were undertaken out of an obligation to national and international guidelines rather than as a conscious or rational choice.

In the following weeks, I listened again to recordings of the focus groups, and it became immediately evident to me that there was a recognisable opportunity for a general improvement of the collection and delivery of measurement dat within the NHS. Reviewing the options of using 3D scanners to assist and alleviate that need, I explored the options for using scanners to address what I believed was the key issue, namely the measurement processes and applications themselves. In short, I wondered if it would be possible to measure automatically as opposed to manually and, if so, what would be the best and most simple means of measuring the human body in healthcare?

Using my 15 years' experience of applying research solutions to solve difficult marketing and product issues for other clients, I ascertained that a simple and easily understood measurement, like BMI, was a pre-requisite. Although there was an undoubted acceptance and recognition from the respondents interviewed that 3D scanners could offer a new dimension to existing measurement techniques, the realisation was that an 'add on' to an existing measurement was in itself a barrier to adoption. Manual measurement, being cost-free and quick and simple to use, effectively meant that usage would in practice not change in the patient care pathway. However, 3D scanning, if adopted as an 'extra' element to BMI, would have the

4.3 The leaflet used by Select to recruit 11 000 participants for the SizeUK project.

potential knock on effect of increasing waiting times and costs to the NHS, both immediate barriers to adoption on a large scale.

We therefore explored the possibility of creating a completely 'new' measurement, to effectively enhance and possibly replace rather than augment the existing manual measurements, and considered what might be the best option. During this evaluation we explored all options for measurement of the human body using a 3D scan and as a result theorised that measurement of volume, as opposed to linear measurement as used for apparel manufacture and design, was probably the best option for creation of a new measurement. Effectively, we decided that to measure the body in reverse to standard applications in retail (where external measurements are paramount) would be the best option for healthcare. In particular, I assessed that adapting 3D scanners to measure part volumes, and the abdominal volume, of a patient's body would provide a measure of weight distribution, a key element to establishing genuine health risk.

We envisaged that measuring internal volume by body part could measure partial weight distribution, and therefore, for the first time, take the individual body shape of a patient into account when assessing their health risk (Fig. 4.2). The idea for BVI was born and I immediately realised the potential impact of this were it to eventually be developed as a new measurement of obesity to enhance and possibly replace manual methods of data collection. In order to simplify the understanding of the measurement, I decided it had to be assigned a designation to emphasise both the link and the difference between the old and the new; therefore I decided on the term 'Body Volume Index' to define and describe the potential new measurement. The abbreviation, BVI, was a simple and effective comparator to the abbreviation for the Body Mass Index – BMI.

I continued to work on calculations and software options for BVI to ascertain how best the internal 'space' of the 3D scan could be filled in to ensure that body composition was both accurate and appropriate to healthcare. Development continued throughout 2000 and early 2001, but was delayed due to the commissioning of the M&S Bra Survey in 2000 and the SizeUK project in 2001.

However, I continued to ponder how to turn my idea into a reality and the best means to achieve this.

Therefore, when our commitment on SizeUK (Fig 4.3) was completed in April 2004, the Chair of the Association for the Study of Obesity, following a meeting and request from me, sent 350 letters to their members across the world inviting potential collaborators to join us in the research and development of BVI. Thus commenced the preliminary research programme for BVI, culminating in the public launch of The Body Benchmark Study on 23 March 2007 at The Institute of Public Health in London to formally announce BVI as a potential new standard for measurement of obesity and health risk with a team of internationally recognised healthcare experts (Fig. 4.4).

4.2.4 Additional research and validation

In 2006 and 2007 we undertook qualitative research with healthcare practitioners in many disciplines which indicated that a 3D scan of the same patient taken for their BVI could potentially be used for review of anaesthetic dosage pre-operation, posture and osteoporosis issues, artificial limb construction and physiotherapy, in addition to general health risks focussed on central (abdominal) obesity.

Further qualitative research undertaken in 2006 also highlighted that waiting times for operations could be reduced and more effective scheduling achieved by using BVI scans at pre-operative level. BMI is the diagnostic reference point for the scheduling of patients' pre-op on a daily basis, but using BVI offers the potential for patient scans to be assessed

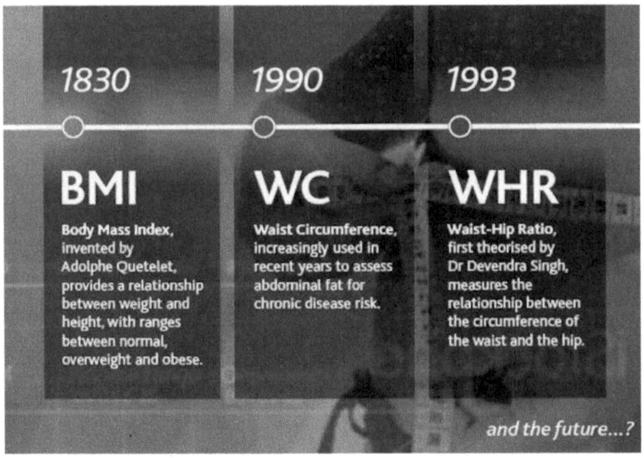

4.4 The leaflet used to promote The Body Benchmark Study highlighting the current manual obesity measures. The public launch date for BVI was on the reverse.

by an anaesthetist in advance of theatre. Weight distribution (measured by BVI) might be a key factor in deciding the length of time an adult patient might require in theatre and therefore which patients should be operated on first.

The ease of the scanning process is equally relevant with children as with adults. Between March 2009 and August 2010, Select Research scanned over 2500 boys and girls aged 4–17 on Shape GB (Fig. 4.5) with no adverse complaints from children or their parents.

This might be significant for a longer term review of the current National Child Measurement Programme (NCMP) in the UK and childhood obesity criteria themselves. The use of BVI may help alleviate some of the well publicised issues in the UK media regarding letters sent to parents about their supposedly obese children, where BMI has been incorrectly used as the reference point to determine whether a child is overweight or not.

Additional research and validation (as yet not in the public domain) compares BVI with current techniques for body composition analysis, such as MRI, DEXA, and the correlation between body volume and cardiovascular health risk. The most comprehensive data of real body composition statistics in the world will be used exclusively as a baseline tool within BVI to ensure that empirical data is used as the cornerstone of the body composition element of BVI. Further validation, using CT (computerised tomography), MRI, DEXA and Bio-Impedance, is currently being undertaken by the five healthcare collaborators in various independent and collaborative trials using the five 3D scanners set up for The Body Benchmark Study.

4.5 The collaborators on Shape GB – The National Childrenswear Survey, subsequently to include additional retailers, M&S and Tesco.

4.2.5 3D scanners, BVI and online developments

Five Tc2 NX16 scanners were used as a baseline data collection tool as we were familiar with Tc2 scanners from our work on SizeUK, but these required sophisticated software to analyse part volume using the body composition data available to us; so new BVI software was developed to allow for segmental analysis of the human body and calculate body composition by body part to create a health risk indicator. Health data including age, gender and ethnic origin, and surrogate markers for cardiovascular disease such as lipid profile, blood pressure and heart rate, were also included.

Using standard data capture through sensors, body volume imaging is non-invasive and creates an exact 'virtual' image of a person's shape. Up to 200 separate linear data measurements can also be extracted. The BVI scanner is described as such because it measures the 'inside' or the 'volume' of the body, in contrast to the scanning applications historically used in the retail field, which measure the outside of the body for clothing.

The scanner can accurately measure people up to 205 kg in weight or up to a 150 cm (60") waist circumference, so more than appropriate for all those except the most obese of patients.

The process is completely harmless, and the quality of data far supersedes the use of manual techniques where operator error and inconsistency produce greater variation of data. Validation by Mayo Clinic on the data output of the scanner[7] has indicated that in a direct comparison between manual and automatic measurement, there is less deviation from the BVI scanner measurements than from manual measurements of the same.

4.2.6 The significance of 'national' and 'benchmark' as reference criteria in sizing

The use of the words 'national' and 'benchmark' in the seven large sizing surveys we have managed are synonymous with the purpose of the surveys and the data sampling needs of the clients we were commissioned by.

SizeUK in 2001 and Shape GB in 2009 were both sampled by us to ensure they were broadly representative of the national population, and the experience gleaned from this work and my 25 years' experience in market research enabled us to establish the most suitable reference for BVI.
Essentially, BVI as an application has two reference points:

- **The Patient** – BVI is referenced to the average part volume distribution of the patient's own peer group based on age, ethnicity and gender. They are compared to their own peer group as represented by that time period in their lives by age.
- **The Population** – BVI records are effectively stored anonymously for public health purposes with a reference number that can identify the individual statistics for BMI, waist circumference, waist to hip ratio, BVI, ethnicity, gender and age. This allows for multi-faceted data analysis from a single data collection point and a single data source.

The above criteria ensure that BVI can effectively serve both as a one-stop reference point for anthropometric data in healthcare and as an individual patient record.

4.3 Research on the use of the BVI

Measuring volumes or part-volumes in itself is insufficient to deliver a meaningful measurement for healthcare and so the measurement and meaningfulness of 'body volume' itself was the key requirement for research. The research study was designed to measure the volume, the segmented volumes and the contents within the 'spaces' of those volumes of the human body.

4.3.1 Research on BVI prior to the Body Benchmark Study

A major new study, The Body Benchmark Study, was launched in March 2007 at the Institute of Public Health to create the BVI, the world's first computer based anthropometric measurement for healthcare using 3D scanners that allows assessment of obesity based on body volume, body composition and body shape. Prior to that, over four years of preliminary work had been undertaken following the letter by the Association for the Study of Obesity in September 2004.

Patients were consulted during extensive qualitative research undertaken by Select Research in 2006, and their insights were used to formulate and create the delivery structure of the BVI application. Research was conducted amongst 85 patients using a series of focus groups and in-depth interviews in a viewing facility, and the sample profile covered all ages, both genders and different socio-economic groupings.

This research highlighted that patients acknowledged and understood the use of BMI as a measure of obesity, but were unaware of its definition and therefore unappreciative of its limitations in measuring obesity. However, when presented with a scenario whereby BMI could be replaced by an alternative, the desire from respondents was for a simple and understandable measure, hence the emphasis on abdominal volume as a percentage of overall volume within BVI. All respondents understood the simple notion of something designed to measure 'the more weight around your stomach, so the greater the health risk'.

Further quantitative research was undertaken over 6 months by researchers from the University of Warwick, on the trial in 2006[8] focussing on whether 3D images had assisted individual patients in addressing their own weight and dietary issues. To augment this research, each patient was given an individual consultancy session with a qualified dietician at 0, 3 and 6 month intervals between scans. The results were inconclusive given the small sample size, but it was evident that some patients were motivated to lose more weight by managing their diet more effectively, and the key motivating factor for them had been the use of their previous 3D scans for comparative purposes.

A prototype version of BVI was developed using the initial calculations from 2000 to 2004 and on the principle that abdominal volume, the section of the body identified as being crucial for diagnosis of central obesity, was attributed with an importance factor. The BVI calculus, algorithms and ratios were planned to interact to provide a predictive risk factor based on the level of abdominal volume (as a percentage of the total body) taking at least age, ethnic origin and gender into account; BMI does not accommodate this level of differentiation. Chest volume was also analysed to ensure that people of athletic build (but with a high BMI) are designated with an appropriate BVI figure, and thigh and leg volume were also contributing factors in any BVI value.

4.3.2 The Body Benchmark Validation Study

Despite this extensive work on BVI, we realised that to ensure widespread adoption and acceptance of BVI, healthcare professionals across the world in appropriate disciplines would be required to validate BVI long term. Therefore, The Body Benchmark Study was set up as collaboration between world-renowned healthcare research facilities and hospitals in the US, the UK and mainland Europe. Some of these, such as Heartlands Hospital in the UK and Mayo Clinic in Rochester, Minnesota, USA, have published data, and other healthcare specialists in child obesity, diabetes, body composition and risk assessment throughout the world will be doing so over the next few years.

The main objectives of the Body Benchmark Study were:

- To validate BVI against established manual measurement alternatives such as BMI, waist/hip ratio and waist circumference and volume measurement devices such as The Bodpod.
- To automate data collection of body measurements and allow for national and international statistical analysis of BVI by age, gender, ethnicity and body shape.
- To establish a series of ratios and relationships between part volumes created by a BVI scan, e.g. abdominal volume, and correlate these with surrogate markers for cardiovascular health risk.
- To validate health risk for an individual by taking into account body shape, distribution of body volume and fat, and age, gender and ethnicity.
- To integrate body composition and body scan data by direct means from cadaver analysis for total and body part volumes.

The published results of the Body Benchmark Study were presented at a major media launch broadcast online on 12 October 2010. Details of the launch and the published results can be found at www.bodyvolume.com.

4.3.3 Financial implications of obesity and the costs to the health sector

The costs of obesity and associated health risks to the NHS were expected to rise to £3.6 billion a year by 2010, and HM Government's Foresight report[9] suggested that if current trends continue, almost 9 in 10 adults and two-thirds of children will be obese by 2050, resulting in an overall cost to society of £50 bn/annum.[10] It is anticipated that BVI will assist in reducing some of these costs, by provision of better data and a weight motivational tool to assist patients in lifestyle changes through choice where this is appropriate for the patient.

4.4 Assessing the value of the BVI

The potential of BVI would only become truly apparent when compared to the existing and established methods of measurement in healthcare. The value of BVI, as with any other product or service, can only be judged against what is already in existence and available to use.

4.4.1 Initial results from BVI compared to manual methods

Four trials have been published by Heartlands Hospital (UK) and Mayo Clinic (USA) and further research in four trials has been completed but is not yet in the public domain. Initial results indicate that with BVI, the long-term monitoring of changes in an individual's body shape may encourage individual weight loss and help with the review of national strategies for obesity management.[11]

Individually tailored preventative and diagnostic treatment appropriate to a patient's health needs is possible, as BVI will offer more accurate information about their history of obesity using 3D data and images. Now, research undertaken during the Heartlands Hospital trial on weight motivation[8,12] has established that some individuals were motivated by reference to their own previous scans and that their own age, gender, and ethnicity were key factors in establishing their own perception of weight and the need to address that through nutritional and diet changes.

Body Volume Index provides data about distribution of body volume, body fat and the shape of an individual, something BMI can never do (Fig. 4.1). Body Volume Index is more accurate in 'contour' measurement of an individual than BMI or a tape measure can ever be, as it measures the actual skin surface of the waist area, which includes the indent at the small of a person's back where the spine is.

Recent validation by Heartlands NHS Hospital in the UK[8,12] and Mayo Clinic in the US[7,13] has highlighted the benefits of using the BVI Scanner and the better accuracy of measurement when compared to manual measurements of waist and hip.[7] These studies illustrate the benefits of using a scanner for repeatability of measurements in comparison to manual measurement, and in particular where part volumes can highlight where volume is and therefore where weight is located. Volume is defined as the space that an object occupies, and the interior of the human body has components in each body section which are unique in content, but more importantly differ in both volume and weight, which BVI can calculate.

The research has shown that BVI offers many advantages over the more traditional BMI, and that BVI can be used to assess body shape, provide automated anthropometric measurements and track changes in body shape over time. This is something hitherto not possible – using BVI to effectively predict and estimate obesity levels of an individual patient based on their variations in body shape over time, in an easy to understand and digestible format.

The preliminary data from this trial has shown that part body volumes can be tracked in a patient and that changes with a patient, in loss or gain, can be automatically recorded and analysed by the scanner. By using techniques like BVI we can spot obesity problems early and manage them earlier.

4.4.2 How BVI works

By taking and comparing the part volumes of a 3D scan, BVI can:

- automatically measure the changes in a person's body volume and part volume;
- highlight the risks associated with greater or less volume in a given region, for example the abdominal area;

- relate the health risks of a person to their body shape, taking medical history, surrogate markers for cardiovascular disease, e.g. lipid profile, C-reactive protein, age, gender and ethnicity into account;
- compare a patient's BVI against normative data of shape for their age, gender and ethnic grouping.

Data from scans can be compared with normative data or merged to create an 'average' shape of a particular age, gender, ethnic group or even whole population. This has tremendous implications when measuring and tracking national and international obesity levels over time.

4.4.3 The benefits of BVI in measuring body shape

Body Volume Index is accurate, reproducible and not operator dependant. Despite World Health Organisation (WHO) guidelines on performing waist measurements, the reality of the manner in which people are measured manually is highly variable and a real issue where there is real inconsistency in practice.

In contrast, MRI scanners are infinitely more sophisticated in terms of individual assessment and diagnosis of internal organs and possible health factors that would affect a patient. However, MRI as a population based screening tool is counter productive, both cost wise and in terms of speed and convenience for a patient. The BVI scanner is meant to provide an alternative to BMI as a risk assessment mechanism for an individual patient, and be a cost-effective tool for public health purposes. Non-medical administrative staff can operate BVI scanners and, as a first-stage assessment tool, the benefits are enormous.

The BVI system as an alternative to BMI will provide benefits to healthcare in the following areas:

- As an obesity assessment tool, BVI will allow for quick, safe and accurate assessment of body weight, body composition, fat distribution and anthropometric measurement of an individual. This data can then be stored and used for the health risk assessment of individuals and offer clinicians a useful assessment tool to discuss sensitive body shape issues with a patient and use the data as a motivational tool for obesity and other health issues.
- Body Volume Index will provide a patient information sheet, with details of measurement data against normative data and comparison against previous scans for the gender, age and ethnicity of that patient.
- Body Volume Index will provide an automated data collection system for individuals, local, national and international population statistics long term.

4.5 Conclusion

New ideas inevitably take time to become acknowledged, accepted and ultimately to become established. The BVI measurement system does not replace BMI, Waist to Hip Ratio, Waist Circumference, but focuses on volume and the body composition of that volume to add a new dimension to human body measurement in healthcare.

4.5.1 The realism of adopting BVI

To change the worldwide use of BMI as the de facto measurement standard for healthcare and assessment of risk is a huge undertaking, given its longevity, universal acceptance, simplicity and the cultural differences of healthcare, where a high-level evidence base inherently negates adoption of change in the short term.

However, there is a growing realisation and acceptance that BMI, irrespective of its simplicity, simply doesn't provide the outcome on an individual level that it purports to do, thereby diminishing and hindering any long-term potential for the measurement in this day and age. In recent years there has been review and proposed re-alignment of BMI to cater for recognised differences in children, those of different ethnic origin and of different geographical region, but this is a short-term 'papering over the cracks' as opposed to genuine refurbishment.

In the short term these will assist healthcare professionals in their use of BMI in practice, and there have been indications that acceptance of these new revised BMI values, especially the added use of waist circumference, may become accepted within the clinical and regulatory bodies. The use of waist circumference specifically as an enhancement or alternative to BMI is currently being reviewed by regulatory bodies such as the WHO and the Centers for Diseases Control (CDC) amongst others, so awareness of the importance of measuring central obesity is increasing.

In fact, in terms of long-term adoption for BVI, perhaps the most significant step has been increasing acceptance within the healthcare community that measurement of the abdominal area of the body, as indicated by the waist, is deemed to be a more reliable indicator of health risk than BMI alone,[15] and this has been adopted in practice through waist circumference being used as a measurement alongside BMI in major health surveys.[16]

The premise of BVI is that, using 3D technology, we are now able to replicate the body shape of an individual, extract the abdominal volume of that individual, and understand the significance of that volume in relation to the overall weight and the percentage weight of that abdominal volume.

As healthcare develops and re-assesses the long-term needs of patients, it is believed that a logical 'next step' from an acceptance of waist circumference within healthcare is that abdominal weight would provide an even greater indicator of health and differentiate between those patients of different body shape.

Body Mass Index can never realistically be replaced, but that was never the intention. The premise behind BVI has always been to provide an enhanced alternative to BMI and, culturally, healthcare professionals are by nature attracted and ultimately should be engaged with any process or procedure that makes their judgement easier and helps their patients get better. As the use of scanners and the use of BVI become more widespread it is envisaged that adoption will be attained. The CDC, WHO, The Department of Health and the appropriate advisory bodies are already aware of BVI and are broadly supportive of any move to replace BMI through technical means. In conjunction with wider use of scanning and BVI in the healthcare environment, lobbying and appraisal of BVI within the governing bodies for healthcare provision will also be required at some stage in the next few years. This is likely to happen organically with a planned rollout of BVI in 2013 and 2014, but universal adoption will require a wider political and clinical willingness to modernise and improve measurement in healthcare at some point.

4.5.2 The significance of cultural and ethical variations

We have scanned over 28000 men, women and children over the past 15 years, and there are recognised and accepted differences in body shape between people of different ethnic backgrounds.[17]

Whether these are differences in height, lower torso or general body shape, for BMI to work effectively means that these changes in body shape need to be taken into account. Some work has been done to adapt BMI for these reasons, but these are, by nature, quick fixes and a long-term solution is required.

Body Volume Index is a computer based system and relies on the collection and analysis of data to produce an output based on pre-determined criteria. An individual's BVI will be compared to the normative values for their own ethnicity, age and gender. The latter two criteria as reference points for changes in body shape are perhaps self-explanatory, but ethnicity does require further evaluation. The scans used as normative data for BVI will be re-aligned in accordance with broad based ethnic backgrounds (e.g. Afro-Caribbean, Asian, etc.) and then a patient's information entered before being scanned, referenced by their ethnicity. The BVI system will then reproduce the correct values for that individual centred upon the normative data for that individual's age, gender and ethnicity.

In summary, there are differences but these will be automatically assessed and calculated within the BVI system, so post analysis of a one-dimensional measurement such as BMI to take these factors into account isn't necessary.

4.5.3 The implications of BVI for the apparel industry

Body Volume Index is a healthcare application using 3D technology, the majority of which has been used in the apparel industry to produce aggregated measurements or bespoke measurements for individual tailoring and garment production.

Developed by the apparel industry via an initial military application, 3D scanning has indirectly provided benefits for healthcare professionals in allowing technical measurement to be achievable on a reasonably low cost level. Conversely, it is imagined that wider adoption and use of BVI on a wide scale will inherently provide the apparel industry with ancillary benefits, by allowing practitioners and deliverers of BVI to offer patients the option of using their scans in an interactive and online series of applications centred on clothing design and fit. These are available from many companies and practitioners in the clothing and design sectors, and through BVI scanning users can generate income through re-sale of scans to patients for clothing purposes, offering potential global business opportunities for those that sell online.

Much of this is dependent upon a cohesive and reliable infrastructure for those offering BVI to patients, and a suitable set up and charging system is under review at the time of writing and should be in place by late 2013.

4.5.4 How prepared the world is for measuring obesity technically

Culturally over the past twenty years, technology has become the focal point by which most people communicate with others and run their lives on a day to day basis. E-mail, texting, smartphones and the internet have revolutionised the way the world works and, in this age, the use of manual weighing scales, a height measure and a tape measure does seem outdated and in need of re-appraisal.

Adoption of BVI does require commitment to invest in scanning technology rather than the incomparable cost of a tape measure, but as technology itself reduces in cost and becomes more efficient, and the accuracy and versatility of 3D scanning becomes more prevalent, BVI is likely to become something that over time will be more accepted as an alternative to BMI.

4.5.5 Summary and conclusions

Body Volume Index is the world's first dedicated measurement standard for obesity and body shape analysis using computers, capable of assessing populations as well as individuals. Rather than one standard measurement technique for all people, BVI is based on individual body shape and able to divide the 3D body image of an individual into multiple body volumes, so part body composition can be analysed, allowing differentiation between people who have the same BMI.

The conclusion is the first computerised anthropometric measuring system in the world for obesity and initial health risk. Applications will include individual assessment of obesity and related health risks and a fully computerised system for the collection and tracking of individual, local, national and international obesity statistics. With BMI nearly 200 years old now, in the computer age, for individual obesity, a manual 'good guess' is simply not good enough. Reliance upon BMI is not conducive to the good of the patient and healthcare management in the long term.

The Body Benchmark Study has hopefully helped to provide the first step in creating a new measurement that is more appropriate and meaningful for the modern world – where tracking and recording of an individual's body shape and weight distribution will ultimately allow for predictive analysis of that person's potential obesity and diagnostic needs in the future.

In late 2011, Mayo Clinic concluded their findings on twelve prospective part volume indices for BVI. Following four years of research, they have drawn conclusions on the most appropriate 'index' for BVI which best indicates the cardiometabolic risk of an individual patient and the added value that BVI offers over current manual methods of measurement in healthcare. This will provide the next development stage in BVI as a long-term measure of obesity and help BVI to become a significant new point of reference for healthcare.

All those involved in the development of BVI since the start of the century are realistic about the difficulty of changing a culture in healthcare from manual to automatic measurement. However, the PC has gradually replaced the typewriter and the mobile the handset, so time inevitably will favour applications where ease of use, ease of data storage and ease of communication are inherent and integral parts of the design. As an example of progress on BVI, in May 2013 we announced exclusivity for healthcare of $[TC]^2$ scanners for use in healthcare for BVI in hospitals and in July 2013 saw the launch on the www.bodyvolume.com website of the online BVI measurement system, which enables a person to be measured from a webcam and therefore any computer anywhere. There will be tablet and mobile versions to follow, but this significant development will inevitably lead to more widespread use and awareness of BVI across the world. It is a major advancement in the automation of measurement and the application of 3D technology in a new field and now with genuine potential to become a viable alternative to BMI.

4.6 Sources of further information and advice

Visit www.bodyvolume.com for further information on development of the Body Volume Index.

4.7 References

1. Sebo P, Beer-Borst S, Haller DM, Bovier PA (2008) Reliability of doctors' anthropometric measurements to detect obesity. *Preventive Medicine* 06/2008 (4):389–393.
2. Nadas J, Putz Z, Kolev G, Nagy S, Jermendy G (2008) Intraobserver and interobserver variability of measuring waist circumference. *Medical Science Monitor* **14**:CR15–18.
3. Romero-Corral A, Somers VK, Sierra-Johnson J, Thomas RJ, Collazo-Clavell ML, Korinek J, Allison TG, Batsis JA, Sert-Kuniyoshi FH and Lopez-Jimenez

F (2008). Accuracy of body mass index in diagnosing obesity in the adult general population. *International Journal of Obesity* **32** (6):959–966. doi:10.1038/ijo.2008.11. PMID 18283284.

4. Romero-Corral A, Montori VM, Somers VK, Korinek J, Thomas RJ, Allison TG, Mookadam F, Lopez-Jimenez F (2006) (2008) Association of bodyweight with total mortality and with cardiovascular events in coronary artery disease: a systematic review of cohort studies. *Lancet* **368** (9536):666–78, 19 August 2006 .
5. Radtke RA (2006) NX12 Body Measurement System; Assessment of Risk for Photosensitive Seizures, Duke University Medical Center, Durham North Carolina, USA, February 2006.
6. Shape GB The National Childrenswear Survey; www.shapegb.org
7. Korenfeld Y, Ngwa T, Friedman L, Romero-Corral A, Xu L, Albuquerque F, Sert-Kuniyoshi F, Okcay A, Somers VK, Lopez-Jimenez F (2009) Validation of a Novel 3D Body Scanner for Obesity Anthropometric Measurements, *AHA* 26th Princeton Conference on Cerebrovascular Disease, Houston, Texas, USA, March 2009.
8. Boelaert K, Palin S, Field A, Rahim A, Barnes R (2008) The impact of 3D body images on motivating weight loss in overweight individuals. *Endocrine Abstracts* **15**:125.
9. Foresight Tackling Obesities: Future Choices – Project Report. Government Office for Science 2007.
10. Hospital Episode Statistics Online 2005–2006. http://www.hesonline.nhs.uk/Ease/servlet/ContentServer?siteID=1937&categoryID=245.
11. Healthy Weight, Healthy Lives: a cross-government strategy for England (2008) Cross-Government Obesity Unit, Department of Health and Department of Children, Schools and Families, London.
12. Tahrani A, Boelaert K, Barnes R, Palin S, Field A, Redmayne H, Aytok L, Rahim A (2008) Body volume index: time to replace body mass index? *Endocrine Abstracts* **15**:104.
13. Romero-Corral A, Somers VK, Lopez-Jimenez F, Korenfeld Y, Palin S, Boelaert K, Boarin S, Sierra-Johnson J, Rahim A (2008) 3-D body scanners – body volume index: a novel, reproducible and automated anthropometric tool associated with cardiometabolic biomarkers. Conference Abstract. *The Obesity Journal* **16**:S124.
14. Gallagher D, Heymsfield S B, Heo M, Jebb S A, Murgatroyd P R, Sakamoto Y (2000) Healthy percentage body fat ranges: an approach for developing guidelines based on body mass index. *American Journal of Clinical Nutrition* September 2000(72):694–701.
15. Manson JE, Rexrode KM, Carey VJ, Hennekens CH, Walters EE, Colditz GA, Stampfer MJ, Willett WC (1998) Abdominal adiposity and coronary heart disease in women. *The Journal of the American Medical Association; JAMA* **280**(21):1843–8.
16. Hirani, V (2011) Adult anthropometric measures, overweight and obesity. The Health and Social Care Information Centre.
17. Fitzgibbon ML, Blackman LR, Avellone ME (2000) The relationship between body image discrepancy and body mass index across ethnic groups. *The Obesity Journal* DOI: 10.1038/oby.2000.75.

5
Psychological and sociological factors influencing consumers' choice of apparel

M. JULIEN, Université du Québec à Montréal, Canada

DOI: 10.1533/9781782422150.1.78

Abstract: Individual evaluation and choice of clothing are always linked to the lifestyles and ways of thinking in a community. Consumers construct a look from subtle variations that exist around them to make them feel unique. In fact, however, differences in clothing and body decoration are neutralized in the common discourse of modern fashion. This chapter describes and analyses the psycho-sociological factors that influence and determine choices in contemporary clothing: the need to look young, to look marginal, to be authentic, to constantly change appearances, to expose the body, to look sexy, to attract attention, to look sporty, and to be ethical.

Key words: contemporary fashion, sociology of fashion, consumer choice, fashion consumption.

5.1 Introduction

Many individuals do not wish to be identified with fashion trends, arguing that fashion is only for those who lack personality, or who conform to market rules. However, nobody gets away from fashion. Twenty years ago, it would have been unthinkable to go to the office with an unkempt beard, or wearing casual clothing. Everybody follows fashion, though not necessarily the same styles; this is all the more evident, lately, as globalisation, hyperconsumption and hyperindividualism have multiplied the variety of available clothing styles. In our time, consumers no longer perceive their look as an immutable identifier of themselves. They prefer to change looks, and more and more consider their appearance as an endless playground. However, even though styles are multiplying and trends are emerging with ever greater speed, they continue to perpetuate a fundamental quality of fashion by crystallising and reflecting the values of their time.

Individual evaluation and choice of clothing and fashion styles are always linked to the lifestyles and ways of thinking of a community. Consumers construct their looks from subtle variations that make them feel unique, or even marginal. In fact, these distinctive modifications give rise to the belief in uniqueness, individualism and exclusivity, even though hairstyles, clothing and body decorations like jeans, T-shirts, tattoos and piercings are increasingly neutralised through their participation in a common discourse. This chapter describes the following key psycho-sociological factors determining choice in contemporary clothing consumption: the need to look young, to look marginal, to be authentic, to constantly change appearances, to expose the body, to look sexy, to attract attention, to look sporty and to be ethical.

5.2 The need to look young

If social distinction used to reside in the ability to pay for expensive clothing, today it comes from the desire and capacity to appear young and transform the body. As indicated by Gilles Lipovetsky (2006, p. 63), 'consumerism claims to constantly rejuvenate lived experience', which ends up affecting our relations to our bodies. Nowadays, having a youthful appearance has become vitally important. Our bodies must respond to the requirements of performance imposed by the society of consumption – and performance is necessarily synonymous with youth. People simply don't want to age, especially the baby-boomers, who are thus increasingly mimicking the styles of the younger generations. They are dressing like them, getting tattoos and piercings, colouring or dyeing their hair, enhancing their muscles, removing unwanted hair, etc.

As noted by Lipovetsky (2006), in our performative society, being young has become more important than being rich, intelligent, or distinguished. As proof, marketers have coined a neologism naming a new target audience of adults who identify as adolescents: the 'adulescent'. Although this categorisation is arbitrary, it is highly suggestive of the superiority of youth culture and its domination of adult culture in terms of aesthetics, behaviour, and sexuality in our society. Indeed, the general attraction of sexy and sporty fashions has not arisen at random. The trends of the past few decades have valorised the behavioural stereotypes of adolescence: rebellion (appearing as an outsider), play (athletic and theatrical clothing), and erotic allure (hypersexual fashion). Moreover, the aesthetic expressions of youth generations from the last half-century have come to dominate the contemporary closet: the black Perfecto, studded belts, Engineers Boots (reappropriated Doc Martens), denim jeans and T-shirts (1950s Rockers); leggings, miniskirts, minishorts, Repettos and thigh-high boots (1960s); hip-hop style (1970s and 1980s gangstas); piercings, ripped clothes and visible underwear (1970s and 1980s punks); vampire and evil-themed clothes (1980s gothic style); and exuberant street style (1980s Harajuku Girls). Given the original impetus behind many of these contemporary styles, there can be no doubt that they have contributed to the postmodern trend among adults to remain for as long as they can within the indeterminate world of youth.

5.3 The need to look marginal and rebellious

Since the 1950s with the emergence of the Rockers (the first adolescent subculture to openly display insouciance and a willingness to enjoy life's pleasures), the conception of a full, meaningful life has totally changed. So much so that the 'juvenile delinquent' is one of society's most glorified behaviours, described by Ève Boissonnault (1998) as expressing what's on one's mind, having fun, letting oneself be guided by emotions, escaping social constraints, being open to all forms of experimentation, taking risks, not compromising, being different and living in the moment. With the prominence of youth culture, it has become trendy to behave playfully and without restraint. Many critics, like Damien Le Guay (2005), have underlined how disobedience, rudeness, and total disclosure are now interpreted as natural and authentic attitudes. This basically represents a return to a more primal state, before people learn the rules of socially accepted behaviour – in other words, before they become adults.

This continual concern with being disobedient has implications for appearance, leading people to choose particular clothing, hairstyles, accessories and body ornaments formerly reserved for the 'rebellious': fetish clothing (visible and provocative underwear, fishnet stockings, skin-tight clothes with animal prints), stiletto high-heels, piercings and tattoos, all of which have infiltrated mainstream culture through the formerly marginalised worlds of pornography, sado-masochism, prostitution, and gangster culture. For example, leather belts with decorative rivets called 'studs', fat leather bracelets and studded dog collars, worn with pride by girls who innocently follow fashion trends, originated in the sado-masochistic world to define the roles of master and slave between sexual partners (Julien, 2010a).

Moreover, the increased disaffection towards certain religious traditions in western society has been another factor contributing to the popularity of rebellious appearances. The body is part of a long and evolving history; individuals haven't always perceived their bodies in the same way. Individuals have, however, always been caught between two spheres: the material (body) and the spiritual (soul). In ancient Greek philosophy, the body was separate from the soul and considered a burden, something to get rid of. The soul was noble, whilst the body despised. This view persisted, and it wasn't until the end of the nineteenth century that a new place of respect was given to the body, after the discovery of the importance of hygiene and cleanliness in controlling disease.

The exultation of the body continued throughout the twentieth century, in its increasing exhibition and, especially, its transformation into a consumer good. Over the last few decades, the body has acquired an existence in itself, and it has even come to dominate the soul (Tissiers-Debordes, 2005). Indeed, the search for absolution no longer begins with the soul, but with the body.

We would rather be beautiful outside than inside, and the current quest for beauty seems to be tied to the desire to seem rebellious and/or above morality (Julien, 2007). In fact, history shows us that there has been little reciprocity between a beautiful body and a beautiful soul. Similar to the fetish fashion of recent years, which is highly representative of the tendency to display a rebellious inner self opposed to traditional, bourgeois codes, accessories with morbid and occult connotations popular among the young are reflective of a spiritual or religious void.

5.4 The need to be authentic

No doubt inspired by the displacement of religion and the collective questioning of other traditional institutions, the need for authenticity is being expressed more and more in western society, in many different ways (Julien, 2010a). Humanity seems to want to move on and search for that which is more real and natural. Fashion is not outside this current; indeed, it has become one of the primary manifestations of this need.

From the fetishistic to the sporty and the theatrical, contemporary fashion is underpinned by the individual quest for authenticity. Like other types of social organisation that preceded it, the form of democracy that developed over the last two centuries in western society has had a significant influence over mores and behaviours. Democracy has given rise to a certain individual freedom, and has placed great value on personality and individuality. Indeed, fashion can be considered as one of the prime manifestations of this exceptional concern for the individual (Lipovetsky, 1996). Democracy has promoted the adoption of more equal human relations, which has led to a revaluation of moral, religious, and civic codes. The consequent disappearance of prohibitions and the dissolution of hierarchies are thus naturally reflected in today's appearances. Hypersexual fashion inspired by prostitution, casual clothing replacing more traditional styles and ethnic symbols worn out of context all suggest the detachment of the emboldened individual from a previous submission to religious, institutional, or political power.

One common way that this is manifested is in women who cultivate a sexualized appearance hoping to demonstrate their authenticity. Throughout history, men have traditionally had control over what women are able (or unable) to display regarding their bodies. It is equally true that men held control over individual women's lives, which is still true in many communities around the world. Even in North America, where women are said to be totally free, the symbolic tradition of asking for the father's approval of a marriage has not entirely disappeared. This highlights just how recent women's independence actually is. As such, 'it's said that in dressing in a sexy manner, women are announcing that only they can decide whether to reveal their body or not, and at the same time claim total freedom to choose their sexual partners' (Julien, 2007, p.31). In other words, by dressing sexily, many women feel that they are being more themselves. Furthermore, the search for authenticity through appearance is also evident in the popular trend of selling or reselling clothing and accessories that reflect the recent, and more reassuring, past (Bak, 2007, exemplified by the explosion of vintage fashion around the globe, as well as the return of Converse All Stars, a must-have on the street (Bak, 2007).

5.5 The need to constantly change appearances

Surprisingly, the quest for authenticity is reflected in various ephemeral corporeal trends and clothing fashions. As emphasised by Serge Tisseron (2007), appearance is no longer a fixed identity marker, but an endless web of games. Individuals try to construct their personalities by experimenting with different looks, so it's no wonder that they reveal inner truths far more than in the past (Amadieu, 2002). Indeed, considering how quickly styles have multiplied, we might think that now, more than ever, individuals feel the need to express who they are. They don't hesitate to mix styles and types, motifs, materials or eras. And even though jeans and T-shirts, the ultimate symbols of democracy and youth, give their closet a normative air, their cuts, colours and ornaments never cease to change. There's no doubt that consumers are now increasingly playing with diversity, paradox, and contradiction. Many people who follow trends enjoy playing different roles and adopting different personalities. Their constant quest for authenticity manifests itself not only in the variety of styles they wear, but also in the great flexibility that these variations exhibit.

According to Lipovetsky (2006), hyperconsumption is partly responsible for a permanent dissatisfaction bred by constant change. At the same time as participating in the elite construction of the self, hyperconsumption is also responsible for providing constantly rejuvenating experiences that lead to change as well as abundance. Women, in particular, are affected by this phenomenon, as they significantly influence 80% of purchasing decisions (Yaccato, 2003). Accustomed to change, they appreciate the diversity of clothing styles and the transformations introduced to their physical appearance.

However, women are not the only ones radically and continually changing their appearance, as young men are increasingly doing so, too. In fact, today's youth are experimenting with a wider diversity of looks than ever before (Julien, 2010a). And knowing no other way of relating to objects than through hyperconsumption, young people are driven not only to transformation, but also to abundance. It is thus not surprising that they are more attuned to bodily alteration than any other previous generation. Their current range of style options seems unlimited (Julien, 2010c): African queen,

American retro, beautiful freak, black metal, boho-chic, chav, tribal-chic, military-chic, cyberpunk, clubber, emo, euro-trash, funk metal, sexy glam, gothic, greebo, grunge, hardo, hip-hop, hippie chic, hispter, kitsch, lolita, metalhead, modern-primitive, nappy, neo-classical, new BCBG, new-wave neo-romantic, pikey, prep, punk, punk-rocker, rapper, rasta, raver, rocker, roller, swag, skater, techno, techno beatnik, townie, trash, urban, wigger, etc.

Even though the search for authenticity manifests itself in consumerism and experimentation with various looks, people ultimately want to be themselves and to have the freedom to express who they are. But freedom is always accompanied by uncertainty, especially as we are now more compelled than ever to interact with others. And this has led to increased visibility. Consequently, skin has become *the* article of clothing which modern people are revealing, as well as modifying and decorating with tattooing, piercing, scarification, fitness, hair removal and cosmetic surgery. And certainly, this trend of exposing the body reflects not only the valorisation of youth, but also a new ethic of authenticity.

5.6 The need to expose the body

A not insignificant influence on fashion lies in our very contemporary need for all to be seen. As theorized by the philosopher Eugène Enriquez (2005, p.49), 'the world is now made to be seen, to be in full view of everyone, not to be understood'. We want to see everything, as much and as far as possible: the bottom of the ocean; direct, immediate violence and death; the interior of the human body; the farthest flung corners of the world; etc. And we want to engage with a fully transparent world. That which is invisible, opaque or difficult to grasp does not interest us. Enriquez (2005) adds that in such a context, nudity – both physical and psychic – has become society's guiding principle. For example, in displaying their bodies as much as possible, adolescents and young people, and also an increasing number of adults, are literally and metaphorically saying they have nothing to hide. This contemporary conceit seems to be motivated by the necessity of making oneself available to others – otherwise, people are suspected of hiding something.

It has become quite fashionable to be completely open and forthcoming to others in our society. Reality television is a good example of this. It promotes the display of one's self entire to an anonymous and voyeuristic audience. In revealing the intimate details of their lives, participants have the impression of being someone. In our current age, 'the exhibition of our private selves seems to prove our individual singularity' (Jeudy, 2007, p.10). Unveiling the self in front of others has thus become a social rite of passage. And moreover, in this age of public confession, those who have committed the most reprehensible acts are the ones who most frequently tell us about their disgraces, perversions and fears. As media specialist Jacques Gautrand

(2002) has outlined, through public exposure, the disgraced have a chance at social rebirth. That is to say, for modern television and internet audiences, it's not 'cool' to be modest, because modesty is perceived as suspicious and anti-social. It's no longer beneficial to be well-respected, nor to be seen as reserved, together and balanced.

For Le Guay (2005), the contemporary desire to be natural, the need to be authentic, the predominance of the visual in our society and the current imperative to be transparent are all interdependent. Therefore, the contagion of sartorial immodesty that some have severely denounced is founded in the obligation to be visible, which is itself connected to the desire to appear authentic. By showing as much of their body as possible, individuals are under the impression that they are coming closer and closer to reaching their natural states. It is no longer clothing that holds attention but rather the body that is subtly hinted at or completely revealed. Skin-tight, faded and/or transparent clothes tend to accentuate body shape and muscle size, and attract attention to certain body parts, especially the buttocks, breast, and genitals. But these fashion trends not only reveal the body, they eroticise it.

Everyone in the west seems to harbour an idealised image of the body which has become the object of a narcissistic and obsessional cult (Gautrand, 2002). In order to love ourselves or even experience pleasure, we feel we need to get as close as possible to this coveted ideal – so much so that our bodies have become life-long renovation projects. We try to control our weight, our hair, our figure, our muscles, and even the aging process itself. In fact, we tend to perceive control of the body, over its pains and weaknesses, as a personal accomplishment. The transformed body (including tattooing and piercing) has thus become an object of pride and gratification. As a sign, too, of social success, the body must be displayed, exposed to the public. Thus, by dressing to expose our bodies, we are looking to others not only for approval of our ability to seduce, but also for recognition of the work we have done on them. Moreover, if we invest significant money into laser hair removal to achieve a perfectly smooth stomach, have a large dragon or butterfly tattooed on our lower back, or pay to have larger, firmer breasts, we certainly want to maximize our investments, using them to impress, be noticed, seduce and/or show off our financial means. Thus does contemporary fashion reflect this new social reality.

5.7 The need to look attractive

As the organisation of corporeal relationships reflects the organisation of relationships to objects and social relations (Baudrillard, 1970), we can also see how consumption can influence people's love lives, and consequently their clothing, in so far as it is used as a tool for seduction.

According to Le Guay (2005), the compulsion to buy, founded on the circulation of goods and need to change merely for change's sake, leads to a diminishing of our romantic lives by encouraging the need to constantly change partners, have multiple sexual experiences and try everything. Gautrand (2002) echoes these sentiments in claiming that the postmodern subject must project a desirable, appetising, and even *edible* image of himself. Indeed, contemporary individuals have transposed consumption behaviour into the sexual sphere. They are insatiable, demanding, impatient, disloyal and motivated by the packaging.

In addition, the increasing frequency with which people change their romantic partners, even later in life, is reinforcing the power and importance of seduction at every age. Consequently, as the ability to seduce has always been connected to beauty and youth – both considered optimal for human reproduction – sex-appeal has become a way to announce that one is sexually active, regardless of age. How then can we be shocked by press titles such as *Fifty and Still Sexy*, or by reality TV shows where hypersexual women in their late fifties are vying with women in their twenties to seduce attractive younger men? It's difficult to deny the link between increasingly sexualised appearances and the expanding imperatives of seduction in our time.

Often inspired by the sartorial codes of prostitution and pornography, contemporary fashion has removed prohibitions by displaying our most intimate body parts. Exposed thongs, midriff-baring shirts, spaghetti straps, super-tight jeans, see-through clothing, high-heels, plunging necklines (even for men), thigh-high boots, miniskirts, leggings, and jeggings: many current fashion trends proclaim the sexual availability of an individual.

Valérie Steele (1997) attributes the success of the eroticisation of clothing to the fact that women see this fashion as a way to assume their femininity whilst also appearing more natural and open. In opting for a provocative style, and adopting tattoos and piercings, women are signifying that they own their bodies, and thus feel more real and free. As an illustration, most women over 40 who get tattoos are in fact recently divorced (Julien, 2010b). These women doubtless have the impression that they are taking back possession of their bodies and recovering their freedom. Marking the body thus becomes proof of their liberation, and a personal statement that they are proud to exhibit.

The sociologist Jean Baudrillard (1970) was certainly not off the mark in stating that women become liberated in proportion to the degree that their bodies are. It seems that the unveiling of a new female body part, which has been occurring steadily since the 1920s, is associated with a new form of liberty: access to public life, a better paying job, control over pregnancy, sexual pleasure, education, independence, access to influential jobs, etc. For example:

1920: Knee-length dresses that expose the legs – the right to go to cafés, drive a car and play sports.
1955: Form-fitting pants that accentuate the legs and buttocks – access to the world of work.
1965: Miniskirts that expose the thighs – the right to abortion and contraception.
1970: Unisex fashion and going braless – sexual liberation and equality of the sexes.
1980: Fine lingerie and the understated sexuality of the *power suit* – the entrance of large numbers of women into universities and the labour market.
1990: Tight-fitting, transparent and/or leather clothes that become part of the body – increasing financial independence and access to more positions of power.

Over the course of the last decade, we have witnessed the exposure of yet another part of the female body, one of the most intimate: the midriff. The display of this part seems to be connected to the increasing freedom among women to have many sexual experiences with multiple partners over the course of their lives. We can see that the younger generation is opting for independence in greater numbers, rather than the more traditional romantic union for life (Chaumier, 2004). In any case, by wearing midriff-baring tops, see-through shirts, and visible underwear, women both young and old are challenging conventions, shocking others, and pushing the limits in announcing their sexual availability. At the same time, they are denigrating bourgeois mores and adhering to rebellious appearances based on authenticity and the value of pleasure. The recent exposed-midriff trend can be seen as one of the most representative of contemporary behaviours because it represents a new social relation founded on accessibility, proximity and transparency. The exposed midriff is also literally representative of contemporary narcissistic behaviour.

5.8 The need to attract attention

Increasingly self-centred, individuals today feel obliged to reassure themselves in terms of the tastes, and the desires, of others. In recent years, a new phenomenon that psychiatrist and psychoanalyst Serge Tisseron (2001) has coined 'extimacy' has gained popularity which, as opposed to intimacy, reflects the need to attract others' interest in an individualistic world where everyone is turned inwards, to their own self. According to Tisseron (2001, p.2) 'most important is not to be valued in the eyes of the other, but to be gratified by their interest'. Harajuku Girls in Japan have exemplified this phenomenon for the last 30 years. Dressed in morbid, manga-inspired, and carnavalesque clothes, these young Lolitas roam the streets in the Harajuku district of Tokyo, stupefying passers-by. Providing inspiration for some of the world's most famous designers, these queens of 'street wear' don't want to be noticed for their intelligence, personality, talent or beauty. Nor do they have a political or philosophical statement; their only intention is to attract attention.

As with reality TV stars, these young Japanese girls are tapping into a form of instantaneous celebrity that Jérôme Béglé (2003, p. 24) has called 'celebrevity', which allows people 'to be stars without having accomplished anything extraordinary'. Celebrevity speaks to a new kind of addiction to the star system whereby people are no longer limited to idolising stars, but instead to take to the stars themselves and become famous. Current fashion trends satisfy precisely this postmodern ambition: theatrical dress, glamorous looks, sunglasses worn at all times, false eyelashes and nails, spectacular hairstyles, flashy body decorations, excessive glitter, etc. One of the most eloquent manifestations of the democratisation of fashion and celebrity is 'street fashion'. Followers of this broad trend claim to be inspired by fashion without really following it, and they create their own looks by drawing on diverse influences from pretty much anywhere. More and more street fashion blogs are appearing, and their goal is to 'humanise fashion by showing real people living real lives' (*Elle Québec*, 2009). But looking at these provocatively dressed 'real people', we have to wonder whether the essential criterion for appearing real and authentic is not originality, or even eccentricity. The same goes for people who dramatically mark their bodies. Self-expression here seems to be confused with performative appearance, and the value of authenticity is confused with the ability to be noticed.

5.9 The need to look sporty

The mass consumption of casual athletic and sporty clothes is one of the most significant changes in the contemporary closet (Herpin, 1986). Comfort has come to prevail over the notion of respectable appearance, long associated with a discreet and appropriate style. The success of athletic clothing worn in public, particularly in the city, can be attributed, as Fabien Ohl (2001, p. 112) writes, to the fact that 'the transgression of athletic appearances opposes bourgeois codes, and reproaches the luxury world symbolised by real brand names'. By wearing shoes and clothing from well-known athletic brands, contemporary individuals are adopting a trendy and desired sartorial style and promoting the idea of performance underlying it, largely because of sponsored professional athletes, rappers and musicians (Ohl, 2001). It thus goes without saying that even though sports seem to be increasingly popular outside the stadium, the signs of athleticism are in fact more common than playing the sports themselves (Ohl, 2001).

Just think about the number of hip-hop fans wearing ball caps, which allow them, consciously or not, to identify with the values they hold in esteem. Based on the headwear worn by baseball players – among the highest paid athletes in the world – it simultaneously symbolizes the Americanisation of culture, social success, youth, pleasure and rebellious behaviour, especially when worn backwards. The international and intergenerational success of the ball cap over the past 25 years unequivocally demonstrates the close link between fashion and social values.

Furthermore, this trend engages with gender issues, as women have certainly long been excluded from the world of sports, and generally from the sphere of social performance. For them, wearing sporty clothes has become another way to promote equality between the sexes and women's emancipation (Moreau, 2008). In addition, because sports have always been associated with the values of authenticity, humanity, and integrity (Cuvillier, 2008), athletic clothing allows women to reflect their attempts at self-expression, as well as their search for liberty, well-being, and independence (Moreau, 2008).

5.10 The need to be ethical

For several years now, consumers have had a growing awareness of ecological and ethical issues when purchasing clothing (Slater, 2003). It seems that so-called green behaviour has replaced the 'philosophy of black' (Blanchard, 2007); you could even say that green is the new black, that saving the planet has replaced the desire to appear marginal and rebellious. These new green trends preach creativity, recycling, local production and moderate consumption. They also advocate materials produced from fibres and textiles whose production minimizes pollution and non-renewable energy consumption. For now, environmentally concerned consumers rely on 'eco-labels' to guide their purchases (Weltrowski, 2010).

Ethical considerations, however, are not limited to the ecological. They also concern the reputation of brands and their influences, the social conscience of managers, the treatment of employees, the symbolism of the clothes (e.g. sexualisation of young girls), and other factors. Moreover, with the development of the principles of individualism and equality, certain consumers are shopping with ethical considerations in mind. For example, in reaction to the dominance of the big mainstream brands, some are refusing to wear designer fashions. Wishing to be more authentic, more 'No Logo', as they claim, they prefer to make or decorate their clothes using their own ideas. As ethical discourse increasingly becomes part of our contemporary social makeup, it is expected that more and more individuals will recognise the value of being 'ethical' when purchasing clothes.

5.11 Conclusion

Today's consumers like to change their appearances and display the extreme permissibility granted to them that was denied their predecessors. They can completely reinvent themselves in a matter of weeks by changing musical tastes, emotional relations and moods, athletic interests, and social groups. As emphasized by Tisseron (2007), appearance is no longer a fixed identity marker, but an endless web of games. Affiliations are more diversified and ephemeral than ever, which renders brand loyalty or adherence to one particular style much more difficult. However, despite the great variety of choice on offer, consumers still perpetuate a fundamental principle of fashion in unconsciously crystallising the social values of their era. Even if we no longer dress to explicitly challenge the rules established by such authorities, the abundant signifiers associated with fashion have created a standard discourse of appearance. Even though they reflect individual choices, they exemplify what Michel Maffesoli (2007) has called an 'invisible force' that fosters a sense of belonging and creates social bonds.

Furthermore, despite the fact that outfits increasingly stand out for what they expose, they are generally based in a static sartorial lexicon that hasn't changed in some time: jeans, T-shirts, Perfectos, visible underwear, tight pants, leggings, etc. Aesthetic preferences in terms of dress and corporeal decoration (piercings, tattoos, radical hairstyles, etc.) are directly tied to the way in which a community sees and understands the world. For example, jeans (the first pants to be worn without suspenders or a belt) and T-shirts (the first undergarments to be worn as outer wear) remain the most popular clothes in the world because they symbolise youth and sexuality (Julien, 2009), both quintessential values in contemporary Western society.

To take another example, it's not too difficult to understand the recent popularity in American high schools of the 'out-of-bed' look: jogging pants, pyjamas, slippers, and tussled hair (*L'Actualité*, 2012). It's also easy to see why certain institutions would ban it, as fashion not only reflects the zeitgeist, but also helps determine behaviour. The out-of bed style positions the bed as the ideal site for living, and the popularity of this style speaks to the current, increasing lack of motivation among youth (Ollhoff, 2010), not to mention the valorisation of exhibiting one's intimacy. Indeed, this style exemplifies the phenomenon of 'extimacy' that can be said to characterize our era (Tisseron, 2001). Can we thus be surprised that the sale of leisurewear in the United States grew by 21% between 2010 and 2011 (*L'Actualité*, 2012)?

Fashion is one of the most telling, indeed revealing, cultural manifestations of our time, indicating the prevailing values of our communities, as well as society at large. The wide diversity of looks, the valorisation of marginal appearances, the popularity of athletic clothing, the international success of luxury designs, and the infatuation with brand names all reflect the importance that our society gives to consumption, sex, celebrity, performance, youth, and pleasure. Even if consumers seem to choose their clothes with full awareness and actively construct their own particular identities through their looks, they remain fundamentally influenced by the society that surrounds them.

5.12 References

Amadieu, J.-F. (2002) *Le poids des apparences*, Paris, O. Jacob.

Bak, J. (2007) *La société mosaïque*, Paris, Dunod.

Baudrillard, J. (1970) *La société de consommation, ses mythes, ses structures*, Paris, SGPP, 'Le Point de la question' 1970.

Béglé, J. (2003) *Célébrièveté*, Paris, Plon.

Blanchard, T. (2007) *Green is the New Black*, London, Hodder & Stoughton.

Boissonnault, E. (1998) Le portrait du masculin dans le cinéma américain de genre 'délinquance juvénile' des années 1950, Master's Communication Thesis, Montreal: Université du Québec à Montréal.

Chaumier, S. (2004) La déliaison amoureuse. De la fusion romantique au désir d'indépendance, Paris, Payot & Rivages.

Cuvillier, D. (2008) *Comprendre les tendances: ceux qui les font et les défont*, Paris, Éditions du Chêne.

Elle Québec (March 2009) 'Ces blogues qui décryptent la mode', http://www.ellequebec.com/mode/style-de-rue/ces-blogues-qui-decryptent-la-mode/s/3033/?ph=21691#ph_21691 [Accessed 22 August 2012].

Enriquez, E. (2005) 'L'idéal type de l'individu hypermoderne: l'individu pervers?', *L'individu postmoderne*, Nicole Aubert (editor), Ramonville-Saint-Agne, Érès, 'Sociologie clinique', 38–57.

Gautrand, J. (2002) *L'empire des écrans*, Paris, Le Pré-aux-Clercs.

Herpin, N. (1986) 'L'habillement, la classe sociale et la mode', *Économie et statistiques*, 188, 35–54.

Jeudy, H.-P. (2007) *L'absence de l'intimité*, Belval, Circé.

Julien, M. (2007) 'La mode hypersexy mise à nu', *Médiane* **2**, 1, Montreal, 27–32.

Julien, M. (2009) 'Le mauvais goût au goût du jour', *L'état du goût*, Bern, Peter Lang, 87–97.

Julien, M. (2010a) 'Looks rebelles, variés et authentiques', *Dictionnaire de l'adolescence et de la jeunesse*, David Le Breton and Daniel Marcelli (eds.), France, Presses Universitaires de France, 557–61.

Julien, M. (2010b) *La mode hypersexualisée*, Montreal, Éditions Sisyphe.

Julien, M. (2010c) 'Looks rebelles, variés et authentiques', *Dictionnaire de l'adolescence et de la jeunesse,* David Le Breton and Daniel Marcelli (dir.), Presses Universitaires de France, 557–561.

Lipovetsky, G. (1996) *L'ère du vide: essai sur l'individualisme contemporain*, Paris, Gallimard.

L'Actualité (2012) 'Les joies du mou', August 2012, 70.

Le Guay, D. (2005) *L'empire de la télé-réalité*, Paris, Presses de la Renaissance.

Lipovetsky, G. (2006) Le Bonheur paradoxal: essai sur la société d'hyperconsommation, Paris, Gallimard, NRF Essais.

Maffesoli, M. (2007) 'Le devenir mode du monde', *Métaphysique de la mode,* Frédéric Monneyron, Virginie Deviller, Jacques Sojcher (eds.), Bruxelles, Amis de la Revue de l'Université de Bruxelles, 81–6.

Moreau, L. (2008) Publicité au féminin: étude des stratégies communicationnelles des annonces-magazines de vêtements de plein air pour femmes, Master's Communication Thesis, Montreal: Université du Québec à Montréal.

Ohl, F. (2001) 'Les usages sociaux des objets: paraître sportif en ville', *Loisir et société/ Society and Leisure*, vol. 24, 1, Sainte-Foy, Presses de l'Université du Québec, 111–35.

Ollhoff, E. (2010) 'La dépression. Épidémiologie', Dictionnaire de l'adolescence et de la jeunesse, David Le Breton and Daniel Marcelli (editors), France, Presses Universitaires de France, 229–33.

Slater, K. (2003) Environmental impact of textiles, Cambridge, Woodhead Publishing Limited.

Steele, V. (1997) *Fétiche: mode, sexe et pouvoir*, Paris, Abbeville.

Tisseron, S. (2007) 'Les nouveaux visages de l'extimité: l'artiste et le délinquant', *Esse, arts + opinions* **58**, 5–7.

Tisseron, S. (2001) *L'intimité surexposée*, Paris, Ramsay.

Tissiers-Desbordes, E. (2005) 'Le corps hypermoderne', *L'individu postmoderne*, Nicole Aubert (ed.), Ramonville-Saint-Agne, Érès, 'Sociologie clinique', 173–97.

Weltrowski, M. (2010) 'L'écologie: un critère à considérer lors de l'achat d'un vêtement?', *Éthique de la mode féminine*, Michel Dion and Mariette Julien (eds.), France, Presses Universitaires de France, 67–80.

Yaccato, J. T. (2003) The 80% Minority: Reaching the Real World of Women Consumers, Canada, Viking.

Part II

Understanding sizing and shapes requirements and choices of particular customer groups

6

Infants and children: understanding sizing, body shapes and apparel requirements for infants and children

N. ZAKARIA, Universiti Teknologi Mara (UiTM), Malaysia

DOI: 10.1533/9781782422150.2.95

Abstract: This study was conducted to develop a standard sizing system for school-aged children in Malaysia. An anthropometric survey was conducted to measure 2035 male and female children, and 50 different body measurements were taken on each one based on the ISO 8859/1989 standard guideline. Multivariate statistical techniques were used to find the key dimensions. Then, cluster and decision tree data mining techniques were used to cluster the groups and classify them according to their own body profile. Finally, it was found that the body sizing system successfully accommodates more than 90% of the sample population with low aggregate loss.

Key terms: anthropometric survey, children's body shapes and size, multivariate analysis, data mining analysis, size designation.

6.1 Introduction

Based on the theory of Maslow's hierarchy of needs, clothing is considered one of the physiological needs, which are the basic needs for humans.[1] In addition, clothing plays many important roles for health and well-being, especially in children's learning and development.[2] The roles of clothing include protection against harsh climate and environment, modesty in social and cultural situations, and a form of adornment and expression of personal taste or style.[3] The most decisive role of clothing is during the early years of human life, which are collectively known as childhood.[4]

Human life is divided into two stages: childhood and adulthood. The legal definition of a child is a person aged between 2 years and 19 years of age.[5] Biologically, a child is anyone in the developmental stage of childhood, between infancy and adulthood.[6] The United Nations Convention on the Rights of the Child defines a child as every human being below the age of 18 years.[7] For this chapter, the data used are based on the study conducted in Malaysia to develop a children's sizing system using anthropometric data.

The data are obtained from samples from elementary, middle and high school children.

Children's clothes need to be functional in order to accommodate growth, provide comfort, promote safety, and foster a sense of independence.[2,8] A few researchers have emphasized that children's clothes first need to be functional. For example, Norum[9] mentioned that regardless of how attractive an item is, how easy the garment is to care for or what value the clothing is to be, it is of no benefit if the garment does not fit the child. In addition, Gautam[4] stated that garment size and seasonally appropriate clothing, with shoes and stockings of correct size, were considered to be comfort features. Besides, several recent studies[10–13] have suggested that better fitting and more comfortable clothing for obese and overweight children can increase participation from a social and psychological perspective.

During childhood and adolescence, physical activity is promoted as a key component of energy balance, and as a lifelong positive health behavior.[14] As a result, children are expected to wear comfortable clothing for various activities in order to be physically, socially, and mentally active and fit.[15] On the other hand, during this period children grow quickly, in different ways and at different rates; therefore, clothing size is significant in giving them the right fit while also allowing room for growth.[16,17] In the UN Convention on the Rights of the Child, Article 27 states that children have the right to a standard of living that is good enough to meet their physical and mental needs with regard to food, clothing, and housing.[18] Kang *et al.* added that when a garment is properly styled according to the right size, it reflects the children's growth and allows free body movements.[8] It is important to provide well-fitting clothes, especially to those that have the greatest potential for growth and body development.[19,20]

Infancy and early adolescence are both periods of rapid physical growth; these two stages are from birth to 12 months old and then between about 9 and 15 years of age.[21,22] Clothing fit is of utmost importance during these two stages of rapid physical growth, hence clothes need to be manufactured according to the body shapes and sizes of the targeted population and figure types.[17,23–25]

Studies have shown that fit is significantly associated with customer satisfaction in shopping for children's clothing.[9,26] Otieno[27] revealed that parents were dissatisfied with garment fit. She also claimed that customer satisfaction regarding children's clothing fit is unique from other measures of customer satisfaction, as the needs of both the parents and the child must be considered. Parents whose children are an irregular size are particularly concerned; sizing then becomes an important issue.[28] Ensuring clothes are comfortable and provide a good fit for children with different degrees of physical, socio-psychological, and cognitive development can be a challenge to all parents and caregivers.[29]

6.2 Understanding different body shapes of infants and children

Children need well-fitting garments. Lee[30] stated that apparel cannot be top quality unless it fits the potential wearers satisfactorily. Other studies showed that children look for fit and size as the most important element when purchasing clothing.[31–34] Teen market consumers were reported to look for qualities in branded clothing that include good fit, look, and style;[35–37] when trying on clothing in the dressing rooms, their decisions to reject or adopt the garments are based three main factors: fit, styling, and appearance.[3,15,39]

Existing apparel companies have worked towards producing clothing according to unique sets of body measurements to fit their own target market, e.g. designing for the plus-sized tweens and teens body range instead of using the general sizing systems.[40–44] Today, it is important to produce clothing according to the body measurements of the target population.[45] Moreover, Tselepis[46] mentioned that of all age groups, adolescents are most likely to spend a lot of time and give special attention to clothing and appearance.

Besides their purchasing power and interest in style, the physical development of the early adolescent (9–13 years) may result in functional and aesthetic problems with fit, which in turn plays a part in expectations of fit.[43,47,48] Fit and size are considered to be two of the important attributes for clothing purchase.[49] Therefore, to ensure that clothing is comfortable and fits children well, their body variations need to be studied. This study of human body dimensions is known as anthropometry, and is elaborated in the next section.

6.2.1 The physical growth and development of children

Growth, maturation and development are major processes that children go through from birth to adulthood.[50,51] Growth refers to increase in the size of the body or specific parts of the body.[16] Development refers to the stages of increment of these traits throughout the childhood lifespan while maturation refers to the stage of complete development.[15] During the development process, there are important changes in body size and proportions. According to Bogin,[52] there are three broad stages in children's development: early childhood, middle childhood, and adolescence; corresponding respectively, as Fig. 6.1 shows, to the infancy stage, childhood and puberty, and adolescence.[52,53]

At puberty, rapid growth occurs in both weight and height. Girls usually start and finish puberty and adolescence earlier than boys (10 to 12 years for girls and about 11 to 14 years for boys).[54] Physical characteristic differences between boys and girls develop during puberty[55] due to the body's hormonal production;[42] this period represents the most important growth spurt.[22] Environmental factors such as nutrition, lifestyle and health appear to be the main causes for differences in growth and maturation between

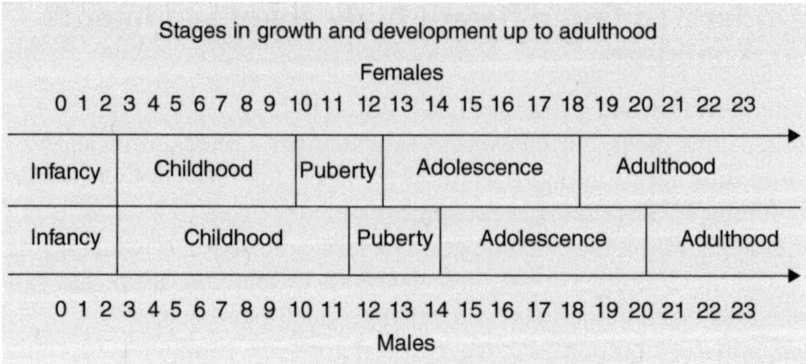

6.1 Stages of growth and development from childhood to adulthood.[53]

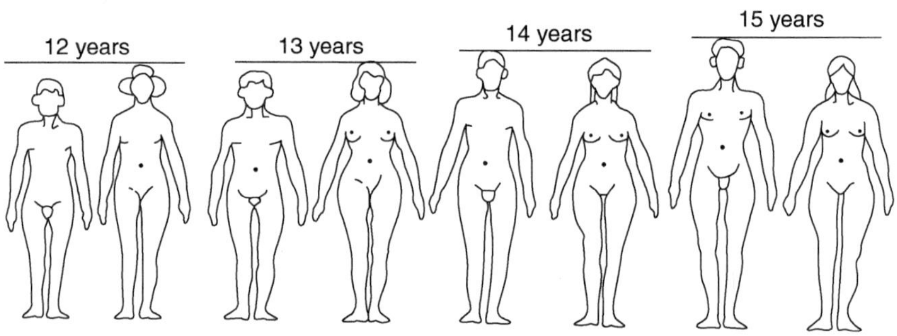

6.2 Differences in height growth in girls and boys. (*Source*: Adapted from Winks.[57])

ethnic and social groups.[42] In conclusion, it could be said that differences in final height are mainly due to prepubertal growth.[56]

The growth spurt produces a rapid increase in both weight and height, but it manifests differently in boys and girls.[17] Ahead of this growth spurt, there are no important differences between boys and girls in weight and height (Fig. 6.2). Girls reach puberty around age 10 and, at the age of 12, are on average taller than boys, who reach puberty around 12. Yet by the age of 14, boys overtake girls in height.[57]

The last stage before adulthood is adolescence (13–17 years old), which is considered the most crucial years for children before they begin their lives as adults.[58] Sexual maturation is also one of the most significant developments during this time.[59] However, sexual development occurs at different ages in different populations, making the comparison of children of the same chronological age in terms of body size and shapes difficult.[60] Growth is generally said to be completed between the ages of 16 and 18, at which time the growing ends of bones fuse.[61]

As mentioned by ISO TC 133, the technical Committee on Sizing Systems and Designations for Clothes, body height is of primary importance in fitting clothes for growing children.[62,63] Children have their own body structure, which is not simply a smaller version of adults. Children's sizing systems are often derived from adult sizing systems, causing a lot of fit problems due to growth variations among children.[64,65] They are also often built around age-related growth spurts for an 'average child'.[66] Because few children are precisely average in all body dimensions, it is difficult for parents to find their child's size. In short, age-related sizes may be an inappropriate basis for a sizing system.

6.3 Sizing requirements for particular populations of children

This chapter is based on real anthropomorphic data acquired in Malaysia on children of both genders from the age of 7–17. For simplicity's sake, only the data for girls aged between 7 and 12 will be used to show the process of the sizing system development (for both upper and lower body) using a data mining methodology.

6.3.1 Body shapes and sizes analysis

Figure 6.3 shows three main stages for the process of body sizing system development. The first stage is the sizing requirements; the second stage is the sizing system development and the last stage is the size designation.

In the first stage, the objective is to understand the different body shapes and sizes. It involved four different steps:

1. fieldwork preparation;
2. anthropometric planning;
3. anthropometric survey;
4. anthropometric analysis.

Anthropometric measurements taken are based on ISO 8559/1989. Table 6.1 lists all the 49 body dimensions inclusive of height and weight, divided into three main groups: vertical length, width length and girth. There are 29 dimensions suitable for upper or whole body and 20 dimensions categorized as lower body dimensions.

6.3.2 Categorical data

The categorical data were first analyzed to show the demographic profiles of the samples. The entire data set was divided into two subsets, male and female. Table 6.2 shows the frequency distribution of the

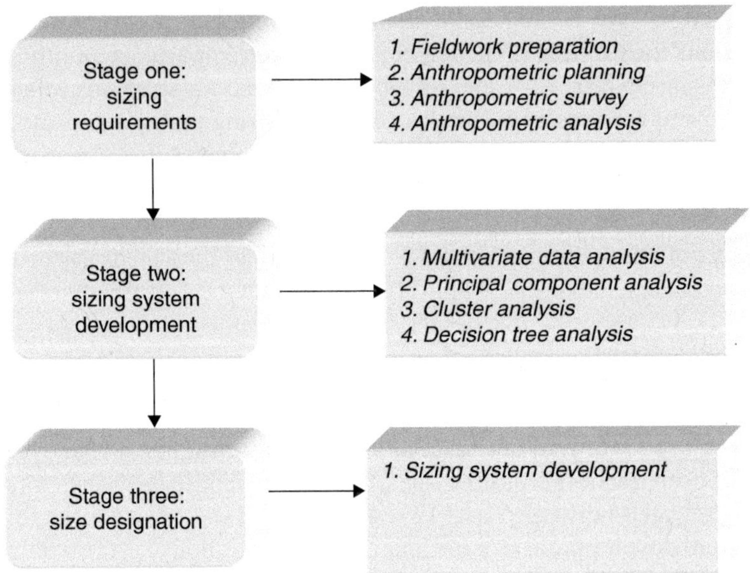

6.3 Stages of sizing development for sample population of children.

Table 6.1 List of body dimensions according to ISO8559/1989

Length (vertical)	Width (vertical)	Girth (horizontal)
Height	Shoulder length	Weight
Under arm length	Shoulder width	Head girth
Scye depth	Back width	Neck girth
Neck shoulder point to breast point	Upper arm length	Neck base girth
Cervical to breast point	Arm length	Chest girth
Neck shoulder to waist	7th cervical to wrist length	Bust girth
Cervical to waist (front)	Hand length	Upper arm girth
Cervical to waist (back)	Foot length	Armscye girth
Cervical height (sitting)		Elbow girth
Trunk length		Wrist girth
Body rise		Hand girth
Cervical to knee hollow		Waist girth
Cervical height		Hip girth
Waist height		Thigh girth
Outside leg length		Mid-thigh girth
Waist to hips		Knee girth
Hip height		Lower knee girth
Crotch		Calf girth
Trunk circumference		Minimum leg girth
Thigh length		Ankle girth
Inside leg length		
Knee height		
Ankle height		

sample population according to age, gender, ethnic group and geographical area. Overall, age 7 has the most subjects and age 17 has the fewest. These numbers are proportional to those found in the global population in Malaysia.

Overall, the proportion of male school-age population in Selangor is slightly higher (51.1%) than the female population (48.9%). Thus, the same percentage is found in the samples totaling 1034 males and 1001 females (33 more male samples than female samples).

In terms of ethnic groups, Table 6.2 shows an almost equal sample distribution (34%, 34% and 32%). The breakdown shows that 688 are Malay, 691 are Chinese and 656 are Indian. As for school setting, 1212 (59.6%) of the sample population come from 17 urban schools while 823 (40.4%) come from 12 rural schools. Therefore, the makeup of the sample has slightly fewer subjects from rural schools than from urban schools.

From the stratum of gender, the number of male subjects from urban schools was 624 while the number from rural schools was 410; 588 females are measured from urban schools and another 413 from rural schools. Thus, the percentages of male and female subjects from rural and urban schools are 51% and 49% respectively.

Table 6.2 Sample population distribution according to four variable body shapes and sizes

Stratum group	Geographical school areas		Ethnic groups		
Age/years	*Urban*	*Rural*	*Malays*	*Chinese*	*Indians*
7	119	106	76	77	72
8	133	85	73	76	69
9	114	96	76	68	66
10	152	67	70	71	78
11	142	72	73	71	70
12	147	66	68	73	72
13	90	74	54	56	54
14	88	75	56	56	51
15	87	67	50	54	50
16	70	65	45	45	45
17	70	50	47	44	29
TOTAL	1212	823	688	691	656
Gender					
Male	624	410	350	356	328
Female	588	413	338	335	328
TOTAL	1212	823	688	691	656
Percentage (%)	60	40	34	34	32

6.3.3 Continuous data

The summary descriptive statistics for selected key body dimensions, namely height, weight, chest girth, waist girth and hip girth for the entire sample size (n=2035) are tabulated in Table 6.3.

As can be seen from Table 6.3, the height distribution of 7–17 years old is from 101 to 184.4 cm with a mean of 141.9 cm, and the standard deviation was 16.8 cm. Their weight ranged from 13.4 to 126.7 kg, while the average weight was 40.8 kg. The smallest chest measures are 35.3 cm, while the largest chests are 134.0 cm. Chest girth showed normal distribution and has a mean measurement of 74.1 cm. The mean waist measurement is 63.8 cm and the range is 113.3 cm. In addition, the smallest waist circumference is 23.5 cm, while the largest waist measurement is 124.5 cm. The mean hip measurement for this sample is 78.2 cm with a range from 32.5 to 131.2 cm. The waist girth distribution is skewed to the left, which means that the waist measurements are mostly scattered in the lower range. Hip distribution is the least skewed and shows an approximately normal distribution of hip measurements among the sample population.

The distribution of key dimensions as shown in this chapter is useful for the estimation of size roll for the sizing system. Overall, the size range for height is 83.3 cm. As for weight, Otieno[67] has mentioned in her study that this dimension is useful only in smaller children. Previous studies clearly suggest that bust and hip girth should be used as key dimensions to replace weight. The reason for this is that both these dimensions (bust and hip girth) are flexible for varying size relationships for upper and lower body.[67,68] Another reason for substitution of weight is that these dimensions (bust and hip girth) are the body dimensions in direct contact with the clothes, while weight is not.[69]

For the upper body, the size ranges for chest and bust girth are 98.7 and 89.0 cm respectively. For the lower body, waist and hip girth measurements indicate size ranges of 101.0 and 98.7 cm respectively. However, these

Table 6.3 Descriptive statistics of selected body dimensions for sizing system

Body dimension	Mean	SD	CV (%)	Min	Max
Height	141.9	16.8	11.8	101.1	184.4
Weight	32.8	17.5	42.9	13.4	126.7
Chest girth	74.1	12.3	16.6	35.3	134.1
Waist girth	63.8	11.9	18.7	23.5	124.5
Hip girth	78.2	13.6	17.4	32.5	131.2

extreme values seem high compared to the size range of the sizing system for Sri Lanka, which was developed recently.[20] For example, for Sri Lankan school children, the height range is 32.5 cm, while for chest, waist and hip girth the size ranges are 43.2, 44.5 and 39.4 cm, respectively. This indicates that the extreme values of the key dimensions may not be suitable for the size range; instead, the percentile value is more suitable. This is supported by previous studies in which the percentile value of key dimensions was used, covering 95% of the population.[70] Thus, for this study, the percentile values are applied to the key dimensions, where the 5th and 95th percentile are used to accommodate 90% of the sample population. This percentile value will be applied to the key dimensions selected in the next stage of this study.

6.3.4 Anthropometric analysis findings

Our analysis of continuous data proves that body dimensions are highly dependent on age and gender. Children aged from 7 to 12 are found to have wide variations of height but children of different ages are also found within the same height measurement. This finding suggests that age is not appropriate as the determining factor for a size system but that body dimensions are. It also shows that no specific height can be associated with a certain age, as humans come in many different sizes and shapes. Thus, this finding supports previous studies, which concluded that it is not appropriate to develop a sizing system for children according to age.[64,67] In addition, the spread of measurement is not only found in height and age but also in girth measurements and length measurements. Therefore, it is essential that the combination of key dimensions comes from girth and length measurements in order to fit the population well, as recommended by many researchers of previous studies.[57,72–74]

6.3.5 Applications of particular body shapes and sizes for sizing system development

The next stage is to develop the sizing system in which to identify key dimensions, then to cluster the sample population and derive how many sizes can be produced to fit the sample population. The continuous data allows the different body shapes available within the sample size population to be visualized.

There are four steps involved in developing the children's sizing system:

key dimensions identification (Section 6.4);
clustering children's body shapes and sizes (Section 6.5);
classification of children's body shapes (Section 6.6);
sizing system development and designation (Section 6.7).

6.4 Key dimensions identifications

This chapter is the first step for sizing analysis. This first step is to identify the key dimensions from the many body dimensions measured. The main objective of this chapter is to reduce the body dimensions which are the variables into the most significant key dimensions. In this chapter, the principal component analysis is the method used to reduce the many body dimensions into a more parsimonious and manageable groups. Key dimensions will be the few significant body dimensions selected to represent parts of the body for sizing system development. In addition, key dimensions are selected to represent upper body garments, lower body garments and whole body garments. By identifying the key dimensions, then these key dimensions will be the variables used to cluster the sample population, for example if for upper body; the key dimensions are bust girth and height then the sample population will be clustered according to their height and bust girth.

6.4.1 Principal component analysis technique to identify key dimensions

There were 49 body dimensions taken from each subject of the sample population. A Principal Component Analysis (PCA) was employed to reduce all the variables into significant components. The data were first divided into two significant variables: gender and age group. Thus, for this chapter, the data for girls aged 7–12 were extracted from the total data and were reduced to components using the PCA method.

6.4.2 The results of components analysis

In general, 49 components were extracted from each sample group in order to explain 100% of the variance in the data. The summarization of variables proved to be very good, as the number of principal components is the same as the number of original variables. This indicates that the original information is not neglected as a variation factor.[75] The body dimensions of each sample group were extracted using PCA and Varimax rotation. This is a common technique applied on anthropometric data to describe variations in the human body in a parsimonious manner by many previous sizing studies.[64,76–80] Table 6.4 shows the principal component analysis (PCA) analysis of the female sample population for age 7–12. From Table 6.4 and scree plot figure, it can be seen that the first 3 components are significant to be retained which accumulated to 86.4%.

According to Pedhazer and Schmelkin,[81] the first few components should extract at least 50% of the variance in order to prove the usefulness of PCA technique. The current study shows that 76% of the variance in the 50 body dimensions for females (age 7 to 12) is explained in component 1. The result indicates variance of more than 50% in the first component, which

Table 6.4 Principal component analysis extraction for females (age 7–12)

Component	Initial eigen values			Rotation sums of squared loadings	
	Total	% of variance	Cumulative %	% of variance	Cumulative %
1	38.1	76.3	76.3	41.9	41.9
2	3.6	7.2	83.5	77.5	77.5
3	1.5	2.9	86.4	8.4	86.4

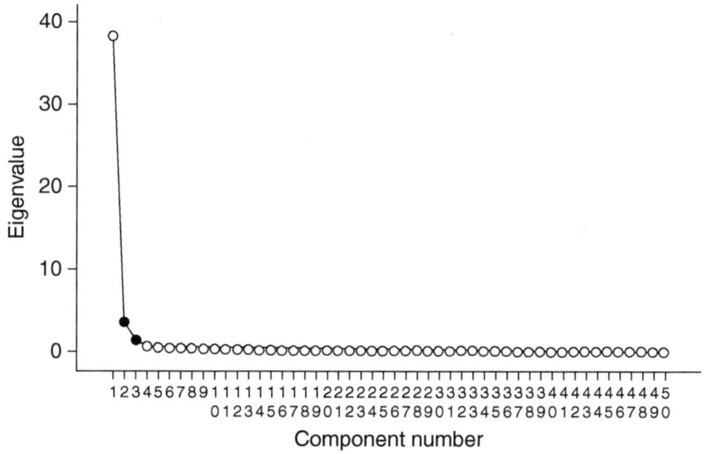

6.4 Scree plot of female sample population (age 7–12).

indicates that PCA technique is an effective method for this study to obtain a parsimonious solution in describing the variations of body shapes in children. Furthermore, according to Hair *et al.*,[82] the extracted factor should be at least 90% of the explained variance in order to show efficiency.

From this finding, 90% of the variance was explained by seven principal components for female samples aged 7 to 12. In contrast, a study conducted by Salusso-Deonier and DeLong[83] showed only 60% of total variance for 15 components, which is considered low. However, in 2006, the same researcher improved her studies by obtaining a 60% total variance accounted for by two components.[84]

In order to reduce the number of components for a more parsimonious solution, the latent root, scree plot and percentage of accumulated variance criteria of retaining components were applied. Table 6.4 shows all the components that have been extracted with an eigen value greater than one, which implies that these components are suitable to be retained. It can be seen that principal component 1 (PC1) has the highest eigen value, which means that this is the principle component of the data set. There are three components (1,2 and 3) for females (age 7–12) showing eigen values greater than one, in which PC1 has a value of 38.1% followed by 3.6% and 1.5% for PC2 and PC3, respectively. The factor rotation value for retained components is observed at 86.4% for females.

For criterion 2, which is the scree plot, the results are shown in Fig. 6.4. An elbow where the curve merges into a straight line can be seen at components 2 and 3 (which are marked by dark gray circles). The component at that merge location is the point to stop. From Fig. 6.4, the elbow is obvious from PC2 and PC3 and thus PC1 to PC3 are retained.

6.4.3 Female samples (age 7–12)

For our female sample, 46 variables were loaded on two components. Out of 46 variables, 30 variables were found to have high factor loadings (≥ 0.75). It was noted that there are 15 variables for upper body and another 15 variables for lower body. The aim of this section is to select the key dimensions. It is apparent from Table 6.5 that most of the variables are highly correlated to the individual components, which can be seen from the high factor loadings. Overall, almost all variables are reduced into two factors.

As can be seen from Table 6.5, for girth upper body dimensions, upper arm girth and bust girth are distinguished as the strongest variables correlated to girth factor with factor loadings of 0.88 and 0.84 respectively. In addition, for length in upper body dimensions, under arm length and inside leg length are noted to have the highest factor loading correlated to length with factor loadings of 0.86 and 0.90 respectively. On the other hand, the highest factor loading for lower body girth is found in waist girth (0.88) and mid-thigh girth (0.87). In addition, for lower body length the two highest factor loadings are found in inside leg length (0.9) and hip height (0.88). However, height is also found to have a high factor loading at 0.85.

In conclusion, for our girls sample population aged 7–12, both length and girth variables are selected to be the key dimensions for a sizing system. Bust girth and waist girth are selected for girth dimensions for upper and lower body respectively. In addition, height is selected as the key dimension for both upper and lower body. Height is selected based on the high factor loading. The current study found that height represents almost 85% (0.85) of the length component for female samples aged 7–12.

From the opinion of other researchers, height must be incorporated in a sizing system.[8,85] In addition, according to James and Stone,[86] height was found to offer an advantage, especially for children, as it could be easily measured in retail shops where a height chart could easily be obtained. Height is also regarded as a better estimator of size than age.[67] Hence, height is the most suitable dimension to cluster the samples into different groups.

Table 6.5 Principal component analysis with Varimax rotation for females (age 7–12)

Body types	Variables	PC 1 (girth)	PC 2 (length)
	Height		0.85
	Weight	0.83	
Upper body (girth)			
	Upper arm girth	0.88	
	Back width	0.75	
	Bust girth	0.84	
	Armscye girth	0.85	
	Neck girth	0.83	
	Elbow girth	0.82	
	Wrist girth	0.78	
	Cervical to breast point	0.80	
	Neck shoulder to breast point	0.77	
	Trunk circumference	0.77	
Lower body (girth)			
	Waist girth	0.88	
	Mid-thigh girth	0.87	
	Thigh girth	0.86	
	Calf girth	0.86	
	Hip girth	0.83	
	Lower knee girth	0.84	
	Knee girth	0.79	
	Minimum leg girth	0.79	
	Crotch	0.85	
Upper body (length)			
	Hand length		0.79
	Upper arm length		0.86
	Cervical to wrist		0.76
	Under arm length		0.85
	Arm length		0.83
Lower body (length)			
	Knee height		0.86
	Inside leg length		0.90
	Hip height		0.88
	Waist height		0.83
	Outer leg length		0.82
	Thigh length		0.84
	Total proportion (%)	41.9	77.5

All the key dimensions mentioned above were finally selected for the classification of the sample population. These key dimensions were selected based on the high relationship with the main body measurements, girth and length. Furthermore, it has been mentioned in previous studies that key dimensions must be convenient to take.[87] This indicates that the selected key dimensions should be the ones that can be measured easily and practically.[38] In addition, it is also stated that if these measurements will be used as the size coding, the customers should be very familiar with the key dimensions.[88] For example, chest girth or bust girth for upper body as compared to upper arm girth.

6.5 Clustering children's body shapes and sizes

A cluster analysis was then used to segment the sample subjects into homogenous groups, which means similar body shapes and sizes. The key dimensions obtained from the PCA methods are then used to group the sample population into different body shapes. The grouping of the children was done in two stages, in which the sample population was first clustered and then individuals classified into their own cluster groups.

Partitioning type of clustering (k-means clusters) was employed to analyze the anthropometric data for females aged 7–12. From the PCA results, the selected key dimensions (length and girth variables) are used as the primary control dimensions, which means that both length and girth dimensions are used to divide the sample population; this differs from many previous studies which used height as the primary control dimension followed by secondary control dimensions of bust, chest or hip girth.[20,77]

In order to create a garment that fits the body well, apparel experts recommend that size development should be designed according to the specific part of the body.[71,89] In addition, Gupta[72] adds 'better fit could be obtained in the garments if the sample is divided separately for the upper and the lower body'. Thus, the clustering of the samples was divided into upper body and lower body. In the previous pages, PCA demonstrated that height and girth are two important components. Therefore, height and bust girth or height and chest girth were used as the primary control dimensions for upper body classification and height and hip girth were used for lower body classification.

In general, the k-means partitioning technique confirms that the grouping of the data into three different clusters is significantly different for each sample group. Similar to previous studies, a cluster analysis technique was successfully used to divide the anthropometric data into three body types, which resulted in distinct groups.[64,90] In addition, the scatter plot of key dimensions for each sample group depicts a clear densely packed distribution of the data separated into well-defined body types. Three distinct body types were observed in the sample population: small size, medium size and large size. The results analysis is presented in the following paragraphs.

6.5.1 Clustering for upper body for female sample size (age 7–12)

The centroids of the clusters for females (age 7 to 12) are depicted in Table 6.6. The majority of the subjects are scattered in Cluster 1 (small size), followed by Cluster 3 (medium) and the fewest in Cluster 2 (large size).

Table 6.6 Cluster groups for upper body: female (age 7–12)

	Cluster 1	Cluster 3	Cluster 2
N	197	194	134
Mean height(cm)	122.5	133.4	145.9
Range (cm)	114.0–131.0	120.5–146.0	132.1–153.7
Mean hip girth (cm)	59.4	67.6	77.1
Range (cm)	53.5–71.8 cm	57.3–83.0	63.0–87.5
Body type	*Small*	*Medium*	*Large*

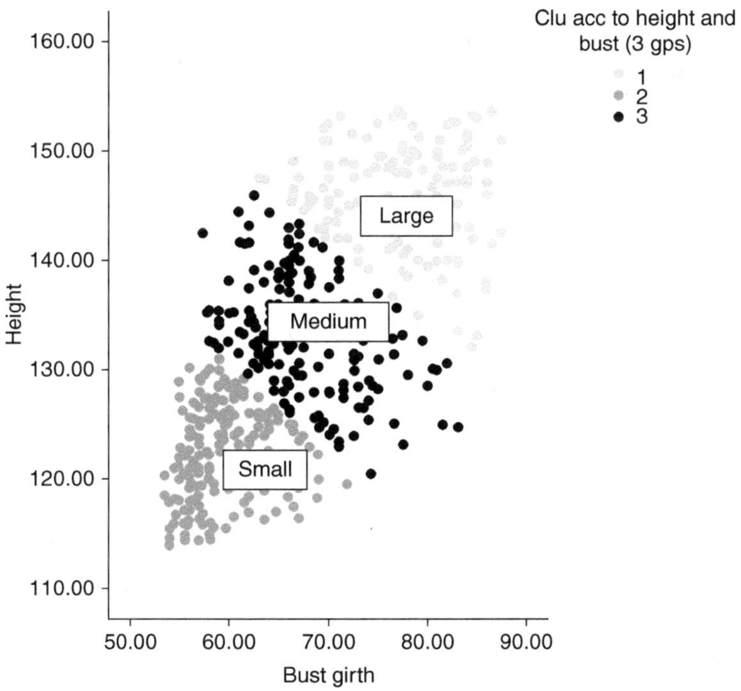

6.5 Scatter plot of cluster groups according to height and bust for females (age 7–12).

Figure 6.5 illustrates the distribution of female samples according to height (y-axis) and bust girth (x-axis). The well-defined clusters are separated into three main body types. Cluster 1, the small size group, has short height and small bust measurements. Cluster 2 samples are tall with big bust. Cluster 3 samples are of average height, and their bust measurements range from small to medium size.

Table 6.7 tabulates the distribution of cluster groups according to age. As shown in the table, the majority are in the small size group (Cluster 1); most in this cluster are 8 years old (62/197). The second highest distribution

Table 6.7 Age vs. cluster groups for upper body: female (age 7–12)

Age	Cluster according to height and bust (3 groups)			Total
	Small	*Medium*	*Large*	
7	*58	14	2	74
8	62	33	1	96
9	42	44	12	98
10	17	*58	29	104
11	13	28	41	82
12	5	17	*49	71
Total	197	194	134	525

*The highest number of students in each group.

Table 6.8 Cluster groups for lower body: female (age 7–12)

	Cluster 1	Cluster 2	Cluster 3
N	176	196	142
Mean height (cm)	122.6	132.4	145.8
Range (cm)	114.0–132.5	119.5–143.3	133.0–154.0
Mean hip girth(cm)	64.39	73.56	82.24
Range (cm)	58.0–74.0	61.0–91.0	70.0–93.5
Body type	*Small*	*Medium*	*Large*

of the sample (n=194) is in the medium size body type (Cluster 3), in which the majority of the sample population are 10 years old. The last group, Cluster 2 (n=134), is the large size children and the majority here are 12 years old.

6.5.2 Clustering for lower body for female sample size (age 7–12)

For the lower body, the female sample size was divided into three distinct clusters as seen in Table 6.8. A majority of the samples are found in Cluster 2 (medium size) followed by Cluster 1, which is the small group. The last group is Cluster 3, the large group. Figure 6.6 shows female samples in three cluster groups divided according to height and hips. In Cluster 1 (small size), samples are mainly found in the short height and narrow hips category. In Cluster 2 (medium size), samples are in the middle of the graph showing they are in the average height category with small to medium size hips. In Cluster 3 (large size), the samples are grouped at the upper right of the graph, showing that they belong to the tall height category with wide hip girth.

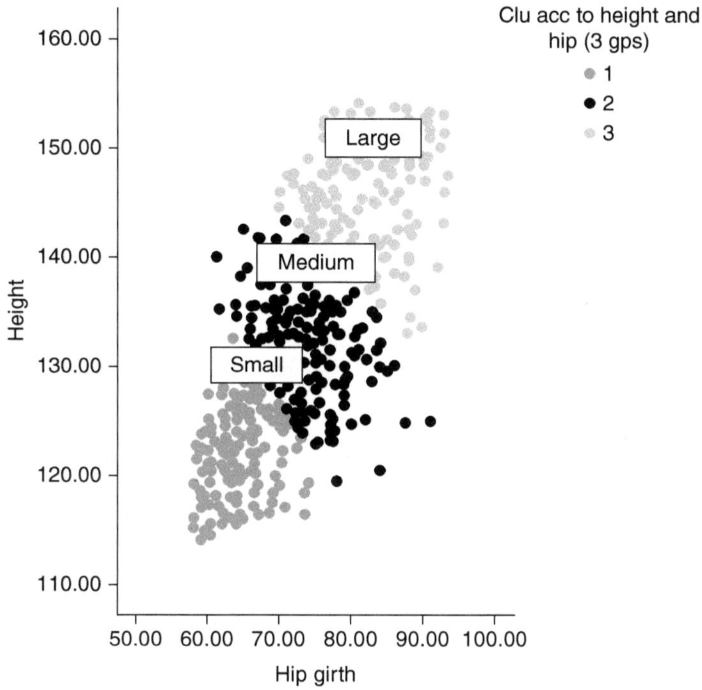

6.6 Scatter plot of cluster groups according to height and hip for females (age 7–12).

Table 6.9 Age vs. cluster groups for lower body: female (age 7–12)

Age	Cluster according to height and hip (3 groups)			Total
	Small	*Medium*	*Large*	
7	53	19	1	73
8	*54	35	2	91
9	33	50	12	95
10	16	*53	35	104
11	14	24	44	82
12	6	15	*48	69
Total	176	196	142	514

*The highest number of students in each group.

Table 6.9 shows that a majority of the sample size (n=196) fall under Cluster 2, medium size. A large group of the samples in this cluster is 10 years old. The second highest distribution of sample size (n=176) is in Cluster 1, small size, wherein the majority is 8 years old. The last cluster group (Cluster 3, n=142) is the large size body type; the majority here are 12 years old.

6.5.3 Summary of cluster sampling technique for sample group (age 7–12)

Figure 6.7 summarizes the final body characteristics extracted via a k-means cluster technique for female and body regions (upper and lower body). This study confirms that there are different cluster trends that describe three different body characteristics in females of the same age group (7–12). The same cluster trend can be seen for both upper and lower body for female samples. For example, in other research it was found that there is a linear growth in all the cluster groups.[64]

In typical sizing research of three clusters, linear trends are observed which are positively correlated; that is, if the height is tall, the girth is also big and vice versa, indicating that the body types are small, medium and large.[20,64,78] The same observation is shown in our female sample (Fig. 6.7). Furthermore, in this age group (7–12), the obvious pattern is that the lower age group belongs to the smaller size body type. For example, the majority of age 7 and 8 are often in small size, 9 and 10 in medium size, and age 11 and 12 always in the large size group. For younger children (age 7–12) their height is differentiated in each of the groups; short, average height and tall. Therefore, by knowing the characteristics of body dimensions, an apparel manufacturer is able to manufacture clothing that fits the human body variations.

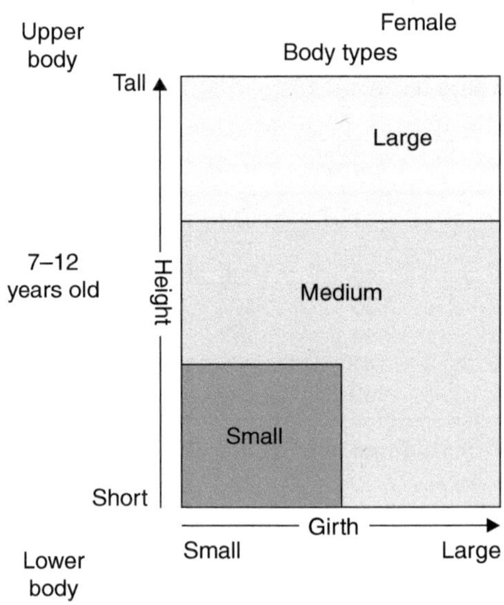

6.7 Body characteristics of females (age 7–12).

6.6 Classification of children's body shapes

The classification and regression tree (CRT) predictive modeling technique employed for this study depicts the rules for separating the data into significant groups based on the three clusters established in the previous section. Moreover, these results verify that the samples are classified into their own cluster groups with high accuracy. It also suggests that the accuracy of classification is excellent in this sample group, with more than 80% accuracy for age 7–12. Hence, this confirms that the sample groups are classified correctly into cluster groups using k-means technique. The profiles of each split are given in the following paragraphs.

For this section, dependent and independent variables are used to classify the cluster groups. The dependent variables are the cluster groups for each individual sample group. The independent variables are the primary control variables (key dimensions) of each sample group specific to either upper body or lower body.

The same observations are seen in all sample groups of both upper and lower body, where the rule of this tree model is as follows: the maximum depth of the tree extends two levels beneath the root node except in lower body of females age 7–12, where it is observed that the maximum depth extends to three levels beneath the root node. The minimum sample at the parent node is 100 and the child node is 50, the same for all sample groups. The same observation is also seen in the predictive variables, where height is selected as the most important variable for all sample groups age 7–12.

6.6.1 Decision tree for female upper body (age 7–12)

As shown in Table 6.10, the decision tree classified the cluster groups with correctness of Cluster 1, Cluster 2 and Cluster 3 as 88.8%, 90.3% and 86.6% respectively. The final overall percentage for this cluster group is 88.4%.

Table 6.10 Regression tree validation, upper body: female (age 7–12)

Observed	Predicted			
Female (7–12)	1	2	3	Percent correct (%)
1	175	0	22	88.8
2	0	121	13	90.3
3	3	23	168	86.6
Overall correctness (%)	33.9	27.4	38.7	88.4

Table 6.11 Profile of upper body types: female (age 7–12)

Node	Body type	Classified rule	*n*
1 and 3	Small	height ≤129.4 cm and bust girth ≤65.3 cm	175
2 and 5	Medium	height >129.4cm and <139.7cm	131
1 and 4	Medium	height <129.4 cm and bust girth >65.3 cm	37
2 and 6	Large	height>139.7 cm	121

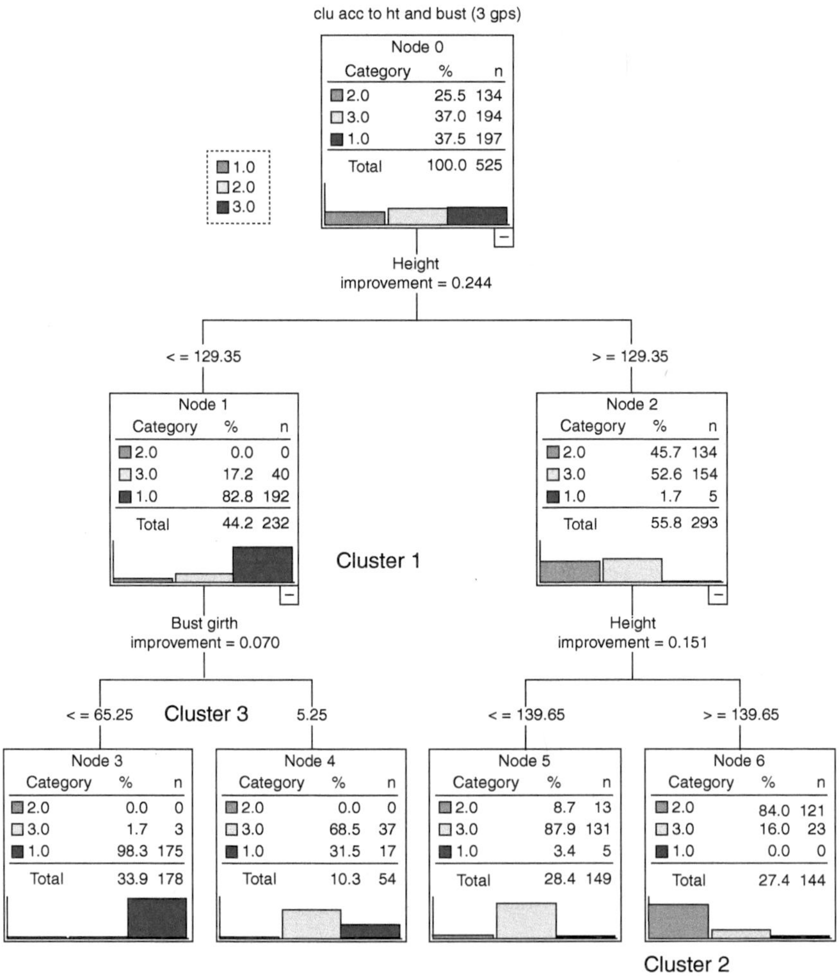

6.8 Decision tree for female upper body (age 7–12).

As seen from Table 6.11 and Fig. 6.8, the samples are first split into two height ranges. One group's height is more than 129.4 cm and the other is less than 129.4 cm. Those who are less than 129.4 cm tall with bust measurement lower than 65.3 cm are most likely to be grouped under the small size (Cluster 1). Those with the same height but with bust greater than 65.3 cm are likely to be grouped under the medium size (Cluster 3). Those samples with height greater than 129.4 cm and less than 139.7 cm are more likely to be grouped under the medium size (Cluster 3). Lastly, those who are taller than 139.7 cm are more likely to be grouped under the large size (Cluster 2). The tree validates the significance of the control variables. The profiles of the clusters shown in Table 6.11.

6.6.2 Decision tree for female lower body (age 7–12)

Decision tree analysis classified these cluster groups with a correctness of 87.4% as a whole. The correctness of classification for Cluster 1 is 99.4%, for Cluster 2 is about 71.9%, and for Cluster 3 is about 90.9%, as shown in Table 6.12. As shown in Fig. 6.9, samples are first divided into two height measurements, greater than and less than 139.7 cm. Those who are more than 139.7 cm tall are most likely to be grouped under the large size, and height seems to be the only significant characteristic for large size people. Those with height less than 139.7 cm and hips less than 67.4 cm are likely to be grouped in the small cluster. Those with height greater than 126.5 cm and less than 139.7 cm with hips greater than 67.4 cm are likely to be clustered under the medium size. Those with height less than 126.5 cm and hips greater than 67.4 cm are also likely to be grouped under small size. Table 6.13 summarizes the profiles.

6.6.3 The classification rules based on gender and age groups

Table 6.14 is the profile of the sample population of females aged 7–12 for upper and lower body. This table depicts the characteristics of each body type in the sizes (small, medium and large). Characteristics are shown in the rules and attributes for each body type. This study implies that the body characteristics of each size vary substantially, which suggests that each rule is needed to accommodate the body variations and in turn provide good fit to the wearer. The same results are shown in previous studies, where the coverage rules confirmed the effectiveness of sizing tables.[91,92]

Table 6.15 shows clearly that in the small size clusters, the height is always less than that of the other two sizes, which means that for the tailoring or

Table 6.12 Regression tree validation, lower body: female (age 7–12)

Observed	Predicted			
Female (7–12)	1	2	3	Percent correct (%)
1	175	1	0	99.4
2	42	141	13	71.9
3	0	13	129	90.9
Overall percentage (%)	42.2	30.2	27.6	87.4

Table 6.13 Profile of lower body types: female (age 7–12)

Node	Body type	Classified rule	*n*
1 and 3	Small	Height ≤139.7 cm and hip ≤67.4 cm	145
1, 4 and 5	Small	Height <126.5 cm and hip >67.4 cm	30
1, 4 and 6	Medium	Height >126.4 cm and ≤139.7 cm and hip >67.4 cm	141
2	Large	Height >139.7 cm	129

Table 6.14 Body types classification for age 7–12

Profile for age 7–12	
Body types	Females
	Upper body
Small	Height <129.4 cm Bust <65.3 cm
Medium	Height >129.4 cm And <139.7 cm Height <129.4 cm Bust >65.3 cm
Large	Height >139.7 cm
	Lower body
Small	Height <139.7 cm Hip <67.4 cm Height <126.5 cm Hip >67.4 cm
Medium	Height >126.5 cm And <139.7 cm Hip >67.4 cm
Large	Height >139.7 cm

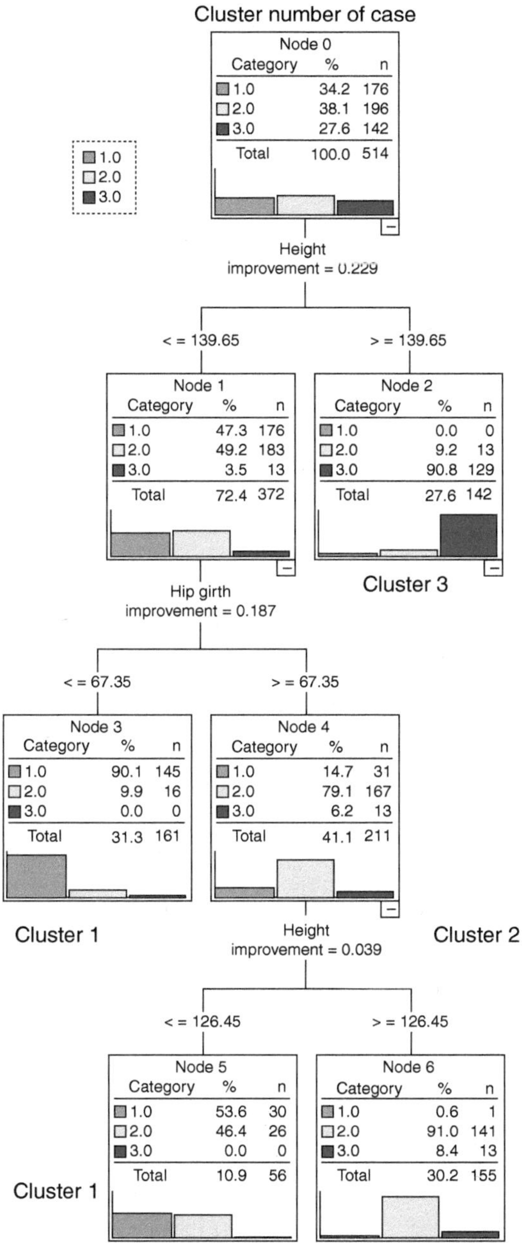

6.9 Decision tree for female lower body (age 7–12).

manufacturing of pants, for example, the length should be shorter as compared to the other two sizes. Note also that for small size males and females age 7–12, the attributes are different: so again, in manufacturing pants for lower body as an example, for male samples height alone is significant, while for

Table 6.15 Small size for upper body: female (age 7–12)

Designation		S114-54	S114-60	S122-54	S122-60
Control dimensions	Height	114–122	114–122	122–130	122–130
	Bust	54–60	60–66	54–60	60–66

female samples both height and hip measurements needed to be considered. Table 6.14 illustrates that height is the most important predictor variable in the younger cluster groups (age 7–12) where height is given as the first variable used to split the sample population. This is due to the fact that children in this age range show rapid changes in height and thus the selection of height could enhance the fit of clothing as these changes in growth are accommodated.[24]

This result is also in agreement with many previous studies, in which height and girth variables were chosen as the most important predictors for body sizes.[8,20,67] Moreover, the previous cluster analyses confirmed that the division of the sample group age 7–12 is proportionate for both height and girth variables, as they show a linear relationship. This indicates that an increase in height predicts an increase in girth. All the significant features found in Table 6.14 provide comprehensive rules for creating the size table. This table is definitely different from such tables in other countries, as it is based on the local sample population, as can be seen by comparing the sizes developed in this table with those developed by previous studies of similar age ranges in other countries.[20,64]

However, it is also noted that the rules of classification differ from one country to another, as it is exclusively built on the said population. It can also differ when different key dimensions are used. This supports the research conducted by previous studies, which verified that the structure of size systems differs substantially across countries and garment types and also within the same shop and same countries.[93–95]

6.7 Sizing system development and designation

In the following pages, the sizing system for each young females group is presented and discussed. All the sizes developed are according to the classification rules in Table 6.14.

The proposed sizing systems are reported and discussed for females age 7–12. The size table is developed according to upper and lower body garments as suggested by the ISO standard.[63,96] The development of a sizing system for this study is based on the range and size interval. Then, to separate the samples into the correct size categories (small, medium and large) the IF/THEN statements are employed on the samples using the profile in Table 6.14.

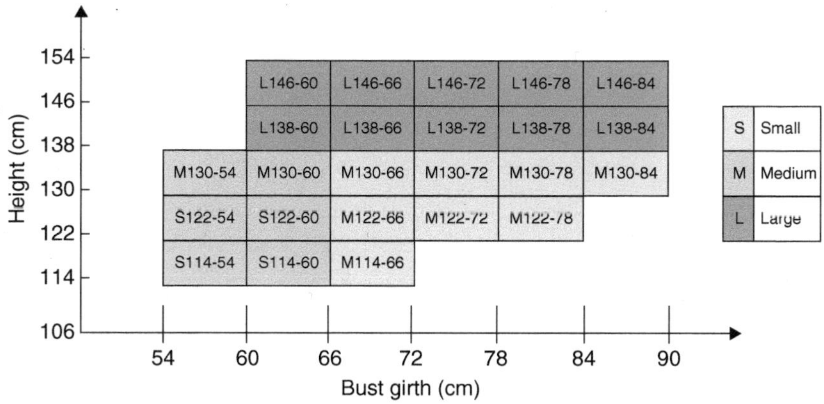

6.10 Distribution graph of bust girth vs. height for female upper body (age 7–12).

6.7.1 Female sizing system for upper body garments (age 7–12)

Figure 6.10 shows the distribution of sizes according to the body types for females aged 7–12. These three body types are produced by a cluster analysis technique and verified by a decision tree analysis. The control variables for upper body are height and bust girth. The size interval for height is 8 cm and for bust girth is 6 cm. The range for height in this population is 114–153.7 cm. This range is divided into six subgroups using an interval of 8 cm. The size interval was chosen based on the best coverage for these samples; in addition, ISO/TC133 commonly uses that as a sizing interval.[63] For the bust, the total range is 31.5 cm from 56 to 87.5 cm; this range was divided into six subgroups using an interval of 6 cm starting at 54 cm. The distribution graph of the key dimensions is illustrated in Fig. 6.10, with bust girth as the X-axis and height as the Y-axis.

Figure 6.10 shows the sample population divided into three sizes: small size, medium size and large size, resulting in 24 sizes. For the small size category, four sizes are provided with height of 114–129.9 cm and 54–65.9 cm bust girth measurement. For the medium size category, ten sizes are provided with height of 114–137.9 cm and bust measurements of 54–89.9 cm. Ten sizes are provided for the last group (large size) for tall children measuring 138–153.9 cm with bust measurement of 60–89.9 cm.

6.7.2 Female sizing system for lower body garments (age 7–12)

In Fig. 6.11, for the lower body, the control variables are height and hips girth. The range for height is 114–153.7 cm and the hips girth range is 58–97 cm.

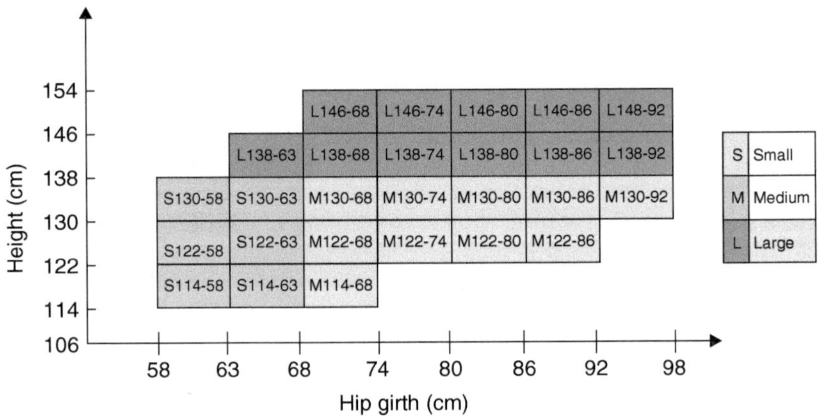

6.11 Distribution graph of height vs. hip girth for female lower body (age 7–12).

The size interval for height is 8 cm, and for the hip girth the size intervals are 5 and 6 cm (5 cm small size samples and 6 cm for medium and large size). These two different size intervals are applied in order to reduce the number of sizes to seven, as compared to eight sizes if using a 5 cm interval.

The samples are classified (Table 6.14) according to the correct sizes as shown in Fig. 6.11. For the small size category, seven sizes are found to accommodate samples with an height of 114–137.9 cm and with a 58–73.9 cm hip girth measurement. For the medium size category, there are nine sizes for children with a height of 122–137.9 cm, and hip measurements of 68–97.9 cm. There are eleven sizes for the large size for samples who are 138–153.9 cm tall with hip measurements of 63–97.9 cm.

6.7.3 Size validation according to gender and age

Several striking results emerge from this section for each sample group. The first is that applying the classified rules from a decision tree analysis resulted in unequal numbers of sizes in each cluster group. These cluster groups reflect the body types for each sample group (small, medium and large). However, these unequal numbers of sizes in each cluster group enables us to arrive at an efficient number of sizes, as shown by the high percentage of total coverage for the sample groups. Thus, this study has proven that unequal numbers of sizes in cluster groups have no adverse effect on building a good sizing system, in contrast to the earlier study, which found that equal numbers of sizes in cluster groups provided the basis for a good new standard sizing system.[97] In addition, the results also demonstrated that the aggregate loss of fit for each cluster group (body type) is accurate, and that

Table 6.16 Medium size for upper body: female (age 7–12)

Designation		M114-66	M122-66	M122-72	M122-78	M130-54	M130-60	M130-66	M130-72	M130-78	M130-84
Control dimensions	Height	114–122	122–130	122–130	122–130	130–138	130–138	130–138	130–138	130–138	130–138
	Bust	66–72	66–72	72–78	78–84	54–60	60–66	66–72	72–78	78–84	84–90

Table 6.17 Large size for upper body: female (age 7–12)

Designation		L138-60	L138-66	L138-72	L138-78	L138-84	L146-60	L146-66	L146-72	L146-78	L146-84
Control dimensions	Height	138–146	138–146	138–146	138–146	138–146	146–154	146–154	146–154	146–154	146–154
	Bust	60–66	66–72	72–78	78–84	84–90	60–66	66–72	72–78	78–84	84–90

Table 6.18 Small size for lower body: female (age 7–12)

Designation		S114-58	S114-63	S114-68	S122-58	S122-63	S130-58	S130-63
Control dimensions	Height	114–122	114–122	114–122	122–130	122–130	130–138	130–138
	Hip	58–63	63–68	68–74	58–63	63–68	63–68	58–63

Table 6.19 Medium size for lower body: female (age 7–12)

Designation		M122-68	M122-74	M122-80	M122-86	M130-68	M130-74	M130-80	M130-86	M130-92
Control dimensions	Height	122–130	122–130	122–130	122–130	130–138	130–138	130–138	130–138	130–138
	Hip	68–74	74–80	80–86	86–92	68–74	74–80	80–86	86–92	92–98

Table 6.20 Large size for lower body: female (age 7–12)

Designation		L138-63	L138-68	L138-74	L138-80	L138-86	L138-92	L146-68	L146-74	L146-80	L146-86	L146-92
Control dimensions	Height	138–146	138–146	138–146	138–146	138–146	146–154	146–154	146–154	146–154	146–154	146–154
	Hip	63–68	68–74	74–80	80–86	86–92	74–80	68–74	74–80	80–86	86–92	92–98

the accuracy of the total size table is good, as seen from the aggregate loss value, which is lower than the ideal value.

6.7.4 Female size designation tables (age 7–12)

Three size designation tables, Tables 6.15, 6.16 and 6.17, are depicted for all the female samples according to small, medium and large body types for upper body and another three size designation tables, Tables 6.18, 6.19 and 6.20, are created for lower body. All sizes are in cm. For Table 6.15, there are four small garment sizes created for females with an height range of 114–130 cm and a bust girth measurement range of 54–66 cm. In addition, from Table 6.16, there are ten different sizes for females with a height of 114–138 cm with a bust girth measurement of 66–90 cm. Lastly, for upper body garments in Table 6.17, there are ten different large sizes created for those with a height of 138–154 cm and a bust girth of 60–90 cm. In total, 24 size tables were created to accommodate 99% of the total sample population based on the 95th percentile value.

There are three main sizes, namely small, medium and large. Table 6.18 shows small sizes in which seven different sizes were developed for females with an height of 114–138 cm and a hip girth of 58–74 cm. For medium sizes (Table 6.19), nine different sizes were created for females within the height of 122–138 cm and with a hip girth of 68–98 cm. Lastly, in Table 6.20, another eleven sizes were created for large sizes with heights of 138–154 cm and a hip girth of 63–98 cm. Overall for lower body, there were a total of 27 sizes created to accommodate 99% of the sample population based on the 95th percentile.

6.8 Future trends

The challenges of developing a good sizing system lie in:

- the data that is being used; and
- the methodology applied.

Clothing sizing systems have been developed since the birth of ready-to-wear clothing; their main aim is to provide a good fit for the clothing. The fit should conform to a specific size range so that everybody in that size range is able to wear the same size. The challenges to creating well-fitting ready-to-wear garments have led to the realization of a number of researches over the years. The first sizing system was created as early as 1941. This means that for the past 70 years, researchers have been working hard to understand the many different body shapes and sizes that exist in every consumer group so that the right sizing system may be created. The challenges are even more important when it comes to creating a sizing system for children, as body growth is a continuous process. Growth among children is so varied that only an in-depth understanding of body shapes can lead to a good sizing system.

Currently, few sizing systems for children are being developed in the world. There also are few national sizing surveys specifically conducted to develop sizing systems for children as compared to national anthropometric surveys for adult sizing systems. The future lies in undertaking more anthropometric studies for children using the highly sophisticated scan technology and artificial intelligence methods to develop a sizing system. The validation of the sizing system can be done using the 'virtual fit try on' to ensure the actual for consumers.

In recent years, the collection of body measurements data using highly sophisticated body scanners has been more widely used for children's anthropometric surveys. In 2009, the UK conducted one of the first large scale projects to measure children, aiming at 6000 boys and girls aged 4 to 17. There were two phases: first measuring children aged 4 to 17, and then measuring infants age 0 to 4. In January 2013, Alvanon, the leader in body shape analysis, announced the body shape reports originating from this survey (http://www.shapegb.org/childrens_sizing_report_launch).

The most important feature for sizing systems for a specific sample population is to accommodate as much of the sample as possible with the smallest possible number of sizes. Using artificial intelligence to develop size systems can contribute to this objective.

6.9 Sources of further information and advice

There are three areas of knowledge to comprehend when developing children's sizing system:

1. children's growth and development;
2. anthropometric study; and
3. sizing system development.

Definitions of children's body shapes and sizes are described in textbooks on child development and growth. There are many books describing the physical growth of children and the different shapes at different phases of growth, such as:

- Life-Span Development, Physical Development and Biological Aging. Santrock, J.W. (2010).
- Garment simulation and animation on children bodies. Magnenat, N. (2005)
- Life-Span Human Development. Sigelman, C.K and Rider, E.A. (2011). Wadsworth Publishing; 7th Edition. ISBN-10: 1111342733.

The understanding of the nature of anthropometric surveys can be found in anthropometric text books. Additional reading in the form of books of ergonomics is recommended to understand the proper way of conducting anthropometric surveys. A number of studies have been realized to try to understand the nature of anthropometric study for children. They may be found in journal articles such as:

- Researchers begin army-wide anthropometric survey. Accetta, D. (2010). Natick Soldier research, development and engineering center (available at: http://www.army.mil/article/47646/).
- Surviving but Not Quite Thriving: Anthropometric Survey of Children Aged 6 to 59 Months in a Rural Western Uganda District. Jilcott, S.B.,Masso, K.L.,Ickes, S.B. and Myhre, S.D. (2007). *Journal of the American Dietetic Association*. Vol.**107** (11).p.1983–8.
- Ergonomics for children. Edited by Lueder, R. and Berg, V. (2008). Details for this study are available at http://www.humanics–es.com/child.
- http://ergonomicsforchildren.com/child-ergonomics.htm

In addition, some good articles about children body sizing system research are available through journal articles and the internet:

- 3-D body scanners capture shape and size of children (23 March 2013) available at: http://www.vision-systems.com/articles/2012/03/3-d-body-scanners-capture-kids-shape-and-size.html.
- Sizing system for functional clothing-uniforms for school children. Zakaria, N. (2011). *Indian Journal of Fibre and Textile Research*. Vol. **36**, December 2011, pp. 348–57.
- Developments of standardized sizing system for the South African children wear market. S Van Huyssteen (2006). Phd dissertation.
- Sizing Systems for Children's Wear in the United States. Su-Jeong H. Shin, Smith & Norris (2004). *ITAA Proceedings 2008* Texas Tech University
- A study of the apparel sizing of children's wears. An analysis of the size increments utilized in children's wear based on an anthropometric survey. Y. Kang, H.S.Choi, and W.H.Do, (2001). *Journal of Korean Home Economics Association* English Ed. vol. **2**, 2001.
- The development of new clothing size charts for female Kenyan children. Part 1: using anthropometric data to create size charts. Otieno,R.B. and Fairhust,C. (2000). *Journal of Textile Institute*. 1754–2340, vol. **91**, pp. 143–152.

6.10 References

1. A.H. Maslow, 'A theory of human motivation' *Psychological Review* vol. **50**, pp. 370–96, 1943.
2. A. Stonehouse, 'Children's clothing in child care' Australia National Children Accreditation Council Inc. (NCAC), 2008.
3. C.P.S. Boden, J. Pilcher and T. Edwards, New Consumers? Children, Fashion and Consumption. Paper presented at 'Knowing Consumers: Actors, Images, Identities in Modern History' Conference, Universitat Bielefeld, Germany, 27–28 February 2004.
4. S. Gautam, 'Concept of General Consideration of Clothing for Preschool Children (2–4 Years) Among Rural Mothers of Palampur Tehsil' *Journal Social Science* vol. **11**, pp. 253–4, 1995.
5. A. Melchiorre, 'At what age...are school-children employed, married and taken to court' Swedish International Development Cooperation Agency, 2nd ed, 2004.
6. Dictionary, 'American Heritage Dictionary' 2007.
7. United Nations High Commissioner For Human Rights, 'Convention on the Rights of the Child' 1989-Article 1. Available at: http://www.ohchr.org/EN/ProfessionalInterest/Pages/CRC.aspx [accessed March 2013].
8. Y. Kang, H.S.Choi, and W.H.Do, 'A study of the apparel sizing of children's wear. An analysis of the size increments utilized in children's wear based on an anthropometric survey,' *Journal of Korean Home Economics Association English ed.* vol. **2**, pp. 95–110, 2001.
9. P.S. Norum, 'Consumer satisfaction with shopping experiences and clothing products in the children's wear market,' *Journal of Consumer Studies and Home Economics* vol. **1**, pp. 331–47, 1995.
10. L.J. Connell, P. Ulrich, K. Simmons, D. Pascoe and D. Bruner. 'Apparel Product Development for Plus-sized Tween and Teen Boys,' Project No. S06-AC03 (Competency: Management Systems), 2006.
11. K. Carol, M. Alexander and V. Spencer, 'Exercise clothing for children in a weight-management program' *Journal of Family and Consumer Sciences* vol. **99**, pp. 68–72, 2007.
12. T. Connolly, 'Plus size clothing industry trends: does bigger mean bigger business' in *Special to Fashion Industry Today.com*, 2 May 2008. Available at: http://www.fashionindustrytoday.com/2008/05/02/plus-size-clothing-industry-trends-does-bigger-mean-bigger-business/ [accessed June 2013].
13. L.J. Chen, K.R. Fox and A.M. Haase, 'Body shape dissatisfaction and obesity among Taiwanese adolescents' *Asia Pacific Journal Clinical Nutrition* vol. **17**, pp. 457–60, 2008.
14. H.W. Kohl and K.E. Hobbs, 'Development of physical activity behaviors among children and adolescents' *Pediatrics* vol. **101**, pp. 549–54, 1998.
15. J. McNeal, 'Children as consumers' *American Demographics* vol. **12**, pp. 36–9, 1990.
16. T.J. Eveleth, *Worldwide variation in human growth, 2nd ed.*, Cambridge: Cambridge University Press, 1990.
17. S. Stinson, 'Growth variation: biological and cultural factors' In Stinson, S., Bogin, B., Huss-Ashmore, R. and O'Rouke, D., *Human biology: an evolutionary and biocultural perspective*, 2nd ed., John Wiley & Sons, pp. 425–63, 2000.

18. UNICEF. *The State of the World's Children 1998*, New York: Oxford University Press, 1998.
19. J.H. Yoon, 'A study on the apparel sizing for children's upper garment.' *Korean Society of Costume* vol. **3**, pp. 307–21, 1995.
20. A.S. Ariadurai, T.P.G. Nilusha and M.R. Dissanayake, 'An anthropometric study on Sri Lankan school children for developing clothing sizes' *Journal of Social Science* vol. **19**, pp. 51–6, 2009.
21. F. Bellisle, 'Effects of diet on behaviour and cognition in children.' *British Journal of Nutrition* vol. **92**, pp. S227-32, 2004.
22. J.M. Tanner, *Fetus into Man*, Cambridge: Harvard University Press, 1978.
23. J.A. Roebuck, 'Anthropometric methods: designing to fit the human body', Monographs in Human Factors and ergonomics, Human Factors and Ergonomics society, Santa Monica, CA, 1995.
24. S.R. Daniels, 'Body shape in children and adolescents' *Journal of Pediatrics.* vol. **152**, p. 45, 2008.
25. P.T. Slee, *Child, adolescent and family development*, 2nd. ed. New York: Cambridge University Press, 2002.
26. E. Bye, K.L. LaBat and M.R. Delong, 'Analysis of body measurement systems for apparel' *Clothing and Textiles Research Journal* vol.**4**(2), pp. 66–79, 2006
27. R. Otieno, 'The role of garment sizing in creation of customer satisfaction: indications from focus groups responses' *Journal of Fashion Marketing & Management* vol. **4**, pp. 325–35, 2000.
28. K. Yeosun, C. Hei-Sun, and H.D. Woei, 'A study of the apparel sizing of children's wear- an analysis of the size increments utilized in children's wear based on an anthropometric survey' *Journal of Korean Home Economics Association English Edition* vol. **2**, pp. 96–110, 2001.
29. S.J.A. Harper, P.J. Dewar, and B.A. Diack, 'The purchase of children's clothing – who has the upper hand' *Journal of Fashion Marketing and Management* vol. **7**, pp. 196–206, 2003.
30. Y.T. Lee, 'Body Dimensions for Apparel' NISTIR 5411, National Institute of Standards and Technology, Gaithersberg, MD, p. 1, 1994.
31. M. Konstanski, A. Fisher and E. Gullone, 'Current conceptualisation of body image dissatisfaction: have we got it wrong?' *Journal of Child Psychology Psychiatry* vol. **45**, pp. 1317–25, 2004.
32. Z. Zhang, G. Ching, G. Ging, J. Moody and W. Liu, 'Investigation of denim wear consumption in China' *Research Journal of Textile and Apparel* vol. **3**, pp. 60–4, 1999.
33. M. Grasson and R. Wright, 'African American mother's needs,search and evaluation of children's clothing' *Journal of Fashion Marketing & Management* vol. **2**, pp. 41–55, 1997.
34. I.J. Grant and S.R. Graeme, 'Buying behaviour of 'tweenage' girls and key societal communicating factors influencing their purchasing of fashion clothing '*Journal of fashion marketing and management* vol. **9**, pp. 450–67, 2005.
35. M. Alexander, L.J. Connell, and A.B. Presley, 'Clothing fit preferences of young female adult consumers' *International Journal of Clothing Science and Technology* vol. **17**, pp. 52–64, 2005.
36. L.D. Cohn, N.E. Adler, C.E. Irwin, S.G. Millstein, S.M. Kegeles and G. Stone, 'Body-figure preferences in male and female adolescents' *Journal of Abnormal Psychology* vol. **96**, pp. 276–9, 1987.

37. P.N. Danziger, What people buy: teen and tween clothing. In *Why people buy things they don't need: understanding and predicting consumer behavior*. Kaplan Publishing; 1 July, Chapter 8, pp. 237–240, 2004.
38. M.J. Chung and M.J. Wang, 'The development of sizing systems for school students.' *In 36th international conference on computers and industrial engineering. taipei, taiwan*, 2006.
39. K.L. LaBat and M.R. DeLong, 'Body cathexis and satisfaction with fit of apparel.' *Clothing and Textiles Research Journal* vol. **8**, pp. 43–8, 1990.
40. Y.S.C. Ming, 'Body shape preference and body satisfaction of Taiwanese and Japanese female college students.' *Psychiatric Research* vol. **133**, pp. 263–71, 2005.
41. D.W. Bunnell, P.J. Cooper, S. Hertz, and I.R. Shenker, 'Body shape concerns among adolescents' *International Journal of Eating Disorders* vol. **11**, pp. 79–83, 1992.
42. L.A. Proos, 'Anthropometry in adolescence -secular trends, adoption, ethnic and environmental differences' *Hormone Research* vol. **39**, pp. 18–24 1993.
43. D.C. Jones, T. Vigfusdottir, and Y. Lee, 'Body Image and the appearance culture among adolescent girls and boys: an examination of friend conversation, peer criticism, appearance magazines and the internalization of appearance ideals.' *Journal of Adolescent Research* vol. **19**, p. 323, 2004.
44. J.K. McCarthy HD, P.M. Emmett and I. Rogers, 'Trends in waist circumferences in young British children: a comparative study.' *International Journal Obesity Related Metabolism Disorder* vol. **29**, pp. 157–62, 2005.
45. M. Alexander, L.J. Connell and A.B. Presley, 'Clothing fit preferences of young female adult consumers' *International Journal of Clothing Science and Technology* vol. **17**, pp. 52–64, 2005.
46. T. Tselepis and H.M. De Klerk, 'Early adolescent girls' expectations about the fit of clothes: A conceptual framework' *Journal of Family Ecology and Consumer Sciences* vol. **32**, pp. 83–94, 2004.
47. M. Suzanne and J. Lee, 'Proximity of clothing to self scale' *Clothing and Textiles Research Journal* vol. **22**, p. 161, 2004.
48. D.C. Jones, T.H. Vigfusdottir and Y.S. Lee, 'Body image and the appearance culture among adolescent girls and boys' *Journal of Adolescent Research* vol. **19**, pp. 851–6, 2004.
49. S. P. Ashdown and L. Dunne, 'A study of automated custom fit: readiness of the technology for the apparel industry' *Clothing and Textiles Research Journal* vol. **24**, p. 121, 2006.
50. S. Sigelman, *Life span of human development*, 4th ed., New York: Thomson learning, 2006.
51. C.B. Malina and B.O. Oded, *growth, maturation, and physical activity*, Champaign, IL: Human Kinetics, 2004.
52. B. Bogin, *Patterns of human growth* 2nd ed., Cambridge: Cambridge University Press, p. 444, 1999.
53. The International Association of Athletics Federations (IAAF), 'Introduction to Coaching Theory: Growth and development' in *IAAF Coaching Theory Manual*, 2009.
54. T.P. Gullota, G.R. Adams and C. Markstrom, *The adolescent experience*, 4th ed., San Diego, CA: Academic Press, 1999.

55. C. Ruff, 'Variation in human body size and shape' *Annual Review of Anthropology* vol. **31**, pp. 211–32, 2002.
56. B. Bogin, P. Smith, A.B. Orden, V.M.I. Silva, and J. Loucky, 'Rapid change in height and body proportions of Maya American children' *American Journal of Human Biology* vol. **14**, pp. 753–61, 2002.
57. J.M. Winks, 'Clothing sizes: International Standardization', The Textile Institute, Redwood Books, Manchester, UK, 1997.
58. S. G. Dudek, *Nutrition essentials for nursing practice*, 4th ed., Philadelphia: Lippincott, 2001.
59. M. Voracek and M. Fisher, 'Shapely centrefolds? Temporal change in body measures: trend analysis,' *BMJ* vol. **325**, pp. 1447–8, 2002.
60. P.T. Slee, '*Child, adolescent and family development*' 2nd ed., Cambridge University Press, pp. 320–31, 2003.
61. J. MacGregor, 'Introduction to the Anatomy and Physiology of Children: A Guide for Students of Nursing, Child Care and Health', Routledge, p. 255, 2008.
62. ISO/TR 10652:1991, 'Standard sizing systems for clothes'.
63. ISO 3637:1977, 'Size designation of clothes – Women's and girls' outerwear garments'.
64. M.J. Chung, H.F. Lin, J.J. Mao, and J. Wang, 'The development of sizing systems for Taiwanese elementary-and high-school students' *International Journal of Industrial Ergonomics* vol. **37**, pp. 707–16, 2007.
65. M. de Onis, A.W. Onyango, E. Borghi, A. Siyam, C. Nishida, and J. Siekmann, 'Development of a WHO growth reference for school-aged children and adolescents' *Bulletin of the World Health Organization* vol. **85**, pp. 660–7, 2007.
66. J. Robotham, 'Children's changing shape' *Brisbane Times*, 2009.
67. R. Otieno and C. Fairhurst, 'The development of new clothing size charts for female Kenyan children. Part I: Using anthropometric data to create size Charts' *Journal of the Textile Institute, 1754–2340* vol. **91**, pp. 143–52, 2000.
68. W.F.F. Kemsley, 'Reviewed work(s): Women's Measurements and Sizes. by Joint Clothing Council, Ltd,' *The Incorporated Statistician* vol. **8**, pp. 143–5, 1958, http://www.jstor.org/pss/2986415.
69. A. Petrova, 'Creating sizing systems' In Ashdown, S.P. (2007) *Sizing in clothing developing effective sizing systems for ready-to-wear clothing'* Cambridge: Woodhead Publishing Limited, pp. 57–82, 2007.
70. A. Beazley, 'Size and fit: Formulation of body measurement tables and sizing systems – Part 2' *Journal of Fashion Marketing and Management* vol. **3**, pp. 260–84, 1998.
71. A. Beazley, 'Size and Fit: The development of size charts for clothing-Part 3' *Journal of Fashion Marketing and Management* vol. **3**, pp. 66–84, 1998.
72. D. Gupta, N. Garg, K. Arora and N. Priyadarshini, 'Developing body measurement charts for garment manufacture based on a linear programming approach' *Journal of Textile and Apparel Technology and Management* vol. **5**, pp. 1–13, 2006.
73. J.C. Yoon, C. Jasper, 'Key Dimensions of women's ready-to-wear apparel: developing a consumer size-labeling system' *Clothing and Textiles Research Journal* vol. **14**, pp. 89–95, 1996
74. S. Ashdown, S. Loker, and C. Adelson, 'Use of body scan data to design sizing systems based on target markets' *National Textile Center Research*

Briefs –Management Systems Competency: NTC Project: S01-CR01, pp. 1–10, 2004.

75. G. Hutcheson and N. Sofroniou, *Introductory statistics using generalized linear models*, Thousand Oaks, CA: Sage Publications, 1999.
76. H.C. Hsu, H.F. Lin and M.J. Wang, 'Development female size charts for facilitating garment production by using data mining.' *Journal of the Chinese Institute of Industrial Engineers*, vol. **24** (3), pp. 245–52, 2007.
77. D. Gupta and B.R. Gangadhar, 'A statistical model for developing body size charts for garments.' *International Journal of Clothing Science and Technology* vol. **16**, pp. 458–69, 2004.
78. C.H. Hsu, H.F. Lin, and M.J. Wang, 'Developing female size charts for facilitating garment production using data mining' *Journal of the Chinese Institute of Industrial Engineers* vol. **24**, pp. 245–51, 2007.
79. C. Hsu, 'Data mining to improve industrial standards and enhance production and marketing: An empirical study in apparel industry' *Expert Systems with Applications* vol. **36**, pp. 4185–91, 2009.
80. C.J. Deonier, M.R. DeLong, B. Frank, and K.R. Krohn, 'A multivariate method of classify in body form variation for sizing women's apparel' *Clothing and Textiles Research Journal* vol. **4**, pp. 38–45, 1985.
81. E.J. Pedhazer and L.P. Schmelkin, *Measurement, design, and analysis: An integrated Approach*, Hillside, NJ: Lawrence Erlbaum Assoc., 1991.
82. J.F. Hair, R.E. Anderson, R.L. Tatham, and W.C. Black, *Multivariate data analysis*, 5th ed., Prentice Hall, Upper Saddle River, 1998.
83. C.J. Salusso-Deonier and M.R. DeLong, 'A multivariate method of classify in body form variation for sizing women's apparel,' *Clothing and Textiles Research Journal* vol. **4**, pp. 38–45, 1982.
84. C.J. Salusso, J.J. Borkowski, N. Reich, and E. Goldsberry, 'An alternative approach to sizing apparel for women 55 and older' *Clothing and Textiles Journal* vol. **24**, pp. 96–109, 2006.
85. W.F.F. Kemsley, 'Women's measurements and sizes, HMSO London', H.M. Stationery Office, 1957.
86. R. James and P. Stone, *Children's wear sizing survey*, Leeds: Clothing and Allied Products Industrial training board, 1984.
87. W. Yu, 'Human anthropometric and sizing system' in *Clothing appearance and fit: Science and technology*, Cambridge: Woodhead Publishing Limited, p.188, 2004.
88. A. Petrova, 'Creating sizing systems' in Ashdown, S.P *Sizing in clothing developing effective sizing systems for ready-to-wear clothing*, Cambridge: Woodhead Publishing Limited, p. 67, 2007
89. J.C. Yoon and C.R. Jasper, 'garment-sizing systems: an international comparison' *International Journal of Clothing Science and Technology* vol. **5**, pp. 28–37, 1993, *Journal of Early Adolescence* vol. **27**, pp. 90–114, 2007.
90. H.C. Hsu, H.F. Lin, and M.J. Wang, 'Developing female size charts for facilitating garment production by using data mining' *Journal of the Chinese Institute of Industrial Engineers* vol. **24**, pp. 245–51, 2007.
91. H.F. Lin, C.H. Hsu, M. J. Wang and Y.C. Lin, 'An application of data mining technique in developing sizing system for army soldiers in Taiwan' *World Scientific and Engineering Academy and Society (WSEAS) Transactions on Computers* vol. **7**, pp. 245–252, 2008

92. H.L. Viktor, E. Paquet, and H. Guo, 'Measuring to Fit: Virtual tailoring through cluster analysis and classification', presented at 17th International European Conference on Machine Learning and the 10th European Conference on Principles and Practice of Knowledge Discovery in Database, September 18–22, Berlin, Germany, 2006.
93. C.Y. Jongsuk, C.R. Jasper, 'Garment-sizing charts: An international comparison' *International Journal of Clothing Science and Technology* vol. **5**, pp. 28–37, 1993.
94. D.Ujević, Brlobašić Šajatović, Blaženka; Hrženjak, Renata; Doležal, Ksenija; Karabegović, Isak; Petrovečki, and Vedran, 'Contribution to Anthropometric Surveys and Principal Comparisons of the Garment Size System' presented at The 86th Textile Institute World Conference, Fashion and Textiles: Heading Towards New Horizons, 18–21 November 2008, Hong Kong, China.
95. D. Ujevi, and I. Karabegovi, 'Anthropometry and the comparison of garment size systems in some European countries' *College of Anthropology* vol. **29**(1), pp. 71–8, 2005.
96. ISO 3636:1977, 'Size designation of clothes -Men's and boys' outerwear garments.'
97. R. Ng, S.P. Ashdown and A. Chan, 'Intelligent size table generation', Proceedings of the Asian Textile Conference (ATC), 9th Asian Textile Conference, Taiwan, 2007.

7

Older consumers: understanding sizing, body shapes and requirements of apparel for this niche

J. McCANN, University of South Wales, Wales, UK

DOI: 10.1533/9781782422150.2.132

Abstract: The author discusses clothing design for active ageing where a major concern is the changing size and shape of the older body. There is little design direction, or time and incentive, for product development teams to consider this emerging, unfamiliar market. Co-design led research indicates that older wearers wish for garment ranges, with fewer products, in less transient styling, with more consistency in sizing, with emphasis on age appropriate fit. In providing styling direction, the author aims to contribute to improving the image of ageing through clothing that is comfortable and flattering to the rapidly growing older market sector.

Key words: age appropriate fit, older body, active ageing, design, older market sector.

7.1 Introduction

This chapter discusses challenges involved in designing clothing for older people, where a major concern is related to the changing size and shape of the ageing body. 'We are now living in a unique historical period. Never before could people expect to live so long. Never before have physical and mental capabilities remained so high into advanced old age' (Huppert, 2003, p. 31). Rising life expectancy and declining family size have together produced a dramatic demographic shift, so that between 1901 and 2003 the proportion of the UK population aged over 50 increased from 15% to 30%, and is projected to rise to 41% by 2031 (Tomassini, 2005). Older people may be segmented as the active ageing, approximately 60–75, and as the fourth age, over 75 or those dependent on care. The author will look at the design issues of relevance to the active ageing, or Baby Boomer Generation, who do not perceive themselves as ill. This generation has been accustomed to making choices in the design of their clothing throughout their lives, since being the centre of attention of the fashion industry in their teenage years in the 1960s. As the fashion industry continues with business as usual, 'consumer research and marketing have failed to realise the relationship between changes in the leading values, views, and behaviours of the marketplace and the New Customer Majority' (Wolfe and Snyder, 2003, p. 9).

In terms of chronological age, certain physiological changes in the older body result in clothing that may be uncomfortable due to inappropriate fit, styling and proportion. User dissatisfaction with fit is a major cause of waste within the garment industry, and of particular interest to the ageing community who feel neglected within the market place. 'While mass-manufacturing and cheap high street stores have provided us with more products to choose from, these choices are more restricted (ask anyone who is different to the "standard" body shape about the difficulty of finding clothes to fit.) The products on sale in our high streets are becoming homogeneous and this lack of choice erodes our individuality and dulls our imagination, limiting our confidence about what clothes can be' (Fletcher, 2008, p. 186). Dame Vivienne Westwood has criticised the rise of 'disposable fashion', warning us that it makes us all look the same and that only 70-year-old women display any sense of style these days. In terms of image, in recent seasons, 'the fashion industry has been prone to portraying bone thin models with long white hair who looked as if they'd been starving on a desert island for two years' (Whitehorn, 2012). Fashion magazines normally discuss ageing in relation to skin care, with older (female) personalities arriving on the cover of French Vogue (No. 932, November 2012).

The lack of awareness of the design needs of the older generation, by predominantly young designers and marketers, and the problem of how to alter processes to address this growing sector, is exacerbated by the globalisation of the textiles and clothing trade. This has led to the general detachment of designers working in the west while mass production is carried out, with long lead times, off shore. 'The prescribed nature of the design process down the path most travelled was an account management method that sought to provide the client with something known and familiar, but expediently constrained designers to design around what they already knew' (Grocott, 2003, p. 85). There is little incentive, design direction, or time, for designers and product managers to consider emerging, unfamiliar markets. Company accountants have gradually passed the costs of technical design development, with the associated knowledge and skills, to locations with lower salary costs. In general, most senior management 'care more about the results that new products produce' and 'less about the processes and tools that generate innovative products and services' (Rhea, 2003, p. 145). Clothing designers, when promoted up the company structure, leave behind practical skills such as pattern cutting and making, while new young designers often have training geared towards 'styling'.

There is great confusion, amongst older consumers, with the unpredictability of commercial garment sizing. As with any age group, they experience little standardisation or synchronisation of sizes between different brands and even between different styles within one brand. This is both in terms of measurement and in the nomenclature of sizes that vary in the size charts from brand to brand. This is exacerbated by the degrees of ease (tolerance) added to body measurements according to the style and mood of the range. In addition, there is the concept of 'vanity sizing' adopted by some brands, said to make customers feel better about themselves and become more inclined to buy. A lack of clear standards with respect to sizing, both within and between different countries, is a challenge for wearers. Research with active ageing participants indicates that both male and female wearers wish for garment ranges, with fewer products, in less transient styling, with more consistency in sizing, with emphasis on appropriate fit for ageing figure types, and with a wider choice of colours and fabrication. Many older people admit to buying more than one garment, in more than one colour, when they find a flattering style!

We need to improve the image of ageing through design that is comfortable and flattering to the size and shape of the rapidly growing active ageing community. In this chapter, a UK design-led collaborative research project 'Design for Ageing Well' is referenced as a case study where capturing the size and shape of older research participants was key to the co-design led research and development of garment prototypes. The aim of this case study was to bring the comfort and easy-care attributes of the great innovations that have taken place in technical and smart textiles throughout the lifetime of the current Baby Boomer Generation, already adopted in performance sportswear, to everyday functional wear for the active ageing. In co-design, the person who will eventually be served through the design process is given the position of 'expert of his/her experience', and plays a large role in knowledge development, idea generation and concept development (Sanders and Stappers, 2008). In this project, size and shape has been of particular relevance, to inform the garment cutting of functional clothing with the incorporation of wearable technologies that demand close body contact.

7.2 Understanding different body shapes of older consumers

In order to design comfortable and flattering styles, designers must be aware of changes to the size and shape of the ageing body, to be addressed throughout the various clothing layers. Physical changes to the ageing body require appropriate design lines and expert cutting to accommodate the changing postures and restricted movements that affect garment fit, balance and proportion, with ease of movement for putting on and taking off garments. Traditional pattern cutting texts provide size charts and pattern block development that normally cater for erect figure types, with a limited degree of articulation. Physical performance capabilities are influenced by the size and shape of the body. Changes in bodily dimensions include age-related shrinkage in height, reduced sensory capabilities, circulatory problems, changing skin sensitivity and complexion and an increased demand for thermal regulation. These factors, as well as the potential incorporation of textile-based wearable technologies, all have a direct relationship to clothing design, cut, and fit and materials selection.

7.2.1 Physical changes related to ageing

The human body goes through one long continuous process of change, from the moment of conception right through to decomposition after death. Inevitably some changes occur with ageing, but lifestyle choices may slow or speed up these changes. How and when changes occur is highly individual, depending on genetic make-up and the damage done to the body by environmental factors and poisons such as accidents, infections, disease, cigarette smoke and alcohol (Macnair, 2011). The human body is made up of fat, lean tissue (muscles and organs), bones, water, and other substances. As we age, the amount and distribution of these materials will change (Langan, 2011). From the third decade the typical adult's body shape changes, especially if people do not keep active. The body becomes broader, features coarsen and cellulite may start to appear, especially in areas with more fat such as the thighs or bottom. However, the most significant physical changes of ageing really begin to appear from about 50 to 60 (Macnair, 2011). The ageing body alters as waists thicken, stomachs expand, shoulders move forward, busts in women lower, and feet become broader as the ratio of body fat to muscle changes.

The tendency to become shorter occurs among all races and both sexes, with height loss related to changes in the bones, muscles and joints. Muscle loss (atrophy) occurs as the liver, kidney, and other organs lose some of their cells. This begins in the twenties but really accelerates after the menopause in women and then after about 60 (Macnair, 2011). Bones may lose some of their minerals and become less dense (a condition called osteopenia or, at its later stage, osteoporosis). Metabolism starts to slow in young adulthood, making weight gain more likely and setting the scene for 'middle aged spread'. Fat tissue may increase towards the centre of the body, including around the abdominal organs, by as much as 30%, with lean body mass decreasing. Men often gain weight until about age 55 and then begin to lose weight. This may be related to a drop in the male sex hormone testosterone. Women usually gain weight until age 65 and then begin to lose weight (Langan, 2011). Weight loss is, in part, caused by a loss of muscle tissue that reduces the amount of water in the body. The loss of muscle mass in the legs, and changes in body shape, can affect balance; as muscle bulk and strength start to fail, simply staying upright and mobile becomes more of a challenge (Langan, 2011). People begin to lose height at a rate of about 1 cm (0.4 inches) every 10 years after age 40, with the average height of 65 to 74 year olds being 5 cm (2 inches) less than the average height of those aged 16 to 24. This height difference is almost double for those aged 75+.

7.2.2 Movement/agility

Physical strength is generally related to body size, with women physically weaker than men at all ages, and with weaker muscles and a poorer power-to-weight ratio. Their disadvantage is even greater on weight-bearing activities such as walking and stair climbing (Huppert, 2003). Particular aspects of physical performance that decline with age and which are important for functioning include mobility, dexterity and the ability to reach and stretch (Metz and Underwood, 2005). Flexibility and an adequate range of movement are essential for many activities, with loss of range, especially in the shoulder joint, common in old age. Arthritis causes swelling and pain in the joints and limited movement and weakness in the arms and hands, resulting in poor dexterity and restricted flexibility while dressing and undressing, and difficulty in getting a firm grip or in making precise finger movements (Metz and Underwood, 2005). However, height loss, muscle strength and weight loss or gain varies from person to person, with diet and exercise playing a large role in these changes.

7.2.3 Sensory considerations

Deterioration of the organs of sense (especially the eyes and inner ear) further increases the risk of falls and accidents as vision, hearing, balance, and co-ordination become less reliable. Hearing loss affects about 30% to 35% of adults between the ages of 65 and 75 years, and up to half of those over 75. Reaction times slow, and when the older person does fall, their bones are often less dense and more likely to break. Changes occur in the organs and tissues as we age. The outer skin layer becomes thinner, and the number of pigment-containing cells decreases, causing ageing skin to appear thinner and paler, with large pigmented spots in sun-exposed areas. The structure of the connective tissue in the skin changes as the production of collagen and natural oils fall, leaving the skin dryer, weaker and less elastic, leading to wrinkles and increased risk of injury. Blood vessels in the skin become more fragile, resulting in small areas of bruising even after minor injury. Wound healing slows and this, combined with loss of fat beneath the top layers of the skin, and decreased activity, can make elderly people especially vulnerable to pressure sores (Macnair, 2011).

7.2.4 Vital signs

Vital signs measurements are liable to change with age and medical conditions. These include body temperature, respiratory rate, heart rate (pulse), and blood pressure; all of which may be observed, measured, and monitored to enable the assessment of the level at which an individual is functioning. Maintaining good blood circulation in the body plays a major role in maintaining good health, as it is the means by which oxygen and nutrients are carried around the body. Improving poor circulation is central to helping stave off a manner of debilitating disorders, from diabetes to heart disease to varicose veins or the inability to think clearly. Preventative lifestyle changes, such as dietary changes, increased exercise, giving up smoking and losing weight, are recommended to ensure good cardiovascular health and help prevent the onset of disease (McCann, 2013).

Accurate regulation of body temperature is essential for optimal function and survival. The elderly are at increased risk of both hypothermia and hyperthermia when exposed to extreme temperatures, as responses to cold stress become impaired with age, with reduced perception of cold. However, the degree of a user's fitness can affect the extent of impairment. When we look at the balance between heat generation and heat exchange, what needs to be achieved is not a constant temperature but rather dynamic equilibrium. As external conditions change, so the body's homeostatic system responds to bring the internal body temperature back to a norm. It is the user's homeostasis that may be enhanced by a layered clothing system. The body can lose heat by convection, radiation, conduction, and evaporation all of which have a direct relationship to clothing performance (Bougourd and McCann, 2009).

7.2.5 The 'feel good factor'

'This new Consumer Majority, not youth and pre-middle-aged adults, is the primary source of today's leading views, values, and behaviours in the marketplace. This historic change in consumer behaviour has made much of what once worked in marketing obsolete. It has changed the rules of marketplace engagement.' Wolfe and Snyder suggest that 'while people are more concerned with authenticity in every aspect of their life, marketing is riddled with inaccuracy and insincerity' (Wolfe and Snyder, 2003, p. xii, p. ix) and that 'this has occurred because for the first time ever, most adults are in the years when the forces of self-actualisation needs exert decisive influences on lifestyle aspirations, buying decisions, and overall consumer behaviour' (Ibid, p. xiii). With respect to addressing ageing design requirements, the physiological capacities, clothing behaviour and individual demands all vary from person to person but, as discussed, diet and exercise play a large role in these changes.

The motivation for older people to engage in healthy exercise will be increased if, at the same time, the activity provides joint or individual pleasurable experiences, an enjoyment of nature, a sense of adventure, education, and entertainment as well as increased fitness and the prevention of ill-health rather than enforced rehabilitation (Best Ager, ISPO, 2010). This demands a design that is pleasurable to wear by virtue of an appropriate blend of form and function, embracing aspects of size, shape, proportion and fit, balanced with appropriate aesthetic style considerations. The author proposes that the design of functional clothing, that is sympathetic to the demands of the ageing body, has the potential to promote independence and well-being for the active ageing within their everyday lives in encouraging participation in healthy exercise.

7.3 Sizing requirements for the older consumer: the design for ageing well project

The aim of the project: 'Design for Ageing Well: Improving the Quality of Life for the Ageing Population Using a Technology Enabled Garment System' was 'To engage in collaborative design practice to produce an innovative *garment layering system* that enhances the comfort and wellbeing and promotes healthy exercise for active ageing through the application of 3D body scanning, innovative smart textiles with embedded technology and novel garment engineering techniques'.

The project involved bringing the comfort attributes of modern fibres and fabrics, within a functional sports type layering system, to address the everyday lifestyle needs of older people and, in particular, to encourage social engagement and participation in walking in both urban and rural environments. In addition, the aim was to introduce assistive wearable technologies to the active ageing, technologies that are normally adopted in garment design for predominantly young athletic figure types, for extreme sport, or as assistive medical devices, with little aesthetic appeal, for the old and infirm. The changing size and shape of the older body was a key consideration in clothing prototype development, especially for the positioning of wearable electronics.

7.3.1 Context

'The idea that older people are now dressing younger is perhaps best exemplified by their widespread adoption of casual dress. The rise of casual dress has been one of the most marked developments in the dress code over the last 20 years; and it now presents the predominant form of dress for most of the population' (Twigg, 2013).

Great innovation has taken place in technical textiles throughout the lifetime of the current Baby Boomer Generation. 'As children in the 1940s and 1950s, the active ageing grew up at a time of great austerity with the global economy impacted by world wars. Most children of that time would have worn hand-me-down clothes, garments their mothers or family members had made for them, all these items would be carefully looked after, be re-sized, passed down to the next child or re-cut to suit another person or purpose. By the 1960s and 1970s they were in their teens and twenties, experiencing the newly developed 'synthetic miracle fibres' (Knowles and Hussey, 2013). Nylon and polyester, waterproof/breathable membranes and elastomeric have all been developed within the lifetime of the active ageing, but with early developments often associated with negative properties such as being harsh to the touch, sweaty, static, and liable to stain and smell through wear. In the 1970s, polyester fleeces emerged (from Calami in Italy and Malden Mills in the USA) and waterproof breathable materials were launched, for example, Gore-Tex. In the early 1980s Nylon was re-launched as 'Tactel' as sportswear became more stylish. Fabric innovation has changed the cut and sizing of garments with the introduction of different manufacturing processes.

Increasingly the popularity of easy fitting styles in jersey and stretch fabrics has meant that basic *'flat' pattern cutting* is now used by a large sector of the retail mass-market, sometimes eliminating darting in creating the shape. The manufacture of this type of clothing is quite different from that of formal clothing, where *'form' pattern cutting* requires blocks that conform more closely to the body form and often involve complex methods of cut and construction dictated by the particular, often traditional, style of the garment (Aldrich, 2008). In terms of design and cut, 'men are especially attracted to conventional pieces with eccentric twists. If this is discrete, it can offer fashion longevity. The majority of men steer away from complicated pieces and impracticality' (Aldrich, 2011, p. 6). The general trend in casual dressing has embraced older males as the mixing of sport and casual wear with formal wear has become an accepted mode of dressing both in terms of the wearing of colour and in garment combinations. It has become conventional to wear polo shirts or tee shirts with a suit, or a formal jacket with jeans' (Ibid, p. 6). Older women also enjoy the comfort attributes of casual easy-care functional materials, but seek flattering as opposed to 'baggy' fit.

7.3.2 Textile innovation driving more tailored sportswear cutting

Recent textile innovation, in tandem with novel garment joining techniques, has introduced a new aesthetic in functional garment engineering, with novel methods of manufacture affecting the cutting techniques to be employed. Early man-made fibres, such as Nylon, were used in waterproof garments, with non-breathable coatings, that had to be cut in loose fitting styles to provide ventilation. As fabrics have become more breathable, and stretch has been added for comfort, leading edge performance clothing for end uses such as active sport has become more tailored and streamlined. 'High-level performance fabrics have had a significant impact in high-cost casual wear and sportswear' and 'it is important to recognise that it is the handle of the fabric that is vital in the creation of garment shape' (Aldrich, 2011, p. 6).

The domination of stretch fabrics has meant an expansion of simple flat pattern cutting techniques, which rely on the stretch of the fabric to create the body shape. However, stretch fabrics married with *form cutting* give quite different effects with bias cutting to add to the drape quality of the fabrics. 'The designer has to consider weight, thickness, shear, drape and stretch' (Aldrich, 2008, p. 5) with some thick pile fabrics constructed with man-made fibres being deceptively light. 'These qualities will affect quite dramatically how a pattern will be cut and how the shape will be realised' (Ibid, 2008, p. 7). Specialist finishes and non-traditional cutting and material joining techniques may be adopted for synthetics and man-made fibre blends. Garment construction ranges from seam-free knit, in base layers, to the 'body mapping' of complex constructions, with many pattern pieces and components, often with heated seam bonding and zip insertion. To date, few of these specialist manufacturing techniques have been seen in designs considered appropriate, in style and shape, for older figure types.

7.3.3 Body scanning and pattern block development

The older research participants, to constitute the user reference group, were selected from walkers (in the 60 to 75 year age group) within a 30 mile radius of the lead university. The process of body scanning was explained to the research participants, with ethical protocols in place. User-engagement began with participants being asked to give details of their height and weight to calculate Body Mass Index (BMI) for potential selection for body scanning. One hundred participants (male and female) were chosen to be body scanned to capture representative size and shape for their age group in relation to SizeUK data. This process informed the final selection of four men and four women as size models (S/M/L/XL) for the development of garment pattern blocks. Basic blocks were generated from the scans (by Sizemic) and made up as toiles (calico prototypes) by the clothing team to be tested on the fit models. The blocks developed represented the starting point for adaptation to articulated blocks for base, mid, soft-shell and outer layer garments. These blocks, fitted to the requirements of the fit models, in collaboration with the wider user group, became the reference point for iterative co-design garment development.

7.3.4 Co-design

A co-design approach was proposed to bring together a breadth of stakeholders to work with older research participants in order to identify, prioritise and address their real, everyday design requirements. The co-design team represented a cross-disciplinary network of academic researchers, with background experience in clothing and textiles, electronics and communications, psychology of behaviour and social gerontology, with independent experts and industry collaborators. Preparatory workshops were necessary for the development of a cross-disciplinary shared language between all stakeholders. Co-design workshops included academic and industry presentations with the demonstration of existing product and fabric innovation, as discussed above. Designers acted as interpreters, translators and visualisers in helping the older research participants to have the confidence to express their design needs and aspirations as well as their limitations (with technology). The iterative co-design investigative stage resulted in a 'range plan' of garment styles, for men and women, to cater for different levels of participation in walking, which were developed, with users, into prototypes.

7.3.5 Cutting for movement

Clothing designed to be 'fit for purpose' requires design lines and detail geared to the technical, aesthetic and cultural demands of the end use as opposed to styling that is driven by transient fashion. Functional wear has quite different requirements from everyday clothing, with key performance features, such as protection, moisture management and cutting for movement, that contribute to overall comfort. In performance sportswear 'the designer must always be conscious that the body is a form. This can be difficult when one has to relate flat pieces of paper to a design that is basically sculptural when it is completed' (Aldrich, 2008, p. 4). A close fit, that follows the silhouette of the body, will promote ease of movement, while larger looser shapes may impede movement. A choice of text books exists for the development and adaptation of standard flat pattern blocks for men's, women's and children's wear, with slight variations in their approach to traditional garment cutting, but little guidance is available to inform the cutting of clothing for functional end use, and for older figure types. Pattern cutting guidance caters predominantly for erect figure types, with blocks that are normally inappropriate in terms of fit and articulation for a range of postures and movement.

In previous work the author, with co-researchers Morsky and Dong, collaborated in exploring cutting for movement, in the development of blocks designed to follow the body's contours in providing ergonomic shaping with enhanced movement. Simple drafting rules were tried and tested for introducing articulation in elbow and knee areas, 'arm lift', adjustable hoods, and within the overall silhouette of the garment. The resulting blocks achieved relatively unrestricted movement, having been manipulated to remove seams from standard positions, such as shoulder and set in sleeve, and subsequently incorporated into new ergonomic seam lines. The blocks were designed to enhance movement while positioning seaming to prevent both friction and ingress of moisture (for outer layer garments). Coordinating design lines were adopted in the development of mid insulation layer and base layer blocks. Articulation may be further enhanced with the application of stretch fabrics and/or in the cutting of certain pattern pieces on the bias of the weave. These functional blocks were further adapted for varied degrees of movement in relation to the demands of a particular end use (McCann *et al.*, 2009).

7.3.6 Co-design prototype development

In Design for Ageing Well, the initial blocks, developed from the body scans, were amended in overall balance, the degree of sleeve pitch, and in articulation for movement, for the co-ordination of garments throughout the layering system. Seaming was manipulated away from the over arm so that subsequent prototype style development would eliminate the rubbing of shoulder seams or ingress of moisture through exposed seams on outer layers. Toiles were made up in representative materials to gauge the degree of fit and tolerances required to enable movement, without friction, between the layers. Independent development was carried out by an expert intimate apparel designer in the development of user specific 'body layers' (to carry textile-based sensors to measure vital signs) as sports-type bras, designed and cut to provide individual levels of support, and as close fitting vests for men. The four bras were styled and engineered to fit the requirements of the representative S/M/L/XL figure types. On-going co-design with participants informed the adaptation of the blocks into style patterns to constitute a comfortable layering system.

Co-design engagement has highlighted recognition of older consumer confusion, with respect to garment sizing, in selecting clothing in retail shops, where they are challenged to find garments that match their own personal preference and at the same time fit the body. It has introduced older people to the benefits of the light-weight, protective, moisture management and thermal regulatory properties of quick drying, easy-care modern fibre and fabric innovations and new production processes for the joining or engineering of garments. It has introduced the functionality of the sports layering system, as a potential platform for wearable technologies, to suit the needs and aspirations of older people. The value of co-design engagement has been recognised in bringing academic researchers and industry stakeholders into close collaboration for enhanced mutual understanding between disciplines and market sectors and, above all, with respect to the design requirements of lay older research participants. During the iterative process, industry partners became increasingly involved in the development of prototype garments, with early findings adopted in commercial ranges.

7.3.7 The need for effective garment design specification

With offshore production the ability to process garments in many variations of style and cloth has been largely due to the adoption of computer aided design (CAD) (Aldrich, 2011). However, from a design perspective, some major challenges result from globalisation due to the detachment of designers, normally in the west, from 'hands-on' involvement in the product development process, often based in the Far East. Designers involved in the generation of design concepts, based on 2D computer generated 'flat' line drawings, now work separately from the pattern cutters and graders, who interpret the flat line drawings, with sometimes limited measurements given, and often with little back up reference in terms of pattern blocks or prototypes. With the pressures of global sourcing, and restricted time lines, the design specification of garments may no longer involve sealing samples, and corresponding patterns, with the result that designs from different manufacturing bases may be interpreted in different ways.

In discussions between the author and design practitioners at the international sports trade event (ISPO, Munich, 2011), an experienced sportswear designer (who wishes to remain anonymous) described the break down in communications as comparable to 'Chinese whispers' as garment designs and specifications are passed down the chain from one sourcing office to the next with no surprises as 'a curved line becomes straight in the process'. One brand's strategy is to introduce common templates for designers to sketch within that, presumably, correspond to pattern blocks at the manufacturing base(s) (McCann, 2012). It has been stated that some factories in Asia prefer not to work with patterns generated by western designers (Lewis, 2012) contributing to the lack of consistency found in garments sourced overseas. Due to this detachment, western designers seldom become involved in designing and cutting 'in the round' in 3D pattern development.

Off shore suppliers have to deal with varying levels of design specification, including sizing information, from brands across Europe and the US. The lack of consistency in the quantity and quality of technical information provided to the manufacturers, seldom with 3D back up, has resulted in one leading garment supplier, to multi-national brands, stating in conversation with a researcher, that 'some companies give me a drawing on the back of an envelope, others give me a three page specification and others may provide a detailed file of 25 pages' (Lewis, 2012). These generic problems, within the youth driven sportswear industry, are exacerbated when catering for the varied design requirements of a breadth of figure types within the wider population. 'Unfortunately, almost nothing from the regular clothing sized lines can be rolled into the plus sized line to reduce its operating costs' (Fasanella, 2011). When embarking on the merging of wearable technologies into the garment layering system, where positioning and fit is critical, the requirement for the effective communication of the garment design and technical specification becomes even more demanding.

In Design for Ageing Well, a product designer and an architect brought their methodologies and skills to merge with clothing design technologists in the specification of the smart, functional, clothing layering system. Flat fashion drawings make reference to body scale measurements, often in limited detail, with patterns seldom created by designers in the west. However, without patterns generated, and a digitised record made, there is no accurate reference from which garments may be made. This would never be the case in product design development or in architectural practice where key dimensions can be drawn in elevation, plan and section and be sent directly to be measured and made to scale. These designers helped to resolve strategies for the clear visual description of the complex garment assembly. They engaged with the body scanning process that informed pattern cutting for older figures, and the subsequent creation of style patterns that were digitised by a pattern technologist. Computer illustrations were produced showing both full garments and enlarged technical details with direct references linked to the digitised 3D style patterns. They visualised the incorporation of soft elements, such as textile-based conductive fibres and ribbons, knitted sensors, and electronic components that have not been within the traditional realm of the fashion designer (McCann, 2012).

7.4 Examples of specific styles for particular body sizes and shapes amongst older consumers: aesthetic needs

Co-design findings from the collaborative research case study that contribute to overall comfort and user satisfaction are discussed as follows beginning with aesthetic concerns.

7.4.1 A positive image of ageing

Fashion styling is normally subject to seasonal changes in silhouette and proportion, as dictated by trend forecasting, often with a focus on the youth market. Older people today are relatively younger, more adventurous and more confident than the previous generation, with the older research participants suggesting that more positive descriptors should be adopted with regard to the ageing, such as experienced, knowledgeable, informed and expert. Designers and marketers need to beware of age discrimination and be aware of mental age (perception) and physical age (chronological age).

7.4.2 Fit

'You learn to appreciate good fit and the reassurance of good fit' (Andrew Crawford, Sizemic, 20 September 12) and this was confirmed as a major consideration in product selection, alongside colour choice. The terminology used was 'Skimming' not 'Baggy' and of a 'Decent' length with, for example, jackets covering the behind. In style development, users proposed a choice of garment lengths to better cater for the variation in height within this age group in terms of body, sleeve and leg hem lines.

They liked the curved hems of performance sportswear jackets; longer at the back to cover the behind and slightly hollowed out at the front for ease of leg movement. Users expressed a need for more inclusive sizing to extend from extra small to extra large sizes. Women, in particular, requested that waists be designed for adjustability, for altering girth, but not 'baggy', possibly fitted through partial elastication.

7.4.3 Greater colour choice

Users welcomed less transient and extreme colour swings than those driven by fashion trend. They were attracted to kinder and more enduring colours, from sober to more characterful tones, considered sympathetic to ageing complexion and changing hair colour. They liked colours influenced by the country side, from deep rich tones to powdered pales, with references to flowers, berries, fruits, lichens, autumn russets, forest greens, with neutral slates and stones, but not beige, and avoiding the current predominance of black within the fashion trade. Both male and female users wished for garment ranges with fewer products, but with more consideration of appropriate fit and proportion for ageing figure types, that you could go back and replace, but with a wider choice of colours and fabrics, in varied textures and blends.

7.4.4 Style lines

The active ageing were found to be more concerned with subtle, non-transient, age appropriate, flattering style lines to address the changing body as opposed to following dramatic swings in fashion. In general, there was agreement (with women being more precise) in terms of style lines that would profile or 'skim' the body, being not too tight or too 'baggy'. There was a preference for softer organic shapes, as opposed to angular seaming, with style lines developed to be sympathetic to the range of figure types of the individual end users. Patterns were cut with flattering side panels, to accommodate the varied shape requirements from bust/chest to waist to hip that, in turn, enabled the elimination of side seams, and any associated friction in walking. Both men and women suggested subtle shoulder emphasis, to take attention away from their less defined older waistlines, in fluid contours to accommodate the ergonomics of movement and performance requirements of the body, with optional fit at waistlines through full or partial draw cords. Once the style lines were determined, the patterns for the garment layers were refined to move in tandem and without friction between the layers.

7.4.5 Less transient styling/fewer styles

Participant feedback indicated user resignation to current ranges often looking very similar and with size and fit that is unpredictable. In general, the demand was for fewer and less transitory styles to be available in a wider size range and greater choice of colours and fabrication. Many older users admitted to buying more than one garment, in more than one colour, when they find a flattering style.

7.4.6 Fabrication/handle/drape

The co-design process revealed that the positive attributes of modern fibres and materials were not initially well understood by the active ageing. Since the early bad image of synthetics, some considered natural fibres, such as conventional cotton and wool, 'good', while synthetic and man-made fibres are considered 'bad'. By means of the co-design process the research participants were introduced to the positive attributes of the fibres and fabrics that have potential to enhance overall comfort within the clothing layering system. The physiological needs of older users must be understood in order to interpret their needs into cutting lines related to the placement of appropriate materials around the body to offer ease of movement in tandem with attributes such as moisture management and thermal regulation. Articulated cutting may be further enhanced through the placement of varied fabric constructions and the use of the direction of stretch of a knit, the use of bias, or any quality with elastomeric content. Fibres and materials of particular interest were Tencel for base layers, due to its natural type handle, tailored fleeces and heated gilets within the mid layer, and the relatively softer and less noisy properties of Soft-Shell materials for outer layers.

7.4.7 Design detail

There was general agreement that neck lines should be designed with collars that were suitably protective when fastened but adaptable to look good when open for more casual wear. There was detailed discussion on hoods, to ensure adjustability and good visibility, as well as easy adjustability for cuffs and subtle adjustment for waistlines. Walkers were keen to have many pockets for specific items. Users were happy to adopt novel manufacturing techniques, such as the 'clean' look of the heat bonding of seams and zip insertions, which they had previously been unaware of. There was also interest in seam-free knit technology, with enthusiastic involvement in a Santoni knit workshop to progress development of seam-free base layer garments engineered to offer less restricting fit, in a Tencel/wool blend, to be knitted in industry. There was general realisation that functional clothing could be designed to be adaptable to a range of activities, both urban and more extreme. There was a general dislike of brand logos – however discreet! The placement of wearable technologies, including devices, their user interfaces and batteries, was to be as discreet as possible, with textile-based electronic cables and soft switches carefully considered within garment assembly.

7.5 Examples of specific styles for particular body sizes and shapes amongst older consumers: needs of the body

Clothing design that is fit-for-purpose should look attractive to the end-user as well as function in relation to their physiological demands. The designer must balance aesthetic considerations with technical design requirements related to the changing older body. Co-design findings from the collaborative research related to the needs of the body are as follows.

7.5.1 Psychological 'feel good factor'

The developed clothing layering system has resulted in enhanced human comfort in addressing the clothing needs of the ageing body in relation to size and shape, movement, dexterity, thermal regulation and protection from the environment. Aesthetic factors that contribute to the design of the garment also affect the wearer's sense of well-being with design that is flattering in terms of fit and proportion, in fabric with attractive handle, and in colour that is sympathetic to user's hair colour and complexion. Fabric that is noisy or Velcro (hook and loop fastening) that catches on clothing can interfere with peace of mind.

7.5.2 Size and shape

There was unanimous confusion around the topic of the commercial sizing of garments, with little standardisation or synchronisation of sizes between different brands and even between different garment styles within one brand. This is both in terms of measurement and in the nomenclature of sizes, exacerbated by the degrees of ease (tolerance) added to body measurements (that vary in the size charts from brand to brand) according to the style and mood of the range. For example, casual wear may allow for more layers to be worn under a jacket by comparison with the more tailored and articulated fit of light-weight protective garments designed for extreme end use. In addition, the concept of 'vanity sizing' adopted by some brands, and the lack of clear standards with respect to sizing, both within and between different countries, is an on-going challenge.

7.5.3 Cutting for movement

Users appreciated the results of 'cutting in the round' to address the ergonomics of movement associated with the activity of walking, as well as catering for the seated figure. Walkers have to bend and stretch with ease of movement in adjusting footwear and putting on and taking off various layers of clothing. Arm lift will also cut down restriction in movement when holding onto the rail in public transport or putting luggage on a rack! Users benefited from the technical attributes of performance clothing in catering for extreme arm lift, elbow articulation, and sophisticated hood design.

7.5.4 Moisture management/thermal regulation

The physiological needs of the end user, and the likely range of environments, must be understood in order to interpret these needs into 'body mapping' or the placement of appropriate materials around the body in terms of moisture management, support and protection. Prior to the co-design process, older users were unaware of the concept and terminology of the layering system, having often worn cotton as a base layer but, once aware, they were happy to adopt wicking materials in more tailored shapes.

7.5.5 Multi-functionality

Some valued multifunctional styling in garments, to encompass both countryside and urban walking, cut with sufficient ease to accommodate inner layers suitable for everyday life in garments considered 'a decent length' (covering the behind!) for the natural outdoors and with longer lines for urban walking. Those involved in more extreme hill walking required greater protection in more streamlined garments with co-ordinating functional inner layers. Some participants engaged in activities that spanned several categories of walking, requiring more activity-specific clothing.

7.5.6 Senses

Materials selection in garment features and trims should address the diminished tactile sensitivity of elderly wearers, with changes in texture used for tactile pleasure as well as for practical end use and safety. In terms of putting on and taking off garments, choosing the right fastening for a garment can significantly alter the effectiveness of the item and the layering system as a whole. Users were anxious to avoid limited visibility from poor hood design, requesting the wiring of brims and easy hood adjustment – even with gloves on. They also collaborated on the positioning of and easy access to pockets and wanted to avoid hoods in collars.

7.6 Future trends

Design for Ageing Well confirmed on-going confusion in relation to garment *size and measurement* within the lay active ageing community. Research revealed that, in terms of functional clothing, the general public has little awareness of the complexity of the global textile and clothing production chain with the minimums and lead times involved. Older consumers have little confidence in *size labels* and cannot understand why it is so difficult to find flattering styles, in appropriate colours, in garments that fit. Through co-design engagement, older research participants gained *awareness of the function of the layering system* and the sophistication of *value added smart materials*, and their attitudes to the value of clothing altered. Being retired, with more time to research purchases, there was general agreement that, with their enhanced understanding, they would pay more for *value added product*, with comfortable fit, if it offered the prospect of *longevity of use*.

There was a clear lack of confidence in knowing where to find products within the retail environment and how to understand point of sale information. There is great suspicion with regard to the claims of brands, and this age group resent wearing logos, however discreet, requesting '*No brand logos!*' In co-design it became clear that the different levels of participation in walking should be categorised and correspond to *fit guidance* in respect to styling, with *clear pictograms* to describe the degree of fit for different end uses.

This relatively 'slower' co-design process (by comparison with the transient fashion cycle) fosters greater consideration of functionality, aesthetics and cultural requirements, in enhancing the overall comfort in garments designed to be fit-for-purpose. In particular, size and shape are considered key to greater user satisfaction and longevity of use. However, the case study project prototypes were cut to fit specific end users, as fit models, and did not represent commercial sizing. Emerging technological advancement in CAD, with the link between body scanning and virtual mannequins, is now contributing to a small movement back towards *customised clothing* in order to provide consumers with options and to attract market share. This may be further linked to individual items being embellished through digital processes such as printing and embroidery.

Mass customisation requires new business strategies for the adoption of more flexible manufacturing systems, with computer integrated manufacturing tools and techniques, and with organisational changes in terms of flexible culture and the empowering of employees by assigning more responsibilities (Bhatia and Asai, 2007). In adopting these kinds of technologies 'the individual is no longer at the end of the clothing chain, but instead forms the start of it' (Daanen and Hong, 2007, p. 16).

Ultimately, a more responsible design approach for the development of better fitting garments, in materials and colours sympathetic to the rapidly growing active ageing community, has the potential to cut down waste, in value-added more sustainable product development. The changing demographic, embracing the rapidly growing active ageing community, could be an ideal market to address by manufacturers closer to 'home'. This consumer is already worried by transient, disposable fashion and anxious that, if they find something that suits them, they may not be able to replace it. Slower, value-added design would be appropriate for both the consumer and for more considered and predictable production. Fewer styles in better fitting product could be personalised with a choice of lengths, in more enduring colours sympathetic to the older complexion, and in fabrics that will not easily date, with meaningful accessories selected by the individual wearer. Operators could build up skills in shorter runs, of known styles, in manufacture that focuses on quality, appropriate size, shape and fit.

7.7 Sources of further information and advice

- http://www.sizemic.eu/
- http://www.size.org/
- www.Bodymetrics.com/
- www.patagonia.com/us/includes/product_size_pop_up.jsp?OPTION=HOWTOMEASURE_SIZE_CHARTS_HANDLER

- www.myseniorhealthcare.com
- www.sizemic.eu/products-and-services/pattern-development.html
- (2009) Effects of aging on your body, Available at: http://www.myseniorhealthcare.com/Effects-Of-Aging-On-Your-Body.html (Accessed: 4th January 2011).

7.8 References

Aldrich, W. (2008) *Metric pattern cutting for women's wear*, 5th edn., Oxford: Blackwell.

Aldrich, W. (2011) *Metric pattern cutting for menswear*, 5th edn., Chichester: Wiley.

Bhatia, A. and Asai, R.G. (2007) Mass customisation in apparel & footwear industry – Today's strategy, future's necessity. WIPRO Applying Thought, http://www.wipro.com/documents/resource-center/library/mass_customization.pdf (accessed 2 March 2013).

Bougourd, J. and McCann, J. (2009) 'Factors affecting the design of cold weather performance clothing', in Williams, J.T. (ed.) *Textiles for cold weather apparel.* Cambridge: Woodhead Publishing Limited.

Daanen, H. and Hong, S. (2007) 'Made-to-measure pattern development based on 3D whole body scans', *International Journal of Clothing Science and Technology,* **20**(1), pp. 15–25.

Fasanella, K. (2011) Apparently, anyone who wears clothes is a sizing expert, http://www.vanitysizing.com/archive/apparently-anyone-who-wears-clothes-is-a-sizing-expert/ (accessed 9 September 2012).

Fletcher, K. (2008) *Sustainable fashion and textiles: design journeys*. London: Earthscan.

Grocott, L. (2003) 'Speculation, serendipity and studio anybody', in Laurel, B. (ed.) *Design research: methods and perspectives*. Massachusetts Institute of Technology, pp. 83–93.

Huppert, F. (2003) 'Designing for older users', in Clarkson, P. J., Coleman, R., Keates, S., Lebbon, C. (eds) *Inclusive design: design for the whole population.* University of Cambridge. pp. 30–49.

Knowles, V. and Hussey, C. (2013) 'Sustainability – understanding the impact of choices when developing active wear garments' in McCann, J. and Bryson, D. (eds) *Textile-led design for the active ageing.* Cambridge: Woodhead Publishing Limited. (In progress)

Langan, M. (2011) Aging changes in body shape, http://www.nlm.nih.gov/medlineplus/ency/article/003998.htm (accessed 9 September 2012).

Lewis, C.M. (2012) 'The Development of a Methodology for the Design of Functional Garments from Initial Concept to Product Launch'. Unpublished thesis (M.Phil), University of Wales Newport.

Macnair, T. (2011) The Physical changes of Ageing, http://www.bbc.co.uk/health/support/elderly/physical_changes_ageing.shtml (accessed 12 December 2012).

McCann, J. (2012) 'Translating the hybrid methodologies and practical outputs of smart textile-oriented research, in clothing for the growing ageing market, for the benefit of all stakeholders', in *The Textile Institute 88th World Conference 2012: Conference Proceedings*, Textile Institute, Bridging Innovation, Research and Enterprise, Shah Alam, Malaysia, 15–17 May 2012.

McCann, J. (2013) 'Smart protective textiles for older people', in Chapman, R.A. (ed.) *Smart textiles for protection*. Cambridge: Woodhead Publishing Limited, pp. 224–75.

McCann, J., Morsky, S. and Dong, X. (2009) 'Garment construction: cutting and placing of materials', in McCann, J. and Bryson, D. (ed.) *Smart clothes and wearable technology.* Cambridge: Woodhead Publishing Limited.

Metz, D. and Underwood, M. (2005) Older richer fitter: Identifying the consumer needs of Britain's ageing population, England: Age Concern Books.

Rhea, D. (2003) 'Bringing Clarity to the 'Fuzzy Front End': A Predictable Process for Innovation', in Laurel, B. (ed) *Design research: methods and perspectives*. Cambridge, Massachusetts: MIT Press, pp 145–54.

Sanders, E. B-N. and Stappers, P.J. (2008) 'Co-creation and the new landscapes of design', *CoDesign,* **4**(1), pp. 5–18.

Tomassini, C. (2005) Focus on older people: Demographic profile, www.ons.gov.uk/.../focus-on-older-people/2005.../focus-on-older-people (accessed 23 December 2012).

Twigg, J. (2013) 'Clothing: identity, embodiment and age', in McCann, J. and Bryson. D. (eds) *Textile-led design for the active ageing*. Cambridge: Woodhead Publishing Limited (In progress).

Whitehorn, K. (2012) 'Vivienne Westwood may think women are stylish at 70 – but where's the help?', *The Guardian,* 20th February, http://www.guardian.co.uk/commentisfree/2012/feb/20/vivienne-westwood-stylish-70/

Wolfe, D.B. and Snyder, R.E. (2003) *Ageless marketing*. Chicago: Dearborn.

8 Overweight and obese consumers: shape and sizing to design apparel that fits this specific market

L. M. BOORADY, SUNY – Buffalo State, USA

DOI: 10.1533/9781782422150.2.153

Abstract: Key issues regarding the design of apparel for the overweight and obese population are examined with reference to how the body changes shape as weight is gained, understanding of the different body shapes of the plus sized consumer and sizing requirements for plus sizes. Most importantly, plus size customers are found to be similar to missy sized customers in that they both want stylish and fashionable clothing that fits. Future trends in plus sizes are discussed.

Key words: plus size, body shape, grading, obese, overweight.

8.1 Introduction

The World Health Organization (WHO) (2012), stated that in 2008 there were more than 1.4 billion adults (20 years or older) worldwide who were overweight. Globally, there are 200 million men and 300 million women who are obese, or about 10% of the world's population (WHO). In the US, 69% of the people are classified as overweight, with 34% of them considered obese (US Department of Health and Human Services, 2010). Since most of the people in the US are overweight, and the majority of these are women, one could assume that the plus size industry is a large portion of the retail market. However, plus size clothing makes up only 18% of the total revenue in the female clothing market (Bellefante, 2010). The assumed cause of this low number is the lack of offering in plus sized clothing.

Plus size is defined as any size above a normal size – but what is normal? According to SizeUSA (Textile/Clothing Technology Corporation, 2004), the average American female is 5'3.5" (161.3 cm) in height and weighs about 163 pounds (73.94 kg). She has a bust measure of 42.75" (108.6 cm) and hips which measure 44.25" (112.4 cm). Compare this to the ASTM D6960-04 (2004) chart for women's plus size figure types, and this 'average' woman would be between a size 16W and 18W based on circumference measurements, yet she would be considered a petite based on height. Plus sizes are the new normal for American woman and many women around the world.

This chapter looks at the issues regarding the fit and design of apparel for women in the plus size range. The main issues that surfaced include understanding how women gain weight and where this extra weight settles on the body frame, as well as a need to understand the data available in order to size garments to fit this population. While minimal research has been undertaken about the plus woman and her proportions, I was able to speak to three industry professionals whose experience in plus clothing spanned between 10 and 23 years. The first (which we shall refer to as Interviewee A from this point on) has over 23 years of experience in sample-making, patternmaking and technical design for the plus market. The second (Interviewee B) is currently the manager of technical design and apparel quality at a well-known conglomerate within the apparel industry and has worked in plus sizes for over ten years. The third (Interviewee C) works as a consultant in fashion technologies specializing in body scan data analysis for the apparel industry.

Plus size fashions started in the 1920s, when Lena Bryant noticed a lack of clothing for the 'stouter' woman and set out to make clothing to fit this section of the population. Bryant took the measurements of 4500 women from her already successful retail business and analyzed them with an additional 200000 measurements from an insurance company to better understand the female proportions of this target market (Mahoney, 1952). From this data, Bryant determined that there were three shapes represented in plus women. Business paperwork accidently changed Lena to Lane, and her business was a success from the start, earning over $5 million in 1923 (Mahoney, 1952). Lane Bryant has been bought and sold numerous times over the years, and is currently in the Ascena Retail Group's stable of brands. Lane Bryant's grandsons remain in the plus size business as owners of Fashion to Figure, an online as well as brick-and-mortar retailer. To understand plus sizes, one must first understand how and why they differ from other size charts; that a plus size is not merely a larger 'small size'.

8.2 Understanding different body shapes of obese consumers

Body fat is not predictable in the general population; everybody gains or loses fat in a variety of areas around the body. Men gain weight differently than women. Figure 8.1 shows the various locations on a female body where fat deposits, though not all in an equal manner. The medical field identifies two major shapes of fat deposits – android and gynoid, more commonly known as apple and pear shapes, respectively (Enzi, 1986). In the apple shape are people who tend to gain weight mostly in the abdomen area, while

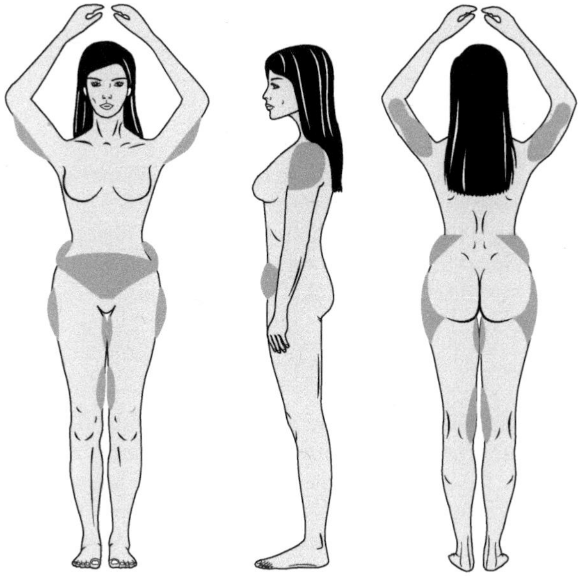

8.1 Fat distribution on females. Fat is usually stored in the low abdomen, buttocks, thighs and hips on women as well as the upper arms. Fat storage can change as a woman ages and fat may migrate from the hip region to the waist.

pears tend to gain weight below the waist in the hip and thigh areas. Some people may gain more in their upper arms than other people with the same bust or waist measure – there is no correlation between measurements.

The Body Shape Assessment Scale or BSAS© (Connell *et al.*, 2006) is a tool developed for the specific purpose of analyzing female body shapes for the development of apparel sizing that will better fit the target population. The analysis of the body shapes used in the Connell *et al.* (2006) study stated that not all women who wear the same size can be assumed to be the same body shape. Plus size women made up approximately 25% of the participant population in the development of the BSAS©. However, as the researchers themselves stated, more research would have to be done to develop a proper body analysis scale suitable for this population. One reason was that the scale may not have all the variations needed for a larger body shape, such as the location of the bust, which does not vary as dramatically in the misses sized population as it may in the plus population.

The amount or proportion of body fat a person has determines how overweight they are. Body fat changes due to weight gain or loss and can be different due to the overall total fat percentage versus muscle within the body. Age and gender also play a role in differing amounts and proportion of body fat for each individual. Stature has no correlation to body fat, but weight does have a correlation (Siervogel *et al.*, 1982). Women show a greater variation in locations and thicknesses of body fat than do men, particularly as they age (Siervogel *et al.*)

Interviewee A stated:

> You have a huge variety of women out there and you must make your clothing to fit a variety of body shapes. Plus women carry their weight in different places. Women have a hard time finding a pair of pants that fit because we vary so much in terms of shape – the same hip size for a customer with a small bust or a large bust means variations. (Personal communication, July 2012)

Interviewee C tells us:

> The difficulty in extended sizes is that the shapes are so individualized – the distribution of soft tissue is unique and hard to generalize across a large demographic. They may have to settle when they buy off the rack and have things altered. The industry is coming as close as you can and still may not be able to satisfy everybody. We accomplish this by looking at the data, where the biggest buckets of general shapes appear – pear and apple are the most recognizable shapes. So these shapes are what we are looking at in our earliest round. The inverted triangle shape is also showing itself in the data. Data analysis is done using SizeUSA and other inputs. (Personal communication, August 2012)

Aging also changes our body shape. Repositioning of fat occurs as we age (Enzi, 1986). Specifically, women's bodies have a tendency to shift the locations of fat from the hips to the waist with no change in actual weight or BMI (Kuk *et al.*, 2009). While the difference in waist measurement is small (1.6" [4 cm] over 9 years for women 60 years of age), it still may be enough to alter the apparel size.

Interviewee B, who works for a company where all clothing is also sold in misses sizes (contrary to the norm in the industry) and where styles must therefore work across a larger demographic, states:

> You need to understand how the body grows and really understand your customer and their body shape based on demographics. A 20 year old is a different body shape than a 50 year old. The same concept applies to missy as well. The body changes as you age – you lose a defined waist, the curvature of spine changes (hunch over), you have tissue loss in the rear (buttock drops and lowers), the breast shape also drops due to tissue loss in breast and the breasts are less firm. (Personal communication, July 2012)

Interviewee C has also seen the changes due to age:

> We are all influenced by the aging process by both shape and posture and this is even more true in the plus size category....as they age their needs change

> further. Plus is not just a size, it is a stage of life, and the fit and style changes as we age. As gravity impacts the weight and spine, there are further changes and this can also affect your clothing choices. It changes preferences and what you like to wear – look around in a mall and you will see jackets that hike up in the back – because of the wearer's weight and age (spine has curved), the clothing fits differently. (Personal communication, August 2012)

Intervieweee B works within her company to help identify better ways to fit the variety of shapes in her target population:

> There is a huge variety of body shapes in plus. We try to realize that most women are not the same size on the top and bottom. Understand that your fit model is the ideal of your population and we do not fit to clothing too close to her exact shape. Some customers will be curvier, some straighter through the waist – recently we have tried to focus on less shaping in the side seams for our straighter customers. (Personal communication, July 2012)

Plus women also have rolls of fat to contend with, and while some women try to hide their body fat, others embrace it. Alvanon, the maker of cutting edge dress forms based on actual 3D data, has developed plus size forms that include indentations of body fat. Figures 8.2 and 8.3 show a comparison of

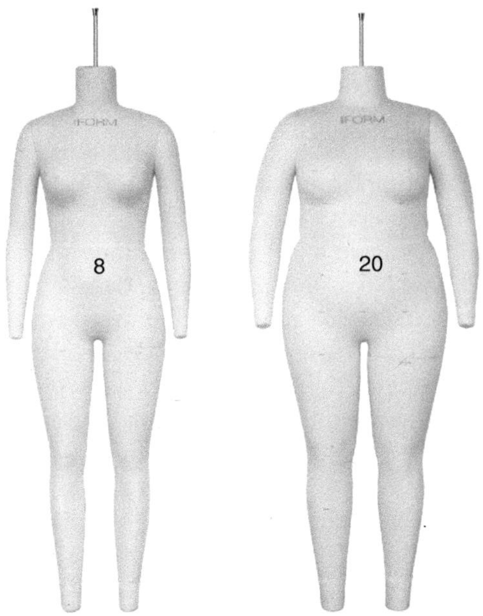

8.2 Front view of Alvanon© forms. This front view shows two dress forms, a misses size 8 and a plus size 20. Note the roll of 'fat' above and below the waist. (*Source*: Image courtesy of Alvanon©.)

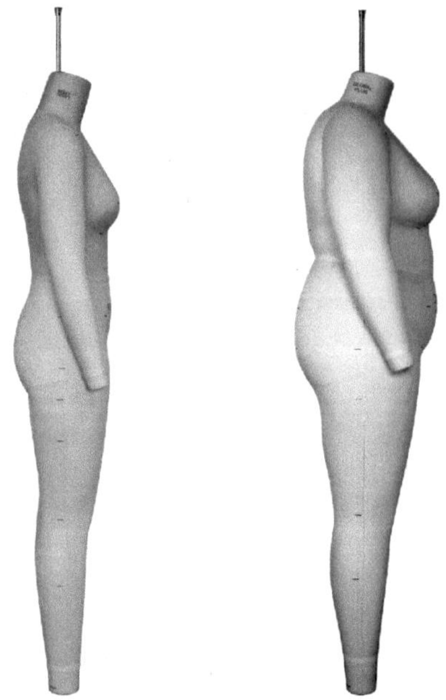

8.3 Side view of Alvanon© forms. The side views of the misses size 8 and plus size 20 forms show a fuller stomach, fuller and lowered bust line and a downward angle from back to front of the waist line for the larger form. (*Source*: Image courtesy of Alvanon©.)

the shape of the plus size 20 dress form developed by Alvanon next to their misses size 8 form. This new line of body forms, launched in 2012, is equivalent to US sizes 16–24, UK sizes 20–28 and EU sizes 46–54. Alvanon (2012) states that these embrace the posture shape and balance of plus women between the ages of 20 and 45 years. Traditional dress forms have an hourglass shape, even in the larger sizes, and they have always assumed a direct proportion between the bust, waist, and hips. Now, fit technicians and designers can see their garments on a static body that looks more like their true customer.

Charming Shoppes conducted a body scan survey on over 50 000 women, and from this data they determined that plus women have three distinct lower body shapes; straight, moderate, and curvy. Each has unique measurements to their body shape and are currently offering pants in these shapes and measurements. They were also able to find three live fit models who fit these shapes, had them body scanned and now use dress forms created based on those scans to better fit their pants (Interviewee B, personal communication). This type of sizing for pants – looking at the hip minus waist measures or 'drop' is also being done in misses' sizes by stores such as JCPenney and L. L. Bean. Figure 8.4 shows each of the three body types based on the 2012 Lane Bryant's catalogue measurements of the three lower body types.

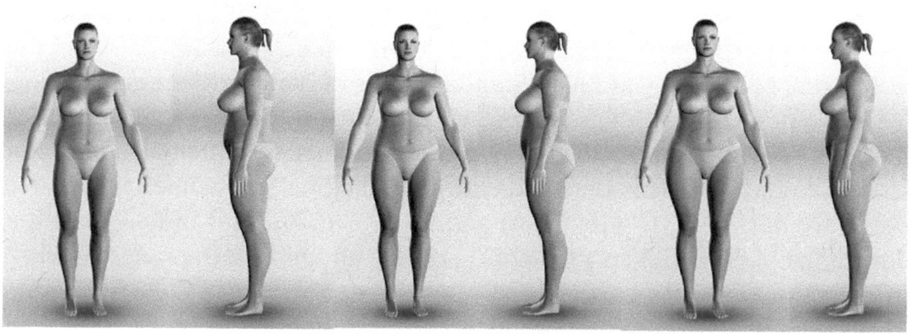

8.4 Straight, moderate and curvy. These avatars show the difference in hip size for the three distinct shapes as determined by Charming Shoppes scanning study. The bust and waist remain the same for all figures, however the hip changes from 41.5″ (straight) to 45.5″ (moderate) and lastly to 49.5″ (curvy).

Interviewee A also brought up another aspect that should be taken into consideration when working in plus – compression garments. She stated that 'compression garments change the shape – more people wear these types of garments today and it is great. They have a smoothing effect on the body yet not as much a compression as a girdle did.' Compression garments do not completely smooth away the rolls of fat but they do alter the softness of the body and provide support. In looking at the overall target population and the best business practices, Interviewee A stated:

> We are absolutely aware of the other body shapes and have dress forms in those different body shapes. We have used different fit models in different body shapes – you can never as a manufacturer survive by making garments for all the different body shapes out there. When you design, be aware that she's out there (women of different shapes) in building your product and assortment. That's how you can be successful. (Personal communication, July 2012)

8.3 Sizing requirements for obese consumers

The Department of Commerce Office of Textiles and Apparel requires that certain labels be attached to garments. Fiber content, care instruction, country of origin and the manufacturer or dealer identity must be on all apparel sold in the US. Size labels are not required. Voluntary Product Standards are developed by groups of interested parties; in the case of apparel sizing it is the American Society for Testing and Materials (ASTM) who develops the standards. The ASTM has committees which work on various standards and these committees are made up of any interested parties, though typically industry professionals make up the bulk of the membership for the apparel sizing committee. However, adhering to the apparel sizing standards is voluntary. There is no regulatory oversight for apparel sizing, indeed size labels are often merely indicators of an approximate size.

The ASTM D 6960-04 (2004) 'Standard table of body measurements relating to women's plus size figure type', which is currently under revision, states that the body measurements for plus sizes are meant for females between sizes 14W and 32W. This chart indicates that the plus size women in this size range are between 5'6" (167.6 cm) and 5'9.375" (176.2 cm) in height with a bust range of 32" (81.3 cm) to 44.5" (113 cm) and hips in the range of 34.5" (87.6 cm) to 47" (119.4 cm).

The ASTM D 6960-04 (2004) chart proportionately grows the plus size body measurements of 2" (5.1 cm) in the bust, waist, and hips while assuming height increases of 0.375" (1 cm) per size. The hip–waist drop remains at a constant 10" (25.4 cm) throughout the size range of 14W – 32W. The bust–waist drop is constant at 8" (20.3 cm). These drop measures represent the traditional hourglass silhouette.

Interviewee C, who is a member of the ASTM committee on apparel sizing, states that 'body scan analysis shows that you cannot grow the body in proportion – think of a copy machine and enlarging something five percent or ten percent. It comes out in proportion – people don't grow that way so we need to rethink how the body grows. We depend on the data to do the grading – it has rocked our world.'

The misses ASTM size chart, D5585-11e1 (2011), has just recently been revised, and this new chart no longer assumes the hourglass silhouette as the only shape of American women. The new chart not only gleaned information from the multiple retailers (and other interested parties) who sit on this committee but also from the SizeUSA body scan data compiled by the Textile Clothing Technology Corporation (2004), whose overall findings included the fact that more women were actually a rectangular shape (less defined waist) than the current size charts assumed. The current ASTM D 5585-11e1 is actually composed of measurements for a straighter figure as well as a curvier figure, which had previously been the only body shape represented. The additional, and updated, information provided by the SizeUSA (2004) data has helped the ASTM committee realize that plus sizing is different than sizing for misses. They are currently working on updating the sizing information for plus sizes. Even though the charts published by ASTM are developed by industry members, adopted and widely referred to, not all retailers follow these standards. Many retailers use their own measurements to cater to their particular target market.

Retailer's size charts for plus must be viewed before choosing a size to purchase, as the measurements vary drastically by company. Lane Bryant, the first retailer to carry plus sizes in the US, states that their size 18W has a 44" (111.8 cm) bust. However, L.L. Bean shows their size 18W with a 45"(114.3 cm)

Table 8.1 Measurements for a size 18W in inches (cm)

Company	Bust	Waist	Hips
Lane Bryant	44 (111.7)	38 (96.5)	47 (119)
LLBean	45.5 (115.6)	39 (99.1)	48.5 (123.2)
Nordstrom	47 (119)	39 (99.1)	48 (121.9)
Monroe and Main	45 (114.3)	39 (99.1)	47 (119)
Fashion to Figure	40 (101.6)	35 (88.9)	42 (106.7)
JC Penney	43.5–45 (110.5–114.3)	36.5–38 (92.7–96.5)	46–47.5 (116.6–120.7)
Avenue	44.5–46 (113–116.8)	37.5–39 (95.3–99.1)	47–48.5 (119–123.2)
Torrid	45 (114.3)	38 (96.5)	47 (119)

bust, Nordstrom's bust measure is 47" (119.4 cm) and the Fashion to Figure bust measure is 40" (101.6 cm). The variations between companies for size 18W can be seen in Table 8.1.

The French mail order company La Redoute, California-based Kiyonna, and the UK's Marks and Spencer all start their plus sizes at 10W, but the measurements for this single size varies drastically. LaRedoute's size 10W has a 37–38.5" (94–97.8 cm) bust measure and Kiyonna's is larger at 41–44" (104.1–111.8 cm) while Marks and Spencer's is a slim 33.5" (85.1 cm). While it is not unusual for size numbers and measurements not to match between retailers, the variation is much wider for plus sizes.

Charming Shoppes conducted their own body scan survey on their customers over the course of five years, scanning over 50000 women across the US. Age varied among the scan participants in order to obtain data on each of their target markets. Lane Bryant, Fashion Bug and Catherine's were in the Charming Shoppe stable of brands when they were bought out in June 2012, joining retail stores Dress Barn, Maurices and Justice under the Ascena Retail Group umbrella of companies.

Interviewee A tells us that 'We know there are multiple body shapes out there and in the past ASTM just put out one standard shape. There was this hourglass figure that didn't fit most women and as people started to gain weight, this shape changed completely. We keep this in mind when we are building any of our garments.'

Proportional grading, increasing the pattern proportionally in height and width, has been the standard method of grading in the industry for decades. However, there are numerous studies which state the inaccuracies of proportional grading (Schofield and LaBat, 2005a,b; Petrova and Ashdown, 2008). Interviewee C goes on to comment about the standard 2" (5.1 cm) grade interval and the standard hourglass shape assumed for plus women, and talks about the difficulty in grading for plus sizes:

> That's [the standard, proportional grading] not happening anymore. They [plus] are even more different than missy is. In all sections of the population everyone is thicker than we were 50 years ago. The waist is where you first put on weight. The waist grows in greater proportion than the bust and hips. Then we look at the hip and thigh grades when we do pear shapes and adjust them more. This is a very simplified view of what we are doing.
>
> The bust also varies to waist – there is so much work to do that you have to start somewhere. Pants are the hardest to fit – the hip-to-waist ratio and getting pants to fit is the greatest need. We have talked about working on the tops but with few exceptions, the bust will be bigger than the waist. In tops it's a matter of styling more than grading in terms of finding the right style that works for your body shape. When you get to sizes 22/24 and above you get a whole different shape and that some with women, especially when they get older, the shape changes. The whole midriff changes are dramatic – we can start to see the differences as early as age 45.

In discussing plus sizing with Interviewee B, she stated that:

> We developed our own sizes through 25 years of selling plus apparel. We used to purchase from other vendors but five years ago we decided to develop our own brand to create a constant fit and style. We implemented our own grading and used customer feedback to tweak it and make it work. We changed sleeve lengths or the grade based on customer feedback.
>
> Grading requires knowledge of how the body grows from size to size – we don't grow proportionately both taller and bigger at the same rate. Shoulder width does not grow at the same proportion as the waist. And growth from back to front differs.
>
> The overall frame shape [skeletal structure] does not change. The frame is set but where does the fat go? You aren't getting any wider in your cross shoulder, you are building mass therefore it cannot be a straight grade up – cross shoulder shouldn't [depending on the style] be the same grade as the bust.

Interviewee C elaborated on her company's decision to move away from the traditional hourglass shape and create new grading rules based on how the body actually grows:

> The grading part – the body grows in ways we didn't understand before – there are skeletal issues that do not change as much – for example, the cross shoulder doesn't grow no matter how much weight you put on but the upper arm gets fuller as weight gets added so the patternmaking is very different. The patterns, the grades, are very different than what we learn in pattern books. In order to get the best fit we are letting the [body scan] data direct us and are not going down one avenue or another based on previous experience. The term we have been using is non-linear grading. We may grade a smaller increment in the smaller sizes and larger increments in the larger sizes but the grading is not the same on all areas. We need to fit the actual body.

The idea that the skeletal frames of bodies do not change but the locations of fat distribution do is seen as important by both of these companies. Grade rules based on this idea are seen as important in properly fitting the plus customer.

However, not all companies which offer larger sizes grade specifically for plus women. Often a company may produce larger sizes but base these sizes off a standard missy size eight or ten, which does not address the specific needs or increases for a true plus woman. This is most often seen in companies which offer only one range of sizes, for example missy sizes four through 20. Interviewee A commented specifically on this industry practice:

> A missy style just graded up to a size 20 doesn't fit well. The changing body shape comes into play. Plus isn't just a larger missy – know how the plus woman grows when grading. Understanding how your customer grows in shape is important. Missy graded up just gives you rounder and longer armholes.

Interviewee C also mentioned the difficulties in fitting the plus customer due to the different qualities of lean muscle tissue versus fat tissue. Fat tissue, being softer and able to be compressed more easily, plays a significant role in fitting plus apparel:

> We have had a lot of conversations about ease that when you are fitting with soft tissue you need to adjust. If you have a grade of four inches and plus needs more compression with an elastic waist to help it hold up so you can't keep the four inch grade. The waist relaxes for elastic bottoms – you can't grade the same as for the hip and expect it to fit. We found that the larger sizes need their bottoms snugger in order for the garment to feel right and stay in place. They feel more secure if the garment is more secure. The soft tissue gives more than toned tissue. Elastic waists help accommodate a variety of shapes.

Interviewee A echoed this sentiment in her comment 'functional ease in plus due to spread – when any woman wears a skirt, we compress when we sit and spread out – we must take that spread into account. If the woman is a plus, she will have more spread than a missy.' On a final note about sizing Interviewee A brought up a problem that is a concern in all size ranges in the apparel industry but could be more prevalent in plus due to the sensitive issue of weight:

> Women – instead of sizing up to fit a garment, she will walk away – we need to get customers to focus on shape not size. Our customers love the different shapes we offer for pants – we have had women who had never worn pants before actually cry because they were so happy to find pants that fit. It is a wakeup call to the industry – customers need help – it is a rallying cry – what are we going to do about it?

8.4 Specific styles for the obese consumer

Thirty years ago, there were very little options for the plus fashion customer. Oversized knit garments and shapeless house dresses were standard wear for people who could not find fashionable clothing in their size. There are companies who have tried to serve this market and then left, such as Liz Claiborne, Sigrid Olsen and Ellen Tracy. Old Navy and Ann Taylor no longer carry plus sizes in their store but do have them available online. There are also companies newly willing to take on the challenge of plus sizes such as Target and upscale designer Elie Tahari. Plus sizes are one of the few areas that are in a growth phase in apparel retailing.

As there are many different types and styles for missy sized women, there are many types of styles and shapes for plus customers. Designers and retailers now recognize that people who are a larger size still have their own personality and style preferences. Not all plus women want to hide their bodies. When asked about specific styles for the plus woman, interviewee A noted that:

> Styles are not good or bad – it should flatter her – it is right whether she is missy or plus. There are a wide range of customers – some women like to dress a bit sexy versus women who like to dress conservatively. Younger plus customers have a specific attitude and conservative plus women vary in attitude as well. Design for the plus size woman is not a challenge if you truly understand the needs of your target market customer. You need to understand the target market, just can't go in and design.

Interviewee B echoed this thought of there being no 'good or bad' styling in her statement:

> For our company, styling is not dependent on size. We carry all styles in all sizes. Ultimately it is up to what your customer wants and will wear. Some plus size customers will wear a halter top in size 4X, some like clothes that are tight fitting, some like them looser fitting. You have to keep in mind that you have a variety of customers each with their own likes and dislikes and try to cater to as many of them as you can.

However, interviewee B also commented on her recent trip to the UK where she saw even more variety of clothing styles for plus sizes than in the US. She said:

> you also see that different countries have different takes on style for plus. The UK has a lot more body acceptance in terms of size and has a lot more mainstream brands which carry plus sizes. Heavy set women wear clothes tighter and shorter than what I expected. I was seeing miniskirts on sizes 16 and up.

While there seems to be no specific designs just for plus women, there certainly are special considerations taken into account when designing for this target market. All the industry professionals noted that deliberate care was taken when using fit models; two of the companies noted that they had recently changed their plus dress forms to the updated Alvanon forms which indicate body fat. When fitting, all companies take into account the larger shapes and the variety of shapes found in the plus market.

When Interviewee A and her team fit a garment they keep style factors in mind:

> Because plus women carry weight in different places and you have a huge variety of women out there, we make our clothing for a variety of body shapes. For us, this is addressed in the styling of the garment. When you do a sheath dress it will fit fewer customers, an A-line style will fit more customers. Hip measurements vary so much – you must put slits in tops to accommodate a wider hip.

Interviewee B stated that they are 'more careful about seaming and shaping and where the seams are located' when they design for their plus customers. Empire style lines must fall below the bust to flatter the figure, not over the bust. All the companies commented that they do not fit too closely in the waist area.

Interviewee A also keeps an eye on fabric and construction when developing a plus garment as larger women tend to put 'more pressure on certain seams'. Using 'appropriate fabric weights that flatter the body and are comfortable' is also taken into consideration. Interviewee A also feels that 'Spandex is God's gift to women, it gives the fabric more movement – any fabric with some stretch is good [for plus sizes]'.

Designing for plus is the same as designing for missy sizes. Interviewee A said:

> Our creative teams travel to other countries, like Milan, and major runway events. The creative teams are inspired by what they see, a good designer will take what they learn and turn it in to "real wear" – clothing for real people. Whether for missy or plus it is the key interpretation of how to turn that [an original runway design] into something that will flatter your target customer. It has to be successful for *your* customer, not one big generic formula. The styles that we offer for our plus customers are a variety of clothing, one is not better than the other – just there are different women with different ideas of what looks good on them. Know your target customer.
>
> We do the same things they do in missy – we go to the same color service with the eye of understanding what translation would be most successful for our company. We use the exact same resources for forecasting that missy does – there are no separate trends for plus.

Plus women want to be as stylish and fashionable as missy size women. Companies recognize this and are creating clothing to meet the needs of this target market. While special consideration must be taken in the construction and placement of seams, in the use of elastic at the waist and different ease allowances when grading, the fact is that plus women are like all other women – different. Different from each other in their style choices and how they like their clothes to fit

8.5 Future trends

People in the US are growing larger according to the Center for Disease Control (CDC) (2012), and this trend has been difficult to stop. The health issues that occur when extra weight is carried on the body are well known. However, losing weight is not easy. There is also the issue of accepting people as they are and a strong movement for self acceptance no matter what your size.

As for what the future may hold for plus women, interviewee B stated 'I would like to think that mass customization is the next new thing but I don't think the average consumer will pay for this. You can add technology to whatever you want but the customer has to like the fit for them to buy it. [But you have to be careful] The [plus] customer in general does not like change.' Interviewee C also looked at technology as a future trend in the plus area:

> There are lots of opportunities especially with 3D scanning and virtual dressing rooms – people who have a hard time finding clothing, they may now have a voice and be able to find more brands that would be suitable for them – "hey given your shape you need to shop these brands." They can find clothing that suits them and fits their shape better.

Another growing target market is plus clothing for the younger set – both children and teens. Interviewee C noted that 'kids today have issues in plus sizes – there is a great need for services in that market.' This statement is backed up by CDC (2012) reports as well as the number of retailers emerging in the kids plus market. Delia's, Justice and Forever 21 all carry plus sizes as well as junior sizes for their younger (tween and teen) target markets.

Interviewee B has looked beyond the US and was recently in China for business. She stated that for the future '…we also need to look more globally – people around the world are getting bigger. China is taller and larger than they have ever been before. They haven't reached our proportions yet but they are running an average size of US 10 or 12. As their economy grows, their people grow.'

Plus sizes are not going away any time soon, and can be a profitable business if one takes the time to understand the target market in terms of fit. Style remains the same no matter what size a woman is – some are flirty, some are conservative, all are unique. Future trends may embrace technology as a method of finding the right size, the right company to purchase from or for customization of clothing in a style unique to you. Future target markets include children and this market is getting younger. People of a variety of nationalities and countries are also trending larger and the plus market is sure to grow overseas in countries which may never have had these sizes previously. While no one can predict the future, one thing is sure – plus sizes are not going away any time soon.

8.6 Sources of further information and advice

It is difficult to find information about plus sizes, however there are runway shows and magazines dedicated to this segment of the market. The following is a sampling of these sources.

The Full Figured Fashion Week (FFFWeek) is produced by DeVoe Signature Events in NYC, and is held every year around June. They use all plus models and show off the latest fashions in a variety of price points. They also hold cocktail parties for their clients and other special events.

Plus size fashion can be hard to find if searching the internet randomly. Some start-ups have culled the world wide web and created useful portals for plus fashion. OneStopPlus.com, FashionFitsMe.com, and Passazz.net are three of the many portals on the internet. They boast impressive findings such as over 700 different retailers on one site. Some portals focus on lingerie for plus women, some on bathing suits, some on professional clothing, so it is worth looking around for exactly what you want.

There are also dozens of plus size fashion bloggers who 'tell it like it is'. If you want to get to know this market better, log in to read the musings of these bloggers and their followers. They provide information such as how they deal with the lack of choice in clothing for their size (some purchase two skirts and sew them together into one in order to have a skirt that fits), their complaints about clothing stores and other issues.

There are also books on plus sized fashion, though they may be considered outdated:

Cooklin, G., 1995. Master Pattern and Grading for Women's Outsizes. London: Blackwell Science.

Deckert, B., 1999. Sewing for Plus Sizes. Newtown, CT: The Taunton Press.

Zangrillo, F. L., 1990. Fashion design for the Plus-size. New York: Fairchild Publications.

Body measurement charts can be found for the plus sizes through ASTM, though I would suggest waiting for the updated ASTM D6960 plus size women's chart to become available.

8.7 References

Alvanon, 2012. Alvaform – Customized. Available at http://www.alvanon.com/M03_S04_T03.html [Accessed 28 July 2012].

American Society for Testing and Materials, (2004). D6960-04: Standard table of body measurements relating to women's plus size figure type, sizes 14W – 32W. ASTM International, West Conshohocken, PA: Author. DOI: 10.1520/D6960-04.

American Society for Testing and Materials, (2012). D5585-11e1: Standard table of body measurements for adult female misses figure type, size range 00 – 20. ASTM International, West Conshohocken, PA: Author. DOI: 10.1520/D5585-11E01.

Bellefante, G., (2010). Plus-Size Wars. *New York Times,* 1 August, MM22.

Center for Disease Control [CDC], (2012). Defining overweight and obesity. Available at http://www.cdc.gov/obesity/adult/defining.html [Accessed 20 August 2012].

Connell, L.J., Ulrich, P.V., Brannon, E.L, Alexander, M. and Presley, A.B., (2006). Body shape assessment scale: Instrument development for analyzing female figures. *Clothing and Textiles Research Journal*, **24**(2), pp.80–95. DOI: 10.1177/0887302X0602400203.

Cooklin, G., 1995. *Master Pattern and Grading for Women's Outsizes*. London: Blackwell Science.

Deckert, B., 1999. *Sewing for Plus Sizes*. Newtown, CT: The Taunton Press.

Enzi, G., Gasparo, M., Biondetti, P. R., Fiore, D., Semisa, M. and Zurlo, F., (1986). Subcutaneous and visceral fat distribution according to sex, age, and overweight, evaluated by computed tomography. *American Journal for Clinical Nutrition*, **44**, pp.739–46. PMID: 3788827.

Kuk, J.L., Saunders, T.J., Davidson, L.E., and Ross, R., (2009). Age-related changes in total and regional fat distribution. *Ageing Research Reviews*, **8**(4), pp.339–48. PMID: 19576300.

Mahoney, T., (1952). *Independent woman.* In Americana Annual. New York: Americana Corporation.

Petrova, A. and Ashdown, S. P., (2008). Three-dimensional body scan data analysis: Body size and shape dependence of ease values for pants' fit. *Clothing and Textiles Research Journal,* **26**(3), pp.227–52. DOI: 10.1177/0887302X07309479.

Schofield, N.A. and LaBat, K. L., (2005a). Exploring the relationships of grading, sizing and anthropometric data. *Clothing and Textiles Research Journal*, **23**(1), pp.13–27. DOI: 10.1177/0887302X0502300102.

Schofield, N.A. and LaBat, K. L., (2005b). Defining and testing the assumptions used in current apparel grading practice. *Clothing and Textiles Research Journal*, (3), pp.135–50. DOI: 10.1177/0887302X0502300301.

Siervogel, R. M., Roche, A. F., Himes, J. H., Chumlea, W. C. and McCammon, R., (1982). Subcutaneous fat distribution in males and females from 1 to 39 years of age. *Journal of Clinical Nutrition*, **36**, pp. 162–71. PMID: 7091026.

Textile/Clothing Technology Corporation [TC]^2, 2004. SizeUSA: The national sizing survey. Cary NC: Author

U.S. Department of Health and Human Services, (2010). Healthy People 2010 Progress Review: Nutrition and Overweight. Available at: http://www.healthypeople.gov/2010/data/2010prog/focus19/default.htm [Accessed 23 August 2012].

World Health Organization, (2012). Obesity and Overweight: Fact Sheet No. 311. Available at http://www.who.int/mediacentre/factsheets/fs311/en/ [Accessed on 23 August 2012].

Zangrillo, F. L., 1990. Fashion design for the Plus-size. New York: Fairchild Publications.

9
Asian and Caucasian: designing apparel for these two different ethnic groups

Y. A. LEE, Iowa State University, USA

DOI: 10.1533/9781782422150.2.169

Abstract: Clothing fit is one of the most important factors that influence consumers' purchase decisions, and has a great impact on purchase returns. To provide satisfactory fit information to consumers, sizing systems are continuously evolving. Technology such as 3D body scanning has been used to collect individual's anthropometric data in various countries (i.e., SizeUK, SizeUSA, SizeKorea, SizeThailand, SizeFrance, SizeBrazil, SizeJapan). Although those measurements provide necessary data for the development of a sizing system, standard measurements alone cannot completely solve the issues of sizing and fit. Other important variables such as ethnicity, body type or shape, body image, and perception of acceptable fit are intertwined. This section will provide an understanding of the rising issues regarding the apparel design for different ethnic groups in relation to meeting their different needs.

Key words: apparel design, body shape, clothing style, culture, ethnic group, fit and sizing, emerging technologies.

9.1 Introduction

These days, clothing design is a sizzling issue in the apparel industry. It has to meet specific individual needs as consumer expectation and satisfaction have become ever more significant to the suppliers of apparel products. Some studies emphasize the importance of determining satisfaction levels with the fit of ready-to-wear garments (Labat and DeLong, 1990). Clothing fit is thus one of the most important factors that influence consumers' purchase decisions and guarantee non-return of a purchase. Satisfactory clothing fit requires an understanding of individual fit preferences, since individuals' perceptions of size and fit are affected by psychological factors related to body size and shape. Therefore, from a global perspective of the apparel industry, understanding the rising issues regarding clothing design for various ethnicities needs to be attained to meet different ethnic groups' specific needs.

First, target market grouping is needed. 'Asian' may be seen as one ethnic group when dividing the world market by continent, yet one can identify further subsets such as East Asian, West Asian, Central Asian, and South Asian according to particular geographical areas or countries. From this grouping of ethnicities, culture related to apparel styles can be considered in designing apparel, and the specific apparel may then appeal to targeted consumers. Different ethnic groups have different body shapes and dimensions as well as different body images; for example, Asian and Caucasian ethnic groups present important differences in body types. One group's body type may be used as a primary guideline facilitating clothing selection. When designing apparel for different ethnic groups, different sizing systems and size labeling methods should be considered.

The ability to offer adequate sizing, and its impact in meeting consumer needs for a satisfactory fit, has long been the focus of the apparel industry. The use of 3D body scanning has often been seen as the panacea for automated fit, size prediction, mass customization, pattern development and even as a magic option for solving fit dissatisfaction (Devarajan and Istook, 2004; Loker, 2007; Bougourd, 2007). Internationally, anthropometric surveys have been conducted to identify optimal fit requirements and sizing systems (Chen *et al.*, 2010). SizeUK (see http://www.shapeanalysis.com/SizeUK%20Stattistics.htm), utilizing both manual and scanning technology focused on integrating scan data, advanced computer aided design (CAD) and commercial applications in production, standards development and virtual shopping (Bougourd, 2007). SizeUSA (see www.sizeusa.com) had similar intentions. However, irrespective of whether manual or non-contact technology is utilized, certain ethical and procedural considerations must be addressed.

With mass customization, for instance, efforts are directed at business technology and consumer satisfaction, of which fit is a critical component. Finding the perfect fit is a challenge to both designers and consumers because of variations in body size and shape, and perception of acceptable fit (LePechoux and Ghosh, 2002). Although women's measurements provide necessary data for the construction of a sizing system, standard measurements alone cannot completely solve the issues of sizing and the related problem of fit. Human measurement approaches have often presented both ethical and methodological issues that have impacted validity and reliability. Since measurement alone cannot meet all sizing needs, and because of the psychological and social implications of size and shape for consumers (Entwistle, 2000), it is crucial to analyze both measurement procedures and consumer views regarding body size and shape evaluation and measurement, which are significant factors of body cathexis, a person's feelings about aspects of his or her body, in assessing fit for apparel products. Consequently, the focus group approach has been recommended for research, especially in evaluating consumer decision making, product development and future planning, and has been used continually to obtain information about consumer attitudes, needs and concerns (Krueger and Casey, 2000).

Over half of the US population is overweight, defined as a body mass index (BMI) greater than or equal to 25, or obese, defined as a BMI greater than or equal to 30. According to the National Center for Health Statistics (2010), among American adults 20 years of age and over, Afro-American females have the absolute highest prevalence of overweight individuals (78%) followed by white Caucasian females (59.6%). In the case of males in the US, 71.7% of black men and 72.9% of white men, a similar percentage fall into this category. Many studies have demonstrated that the high prevalence of overweight people in the black population, especially among black females, might be attributed to low physical activity, fewer educational and financial resources, and cultural influences leading to a de-emphasis of the thin body type in this community. In contrast, whites Caucasians, especially white females, appeared to be more concerned about their body weight. This perception difference in body image may account for a large part of racial differences seen in overweight prevalence.

Wells *et al.* (2008) found no significant ethnic differences in body shape among white, African- and Hispanic Americans. However, Asians' standards for body shape analysis are different from the others. For instance, lower BMI cut-offs for screening the metabolic syndrome have been proposed for Asians in general. As the fashion market becomes ever more global, apparel designers and product developers need to understand these different body shapes of other countries' consumers. For example, Thai adults in both genders had significantly smaller body girths than UK adults after adjusting for age and height. Matching for BMI, and adjusting for height and age, Thais of both genders tend to have similar or greater limb circumferences, but significantly smaller torso circumferences (especially waist and hip) than UK individuals (Wells *et al.*, 2012). Shape-age associations also differed between the populations. Young Thai adults have a significantly slighter physique than white UK adults, with a less central distribution of body weight. However, these differences reduce with age, especially in males (Wells *et al.*, 2012).

Rasband (2006) states that body figures vary from the ideal figure in six characteristic ways; 1) height, 2) bone size or structure, 3) weight, 4) proportional body areas, 5) contour, including weight distribution or figure type, and 6) posture. There are several typical patterns of weight distribution when it comes to overall body contour. Body types can be recognized or identified according to the specific areas on the body where weight tends to accumulate, regardless of height. Body types can be a guideline to creating easier clothing selection for consumers when shopping. Figure 9.1 has been developed by the author to show various figure types identified by Rasband (2006).

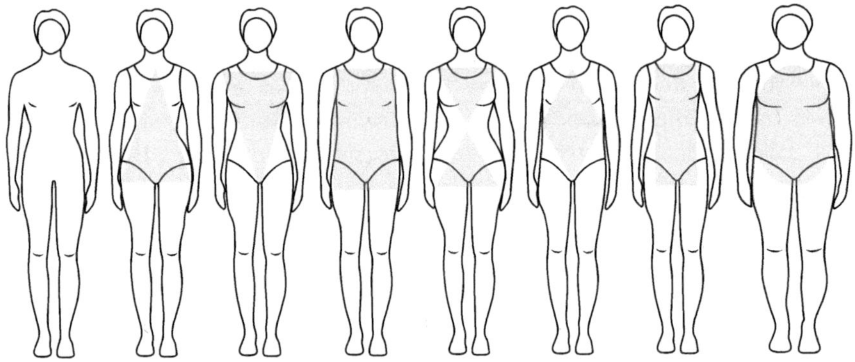

9.1 Selected examples of body type variations.

Table 9.1 shows the summary of body type classifications by various researchers. Bonnie August's Body ID (1981) scale contains eleven proportion designators categorized into three different groups: front view width, side view width, and front view length. These proportions are represented by alphabetic letters, which visually align with the selected landmarks amongst the varying shapes of the body forms. August's decision to utilize the alphabet in contrast to shape names was influenced by the notion that certain shape classifications may seem derogatory to women. These alphabetic representations are A, X, H, V, W, Y, T, b, d, i, and r. The design of this scale allows for frontal categories to be combined with side view categories (Sidberry, 2011). August's (1981) scale is a nominal scale, as it does not contain a ranking. The stimulus used in Bonnie August's Body ID scale (1981) contains body silhouettes with the appropriate alphabetic representation aligning with the landmark points of the body shape. The landmarks used in the formation of this scale include shoulder points, natural waistline, and widest hip point. This scale also uses the shape recognition classification as a basis for grouping unidentified shapes (Sidberry, 2011).

As shown in Table 9.1, Farrell-Beck and Pouliot (1983) used hip shapes for their body type classification. An experimental method was developed to alter pant patterns utilizing graphic somatometry techniques. In the study, they used five figure variations, which are round hip, pear-shaped hip, average hip, weight in front and weight in back (Devarajan and Istook, 2004). The pattern industry developed its figure types and standards to meet the needs of consumers. Patterns were sized using the bust as the single control dimension. The figure types include Young Junior/Teen, Junior Petite, Junior, Miss Petite, Miss, Half-size, and Woman (Armstrong, 1987; Devarajan and Istook, 2004).

Table 9.1 Body type classifications by various researchers

	August (1981)	Farrell-Beck and Pouliot (1983)	Armstrong (1987)	Simmons (2002)	Connell *et al.*, (2006)	Rasband (2006)
Classification	A X H V W Y T b d i r	Round hip Pear-shaped hip Average hip Weight in front Weight in back	Young Junior/ Teen Junior Petite Junior Miss Petite Miss Half-size Woman	Hourglass Bottom hourglass Top hourglass Spoon Rectangle Diamond Oval Triangle Inverted triangle	Hourglass Pear Rectangle Inverted triangle	Ideal Triangular Inverted triangular Rectangular Hourglass Diamond Tubular Rounded
Classification Criteria	Front view width, side view width, and front view length	Hip shapes	Bust	Bust, waist, stomach, and abdomen circumferences	Frontal body shape: shoulder point to shoulder point, the frontal waistline, and the widest point between the waist and crotch line	Bust, hip circumference

The Female Figure Identification Technique scale (Simmons, 2002) utilized circumferential measures as opposed to silhouette. Prior to utilizing a Textile and Clothing Technology Corporation $[TC]^2$ NX-12 3D body scanner, researchers reviewed the existing information regarding female body shape to find commonalities amongst the data. Next, common body shapes were acquired from the existing information and synthesized into five shapes: triangle, which is defined by hips that exceed the shoulder in width; inverted triangle, which is defined by shoulders that exceed the hips in width; rectangle, which is described by a lack of deviation between shoulder, waist, and hip widths; hourglass, which is characterized by equally wide shoulder and hip widths that created a defined waistline; and oval, which is defined by a narrow top and bottom torso, with a wide or full abdomen (Simmons, 2002). Once these five shapes were derived, they were assessed using the body scan data of a sample of 222 women above the age of 18, in combination with mathematical criteria that identified each body shape. The landmarks used in this assessment were bust, waist, stomach, and abdomen circumferences. The results were that more body shape categories were needed, as many subjects failed to be categorized. To overcome this problem, four new categories were developed; these groups were labeled spoon, diamond, bottom hourglass, and top hourglass, as they each resembled the shape of the object after which they were named (Simmons, 2002).

The Body Shape Assessment Scale (BSAS©) developed by Connell *et al.* (2006) contains nine variants assessing frontal body shapes. This scale came about as a result of the assessment of body scans derived from a sample of 42 women between the ages of 20 and 55, in combination with the review and revision of existing body scales. The results of this analysis yielded four prominent frontal body shapes. The BSAS© uses digitally illustrated female body silhouettes to represent each of the four body shapes based on the following landmark locations to differentiate whole body shapes: shoulder point to shoulder point, the frontal waistline, and the widest point between the waist and crotch line, as seen from the front (Connell *et al.*, 2006). Rasband (2006) used eight different body variations, which are: ideal, triangular, inverted triangular, rectangular, hourglass, diamond, tubular, and rounded figure types. Those body types are derived from two main landmarks, bust and hip circumference, which are recognized as the specific areas on the body where weight tends to accumulate, regardless of height (Rasband, 2006). Figure types are a basic guideline used to make clothing selection easier, and people want to look for the figure type most similar to them. In Rasband's book (2006), all figure types but the ideal are identified according to their relative similarity to easily recognized geometrical shapes as viewed from the front.

9.2 Understanding different body shapes: the case of North American college students from Asian and Caucasian ethnic groups

Based on the rapidly growing Asian population in the US during the past decades, and globalization of the fashion market, an interesting study related to body shapes of different ethnicities was conducted. This case study explored the body shape differences between white Caucasian students and Asian students, which impacted the apparel sizing systems of each respective culture and might relate to body image and weight issues. This study also addressed some important issues related to body type, body preference, and weight perceived by different genders and different ethnicities. The results of the study can provide directions about apparel design concepts and marketing strategies for the apparel industry.

A total of 141 college students participated in this study; 27 female and 14 male students were Asian, and 93 female and 7 male students were white Caucasian. The sample consists of 120 female students and 21 male students. Participant's ages ranged from 18 to 32 years. For this study, only 3D scanned images and the demographic information of each participant were obtained. A $[TC]^2$ NX-16 white-light 3D body scanner was used to collect scan images for the original research in one of the Midwestern universities. All subjects participating in the project were voluntary, and were required to sign a form that released all data resultant from the experiment so that it could be used for further research. Before the data collection, each study participant completed a demographics form, including age, gender, ethnicity, education level, height, and weight. The height and weight information was required for the BMI calculation. Every participant wore white, close-fitting clothing and stood straight with a relaxed posture in the 3D body scanner for 10–15 seconds until the scan was completely processed. For the analysis of body shapes of those ethnicities, 3D scan images and the participant's demographic information, which were collected from 2008 to 2010, were used and the data were coded by using SPSS 17. In this study three subsequent categories were explored: (1) body type, (2) body preference, and (3) weight issues.

The study showed significant differences between genders in body type and body image, with insignificant differences between ethnicities (white Caucasian and Asian). The results show that college students' body shape concentrates in several body types or groups, and male students have fewer body shape differences than female students. Particularly, for the female students, rectangular, bottom hourglass, top hourglass, hourglass, triangle, and spoon were the dominant body shapes. The results also support the idea that female students' actual or real body types are far away from their ideal body type while male students mostly have insignificant differences between their

Table 9.2 Body type categorization by human body measurements

Body types	Measurement
Hourglass	If (bust – hips) ≤ 1, then if (hips – bust) < 3.6, then if (bust – waist) ≥ 9 or (hips – waist) ≥ 10
Bottom hourglass	If (hips – bust) ≥ 3.6 and (hips – bust) < 10, then if (hips – waist) ≥ 9, then if (high hip/waist) < 1.193
Top hourglass	If (bust – hips) > 1 and (bust – hips) < 10, then if (bust – waist) ≥ 9
Spoon	If (hips – bust) > 2, then if (hips – waist) ≥ 7, then if (high hip/waist) ≥ 1.193
Triangle	If (hips – bust) ≥ 3.6, then if (hips – waist) < 9
Inverted triangle	If (bust – hips) ≥ 3.6, then if (bust – waist) < 9
Rectangle	If (hips – bust) < 3.6, and (bust – hips) < 3.6, then if (bust – waist) < 9 and (hips – waist) < 10

real body shape and their ideal body shape. An ideal body type was measured by waist-hip ratio (WHR) and BMI. There were no significant differences in real body type, body preference, and weight between the two different ethnicities in this study.

First, our results for both white Caucasian and Asian female students showed the same result in body type prevalence. Table 9.2 shows how to group bodies into seven different body types using the method from Lee *et al.* (2007). Most female students had a rectangle body shape, followed by the bottom hourglass, triangle group, hourglass, and spoon shapes (see Table 9.3). For the male students, the two body types most frequently seen were the rectangle and triangle shapes. Secondly, when it comes to the body preference, both white Caucasian and Asian female students considered that their real body types were significantly different from their ideal body types. In this study the comparison was done through the analysis of the relationship between their current BMI and the ideal BMI. On the other hand, both white Caucasian and Asian male students showed no significant differences between their real body types and ideal body types based on their BMI.

Finally, in the weight analysis, the majority of female and male students belonged to the normal weight category (see Table 9.4). There were no significant correlations between different genders and weights, nor was there any significant relationship between different ethnicities and weight. This result might be because the college students who tend to volunteer to participate in this kind of research usually belong to normal body type and weight categories.

In sum, the case study results show that there were significant differences between different genders in body type and body image, with insignificant differences between white Caucasian and Asian. The results show that college students' body shape concentrates in several body types and

Table 9.3 Comparison of body mass index (BMI) and waist-hip ratio (WHR) between different body types, gender, and ethnicities

			Total	Hourglass	Bottom hourglass	Spoon	Triangle	Rectangle
Female	White Caucasian	N	93	4	29	2	15	43
		% of total	100	4.3	31.2	2.2	16.1	46.2
		Sig of BMI	.000*	.112	.002*	.37	.000*	.000*
		Sig of WHR	.000*	.088	.000*	.392	.000*	.000*
	Asian	N	27	1	5	–	3	18
		% of total	100	3.7	18.5		11.1	66.7
		Sig of BMI	.007*	.009*	.000*		.007*	.018*
		Sig of WHR	.000*	.000*	.000*		.007*	.000*
Male	White Caucasian	N	7	–	–	–	2	5
		% of total	100				28.6	71.4
		Sig of BMI	.694				.509	.322
	Asian	N	14	–	1	–	1	12
		% of total	100		7.1		7.1	85.7
		Sig of BMI	.754		.932		.448	.717

$N = 141$ $^{*}p < .05$

Table 9.4 Weight comparison between white Caucasian and Asian students

Weight comparison between white Caucasian and Asian								
			Under-weight	Normal	Overweight	Obese	Total
Female	WEA	Count	7	66	17	3	93
		% of total	7.5	71.0	18.3	3.2	100
	Asian	Count	4	21	2	–	27
		% of total	14.8	77.8	7.4		100
Male	WEA	Count	1	5	–	1	7
		% of total	14.3	71.4		14.3%	100
	Asian	Count	2	8	4	–	14
		% of total	14.3	57.1	28.6		100

male students have fewer body shape differences than female students. The results also supported the idea that even though the college students were part of one of the largest groups that have normal body types and weight, female students' real body types are far from today's ideal body type, while male students mostly have insignificant differences between their real body shape and their ideal body shape. The fact may stimulate females to change their body shape more than males. It is important to note this result may be due to the fact that there were only a very small number of male participants.

9.3 Sizing requirements for different ethnic populations

To understand characteristics of different ethnic groups in designing apparel, it is important to continuously gather sizing-related features of different ethnic populations such as sizing system, body measurement specification and standards, sizing labeling and real anthropometric data of different target groups (e.g. men, women, children). Although various endeavors to identify sizing standards used in clothing design have been attempted, the apparel industry has not applied a single sizing system due to their differentiated marketing strategy (Simmons *et al.*, 2004). Therefore, those size differences between countries, companies, brands, even product categories in a brand caused problems in understanding sizing systems of the apparel industry at the global level. The apparel industry needs to find a relationship between different sizing systems used in various countries around the world. Converting sizes give the ability to translate a symbol from one system to another according to the encoded size of clothing, but this is just the conventional relationship, and so far has not standardized.

The goals of sizing standards are to ensure consistency and clarity in clothing sizes and size labels offered to the consumer and to fit a large segment of a target population adequately (Labat, 2007; Alexander *et al.,* 2012). Two major types of sizing standard are tables of body measurements and size names or designations. The first provides tables of body measurements for a segment of consumers such as women, men, or children. The second offers standard language for size designations (Labat, 2007). Body measurement standards are based on anthropometric studies on body measurements of a sample representative of a population of people. There are many methods for segmenting the data into sizes, with different methods being used by different researchers and companies.

As the fashion market is growing into a global level, understanding sizing systems used in other countries is important, even though adopting a worldwide clothing sizing standard and body size dimensions is difficult given the huge variation in body dimensions and cultural diversity in shopping and product expectations (Labat, 2007). Designers often change their standard model along with changes in the average target customers' figure, and brands also set their size according to different models (Rasband, 2006). To complicate an already complex situation, types of garment are not sized alike. For example, shirts are sized differently than sweaters, dresses differently than coats, that is, garment style makes a difference (Rasband, 2006).

Some international brands such as H&M and Zara include the size denominations from multiple countries in their garment labels, while others such as Uniqlo (Japanese clothing brand) post Japan and European equivalent charts on the sales floor in their stores. Global brands such as Nike, The North Face, Lacoste and Prada have also developed Asian-specific fits. An increasing number of brands are beginning to understand that the one-size-fits-all approach will limit their growth and success (Miller, 2011). It is difficult to find the balance between having too many sizes versus too few. If a brand has more than 10% of its sales in the smallest or largest size offered, then it needs to add a size or else they risk missing sales (Miller, 2011).

Organizations that develop sizing standards in various countries are currently working on the labeling issue by offering better information and guidelines on the schemes and by crafting more globalized standards. Table 9.5 shows the selected examples of international sizing standards (Labat, 2007; Alexander *et al.*, 2012; ASTM, 2008).

The ISO 3635 standard indicates size designation of clothes, definitions, and body measurement procedures (ASTM, 2008). The ISO TC133 technical committee for sizing systems and designations for clothes even became active again in March of 2011 after a pause of more than 20 years (Miller, 2011). The most recent standard for plus-size figures is ASTM International Standards D6960-04. Updated in 2004, ASTM Standard D6960–04 clarified the body measurements for plus-size women sizes 14W–32W in an effort

Table 9.5 Selected examples of international sizing standards

Country	Sizing standard /Publisher	Size designation/ dimensions
China	GB 1335-81 (1981, updated in 1991)	Anthropometric survey of 14 000 subjects
EU	EN 13402	European standard for labeling clothing sizes based on body dimensions, measured in centimeters
England	BS 3666 (1990)	Women's body measurements
Germany	DOB-Verband (1983)	Nine figure types height by hip
Hungary	Office of Hungarian Standards (1986)	Height and body build
Japan	JIS L 0103/Japanese Standards Association (1990)	Garment measurements, not body measurements
Korea	KS K 0051/Korean Standards Association (1990)	Five height groups
Thailand	Wacoal (1981 & 1987)	Women's preferred sizes
USA	ASTM D6960-04 (2004) ASTM D6240-98 (2006) ASTM D6458-99 (2006)	Average height and more rounded body figures (clarified plus-size women) Body measurement of men size 34–60 Body measurement of boys size 8–14 slim and 8–20 regular

to help reduce consumer confusion and dissatisfaction in regard to apparel sizing (ASTM, 2004). Rather than plus-size measurements being associated with a taller figure, the revised standard D6960-04 reflected plus-size measurements for the average height and more rounded body figures (Keiser and Garner, 2008; Alexander *et al.*, 2012). The most recent standard for men and boys are ASTM International Standards ASTM D6240-98 and ASTM D6458-99 (ASTM, 2008).

The European Union has produced a standard EN 13402 'Size designation of clothes' for labeling clothes sizes, intended to replace existing standards in the member countries. It is based on body dimensions, the metric system (SI), data from new anthropometric studies of the European population performed in the late 1990s, and similar existing international standards (ISO 3635). This standard consists of four parts: terms, definitions and body measurement procedure; primary and secondary dimensions; measurements

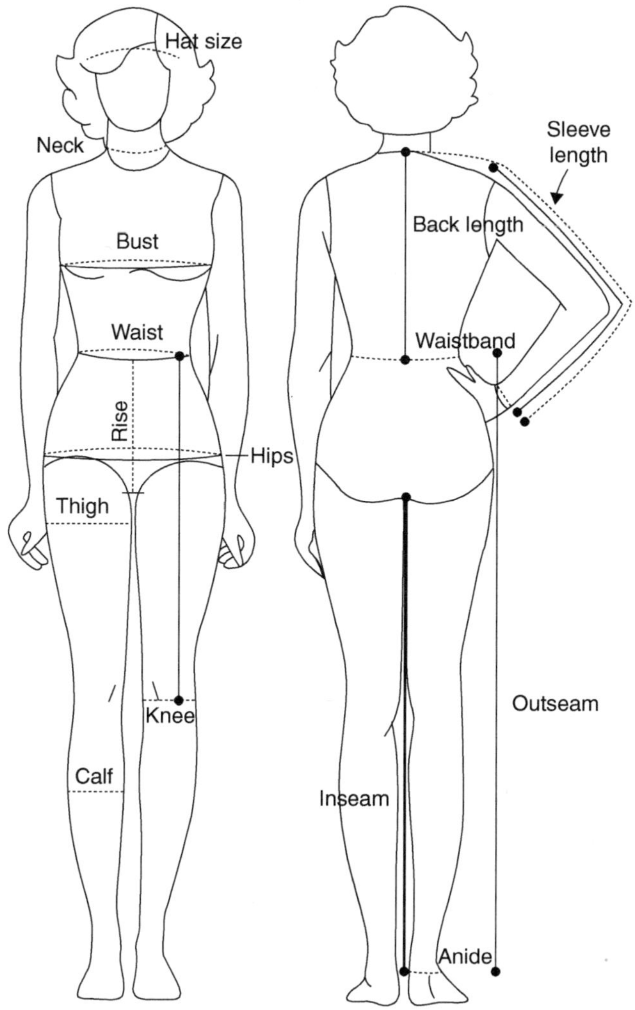

9.2 Key body measurement areas for clothing sizing systems.

and intervals; coding system. This standard, in force in some European countries, combines the dependency code letter not with the known size, but the dimension values of the chest (men) or bust (women) girth, and then with other measurements, such as height. Standard clothing sizes in the US were developed from statistical data in the 1940–50s. It is similar in concept to the EN 13402 European clothing size standard, however now the most common standard is known as the US catalog sizes. Catalogs have departed from the US standard sizes since approximately the 1980s. Companies now may provide the measurements for their sizes, which may vary even among different styles of the same type of garment. According to these catalogs, women's sizes are divided into various types, depending on the overall height and the relative bust and waistlines heights. Figure 9.2 illustrates the key body measurement areas that are used by various countries when making clothing.

In view of widespread globalization, including in the garment industry, there is a need to find a relationship between the different systems of garment sizes used in various countries around the world. Only a properly defined and corresponding correlation creates the possibility for commercial transactions between companies from different countries, and provides correct relationships between imports and exports. Converting sizes would give the ability to translate a symbol from one system to another according to the encoded size of clothing. Theoretically, a relationship can be found between these systems of clothing sizes (see Table 9.6), but this is just the conventional relationship, and so far has not been standardized. Finding a connection is not simple and straightforward because there are fairly significant differences between these systems. They arise not only from the use of different measurement units, but also from different approaches, especially in children's sizes. For children's clothing, Europeans go by height rather than age. Meanwhile, the UK system uses only age to define size. Such thinking is not always correct because children of the same age can have different height (Bogusławska-Bączek, 2010).

Table 9.6 Selected examples of international clothing size conversions

Country	Size													
	Women's clothing													
International sizes	XS		S		M		L		XL		XXL		XXXL	
EU	32	34	36	38	40	42	44	46	48	50	52	54	56	58
UK	–	6	8	10	12	14	16	18	20	22	24	–	–	–
France, Spain, Portugal	34	36	38	40	42	44	46	–	–	–	–	–	–	–
US	–	4	6	8	10	12	14	16	18	20	22	–	–	–
Italy	38	40	42	44	46	48	50	–	–	–	–	–	–	–
Japan	7	9	11	13	15	17	19	21	23	–	–	–	–	–
	Men's clothing													
EU	40	42	43	44	46	48	50	52	54	56	58	60	62	64
–	–	32	34	36	38	40	42	44	46	48	50	52	–	–
US	–	32	34	36	38	40	42	44	46	48	50	52	–	–
	Children's clothing													
EU	28	30	32	34	36	38	40	42						
UK	3	5	7	9	11	13	15	17						
US	1	3	5	7	9	11	13	15						

9.4 Examples of specific styles for different ethnic populations

The degree to which a style is found fashionable is generated by its popularity. With more advanced concepts, social norms also define the style. Whether individual consumers accept an apparel style in a particular group or community is influenced by the normal standards of the group which accepts the style. Different ethnic populations have different cultures rooted in their own history and may have different preferences for their body shape and body image. Satisfaction or dissatisfaction with clothing design may stem from both social norms and cultural ideals of style. Therefore, designers and merchandisers in the fashion industry should consider their specific style preferences of clothing based on their specific social norms and culture. There can be culture inspired styles, body preference driven styles, body shape feature driven styles, and the industry needs to invest in researching those styles according to different ethnicities to meet various consumer needs in the global market.

Since 1982, the average Thai woman aged under 50, has become 5 cm taller and 5.4 kg heavier, while her waistline has ballooned by 9.4 cm, her bottom has grown by 5.1 cm, and her chest has expanded by 5.6 cm. The average Thai man aged under 50 has become 7.2 cm taller and an amazing 11.4 kg heavier. His chest has expanded 13.7 cm, his waistline has increased by 10.2 cm, and his bottom has gained 8.6 cm. The fact that women's waistlines have grown almost twice as much as their chests and bottoms means that the shape of the average Thai woman has also changed. Not only have women become bigger, they have also become more tubular and have less of the traditional hourglass shape. According to the SizeThailand data, SizeThai (TH) established nine standard sizes for men: TH 32, 34, 36, 38, 40, 42, 44, 46, and 48, all based on the measurement of chest and waist. For women, they defined ten sizes: TH 28, 30, 32, 34, 36, 38, 40, 42, 44, and 46, all based on the measurement of chest, waist and bottom. The majority of Thai men are size TH 40 and the majority of Thai women are size TH 36.

9.5 Future trends

In designing for various markets, different strategies need to be considered in the future. With the development of new technologies, consumer's involvement in designing apparel has become crucial. The 3D scanning technology provides detailed digital anthropometric data and some companies match consumers' exact measurements to clothes that will best fit their body shape and make it easier for consumers to shop. Development of comprehensive specifications for sharing apparel product data among all stages of the product lifecycle is also ongoing, to ensure better clothing fit for specific consumers. In the near future, apparel companies will use various sizing data from different countries to create garments that meet the target market's various clothing needs.

The 3D photonic scanning provides detailed digital anthropometric data capable of monitoring between- and within-individual shape variability. The technology merits further application to investigate whether variability in body shape is more sensitive to metabolic risk than BMI within- and between-populations (Wells *et al.*, 2012). The sizing variations are a big contributor to the $194 billion worth of clothing purchases returned in 2010; this amounted to more than 8% of all clothing purchases (Clifford, 2011). The scanners are a modern solution to an old problem. A study of dress sizes in Vogue advertisements from 1922 found that clothing sizes have been irregular for decades. Clifford (2011) stated that a woman with a 32" bust would have worn a size 14 in the 1937 Sears catalog. By 1967, she would have worn a size 8. Plenty of people have tried to address these random sizes. Fitlogic, which is a labeling system, discovered just how difficult it is to change manufacturers' approach to size over the last few years.

For instance, Me-Ality, previously called MyBestFit, (2012) matches consumers' exact measurements to clothes that will best fit their body shape to make consumers shopping easier. The customer steps into a circular booth, fully dressed. A wand rotates around him or her, emitting low-power radio waves that record about 200000 body measurements within 10–15 s. Then, the system matches the customer's measurements to the sizing specifications of apparel brands in its database, eliminating the guesswork and frustration associated with shopping for clothes. The database contains measured clothes from about 135 brands/stores, including Old Navy, Eddie Bauer, Levi's, Nike and Talbots to name a few. Each customer receives a printout of the sizes from each store that ought to fit him/her best. This guide can be filtered by style, price or retailer (see http://www.uniqueltd.com/). This option will eventually open the doors for various ethnic groups to find clothing that fits well into their different body shapes.

The apparel industry has used computer systems to great advantage. It has automated many of its manufacturing processes. However, the manufacturing innovations often stand alone as 'islands of automation'. Integrating the separate automated processes could greatly improve the effectiveness of the entire enterprise. In recent years, the National Institute of Standards and Technology (NIST), under the sponsorship of the Defense Logistics Agency, has been developing the apparel product data exchange standard (APDES). The goal of the project is to develop a comprehensive specification for sharing apparel product data among

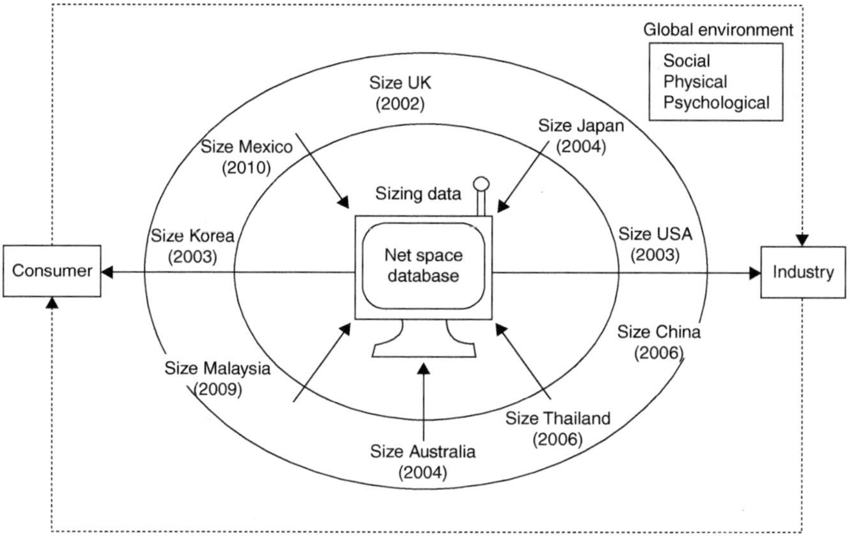

9.3 Technology integration with international sizing systems to meet diverse ethnic group's clothing needs in global environments.

all stages of the product life cycle. The APDES project members at NIST have determined a set of manufacturing data interfaces that could be standardized for the effective integration of the information systems required to operate an apparel manufacturing enterprise. Now, ISO 20685:2005 – 3D scanning methodologies for internationally compatible anthropometric databases – is used to get anthropometric data through data surveys. Figure 9.3 shows a few selected countries that have joined to conduct the national sizing survey using 3D body scanning technology, and illustrates how the databases from each country can possibly be merged and used for diverse target population's clothing needs in the global environments.

In the near future, apparel companies could use various sizing data from different countries to create clothing which meets the target market's needs. The company could save the data, with ID numbers but not with names, and may give collective information to retailers as feedback. Customers will be able to measure their sizes overseas and then send the data to the Net Space database, which will produce 3D drafts and accurately determine sizes for customers to custom make clothes. In summary, this section provided an understanding of the rising issues regarding the apparel design for different ethnic groups in relation to meeting their different needs, and of the way the technology could make it possible to meet these consumer's clothing needs in this dynamic global environment.

9.6 Sources of further information and advice

Websites

- ASTM International: http://www.astm.org.

Offers developed international standards for materials, products, systems and services used in construction, manufacturing and transportation.

- International Organization for Standardization (ISO): http://www.iso.org.

Offers developed and published international standards to ensure products and services are safe, reliable, and of good quality.

- Me-Ality: http://www.me-ality.com/.

Company website specializing in 3D body scanning, offering services such as scanning locations, size matching to brands, and a shopping guide.

- National Center for Health Statistics: http://www.cdc.gov/nchs.

Offers current statistical information related to the health of the American population.

- National Institute of Standards and Technology (NIST): http://www.nist.gov

Non-regulatory US federal agency website promotes advancement of measurement science, standards, and technology in ways that enhance economic security and quality of life.

- SizeKorea: http://sizekorea.kats.go.kr.

Offers size statistics on the Korean population derived from a study utilizing 3D body scanning technology to obtain measurement data.

- SizeThailand: http://www.sizethailand.org.

Offers size statistics on the Thai population derived from a study utilizing 3D body scanning technology to obtain measurement data.

- SizeUK: http://www.shapeanalysis.com/SizeUK%20Stattistics.htm.

Offers size statistics on the UK population derived from a study utilizing 3D body scanning technology to obtain measurement data.

- SizeUSA: http://www.sizeusa.com.

Offers statistically accurate shape and sizing data for the US population obtained through extensive 3D body scan surveys.

- [TC]²: http://www.tc2.com.

A company providing 3D body scanning equipment, software for measurement extraction, avatar creation and virtual fashion, and specializing in technology development and supply chain improvement

Major newspapers that report on fashion

- Apparel magazine: http://apparel.edgl.com.
- Financial Times: http://www.ft.com.
- International Herald Tribune: http://global.nytimes.com.
- New York Times: http://www.nytimes.com.
- The Economist: http://www.economist.com.
- The Telegraph: http://www.telegraph.co.uk.
- Washington Post: http://www.washingtonpost.com.
- Women's Wear Daily: http://www.wwd.com.

9.7 References

Alexander M, Pisut G R and Ivanescu A (2012), 'Investigating women's plus-size body measurements and hip shape variation based on SizeUSA data', *International Journal of Fashion Design, Technology and Education,* **5**(1), 3–12.

Armstrong H J (1987), *Patternmaking for fashion design*, New York, Harper & Row Publishers.

ASTM (2008), *Annual book of ASTM standards 2008: Section 7 textiles*, West Conshohocken, PA, ASTM International.

ASTM (2004), *Annual book of ASTM standards 2004: Section 07.02 textiles (II)*, West Conshohocken, PA, ASTM International.

August B (1981), *The complete Bonnie August dress thin system: 642+ ways to correct figure with clothes*, New York, Rawson, Wade Publishers.

Bogusławska-Bączek M (2010), 'Analysis of the contemporary problem of garment sizes', *The 7th International Conference – TEXSCI 2010.*

Bougourd G (2007), 'Sizing systems, fit models and target markets', in Ashdown S P, *Sizing in clothing: developing effective sizing systems for ready-to-wear clothing*, Cambridge, Woodhead Publishing Limited, 108–51.

Chen, C M, LaBat, K and Bye E (2010), 'Physical characteristics related to bra fit', *Ergonomics*, **53**(4), 514–24.

Clifford S (2011, April 24), One size fits nobody: Seeking a steady 4 or a 10, The New York Times. Available from http://www.nytimes.com/2011/04/25/business/25sizing.html?_r=1 [Accessed 19 April 2012].

Connell L J, Ulrich P V, Brannon E L, Alexander M and Presley A B (2006), 'Body shape assessment scale: Instrument development for analyzing female figures', *Clothing and Textiles Research Journal,* **24**(2), 80–95.

Devarajan P and Istook C L (2004), 'Validation of 'female figure identification technique (FFIT) for apparel' software', *Journal of Textile and Apparel, Technology and Management*, **4**(1), 1–23.

Entwistle J (2000), *The fashioned body: Fashion, dress and modern social theory*, Cambridge, Polity Press.

Farrell-Beck J A and Pouliot C J (1983), 'Pant alteration by graphic somatometry techniques, *Home Economics Research Journal*, **12**(1), 95–105, doi: 10.1177/1077727X8301200112.

Keiser S J and Garner M B (2008), *Beyond design: the synergy of apparel product development*, New York, Fairchild.

Krueger R A and Casey M A (2000), *Focus groups: A practical guide for applied research*, Thousand Oaks, CA, Sage Publications.

Labat K L (2007), 'Sizing standardization', in Ashdown S P, *Sizing in clothing: Development effective sizing systems for ready-to-wear clothing*, Cambridge, Woodhead Publishing Limited, 88–107.

Labat K L and DeLong M (1990), 'Body cathexis and satisfaction with fit of apparel', *Clothing and Textiles Research Journal*, **8**(2), 43–8.

Lee J Y, Istook C L, Nam Y J and Park S M (2007), 'Comparison of body shape between USA and Korean women', *International Journal of Clothing Science and Technology*, **19**(5), 374–91.

LePechoux B and Ghosh T K (2002), 'Standard sizing and fit testing applied to women's hosiery', *Textile Progress,* **32**(1), 1–59.

Loker S (2007), 'Mass customization and sizing', in Ashdown S P, *Sizing in clothing: developing effective sizing systems for ready-to-wear clothing,* Cambridge, Woodhead Publishing Limited, 246–63.

Me-Ality (2012), Official Web site. Available from: http://www.me-ality.com/ [Accessed 26 November 2012].

Miller K (2011), just-style management briefing: International sizing standards expand, just-style. Available from: http://www.just-style.com/management-briefing/international-sizing-standards-expand_id112057.aspx [Accessed 11 April 2012].

Rasband J A and Liechty E L G (2006), *Fabulous fit: Speed fitting and alteration*, New York, Fairchild.

Sidberry P A (2011), Effects of body shape on body cathexis and dress shape preferences of female consumers: A balancing perspective, Unpublished Master's thesis, Auburn, Auburn University.

Simmons K P (2002), *Body shape analysis using three-dimensional body scanning technology*, Unpublished doctoral dissertation, Raleigh, North Carolina State University.

Simmons K, Istook C L and Devarajan P (2004), 'Female figure identification technique (FFIT) for apparel–part I: Describing female shapes, *Journal of Textile and Apparel, Technology and Management*, **4**(1).

Wells J C K, Cole T J, Bruner D and Treleaven P (2008), 'Body shape in American and British adults: Between-country and inter-ethnic comparisons', *International Journal of Obesity,* **32**, 152–9.

Wells J C K, Treleaven P and Charoensiriwath S (2012), 'Body shape by 3-D photonic scanning in Thai and UK adults: comparison of national sizing surveys', *International Journal of Obesity,* **36**, 148–54, doi: 10.1038/ijo.2011.51.

10
Males: understanding sizing requirements for male apparel

J. BELLEMARE, Université du Québec à Montréal, Canada

DOI: 10.1533/9781782422150.2.189

Abstract: This chapter presents a look at the male apparel industry with respect to product development, design and production of clothing and accessories. The chapter is divided into four sections, starting with a comparison of key issues in order to better understand the mass production of menswear, followed by a study of men's standard sizing for the mass apparel industry. The second section examines different body shapes using case studies, along with the identification of different morphologies associated with fit using a cluster-type approach. The third section discusses the sizing requirements for male apparel and provides some examples of sizing systems. To conclude, some future trends are discussed as ways to further our research against the background of globalisation and new web-based approaches to collaboration.

Key words: male apparel body shapes, mass customisation, fitthinking, co-fitting, eSizing.

10.1 Introduction

This chapter presents a look at the male apparel industry with respect to product development, design and production of clothing and accessories. It will specifically explore the evolution of men's fashion to better understand the impact of body shape and size, as well as the major influences associated with the idea of 'fit' in the industry. The chapter is divided into four sections, starting with a comparison of key issues in order to better understand the mass production of menswear, followed by a study of men's standard sizing for the mass apparel industry. The second section examines different body shapes using case studies, along with the identification of different morphologies associated with fit using a cluster-type approach. This will help us recognize the importance of adapting a new measurement chart for a globalized context as it becomes increasingly difficult to meet the mass demand for different fit specifications. The third section discusses the sizing requirements for male apparel and provides some examples of sizing systems. From this point of view, the issues raised in this chapter aim to better comprehend the industry's perspective, in addition to the consumer's perspective. To conclude, some future trends will be discussed as ways to further our research against the background of globalisation and new web-based approaches to collaboration.

10.2 Key issues for comparison

Anthropometry, which refers to the process of measuring different body dimensions and features (Sun and Su, 2009), is playing an increasingly important role in ergonomics, as well as in clothing design. One of the key points of comparison the study of ergonomics has introduced to the clothing world is the notion of 'fit', defined in French by Serwatka (2008) as '*le bien allé*' and by Bougourd (2007) as a relation between size, contour and shape that perfectly matches the shape of a human body.

10.2.1 Fit

When we speak of fit, we also follow Chen (2006), who writes of the dimensions of a garment adjusted for a specific size. According to Ashdown (2007), the greater the increase in height and weight, the more problematic this adjustment becomes due to a lack of proportion. Otieno, Harrow and Greenwood (2005) explain that the industry is currently facing a major transformation in male morphology caused primarily by an increase in the variety of weights and statures. The greater the difference among morphologies, the more difficult it is to find the right fit.

For both manufacturers and consumers, finding the perfect fit comes from knowledge backed up by experience. Thus, Rasband and Liechty (2006) suggest that the first step toward achieving it is to familiarize oneself with six characteristic ways one's figure may vary from the average or ideal: bone size or structure, height, weight, proportional body features, contour (including weight distribution) or figure type, and posture.

10.2.2 Bone size and structure

The skeleton changes over time with the development of bone structure; body weight combined with bone size directly influences human height. The maximum height of an adult man is usually attained in his early twenties, and exceptional height variation within a particular population is usually due to gigantism or dwarfism, two medical conditions caused by specific genetic or endocrine abnormalities (Gangong, 2001). Bone mass, comprising the entire human skeleton, accounts for about 15% of total body mass, according to Ruff (2000b). We measure this according to wrist size, following Rasband and Leichty (2006), who define bone size in the following way: 16.5 cm or less is considered small, 16.5 to 19 cm is medium, and more than 19 cm is large for males between 1.65 m and 1.92 m in height. It is also important to point out that bone mass changes with age. According to Ruff (2000a), skeletal volume progressively decreases after age 50, and this naturally has a subsequent and direct effect on body mass, posture and contour.

10.2.3 Height

Over the last 50 years, bone mass has generally increased, leading to an overall increase in the strength of the skeletal structure. According to Faust (2012), since 1970, the average American male has gained 5.5 cm in height and 5.4 kg in weight, measuring today at 1.75 m and 77.4 kg, while Workman (1991) indicates that the average height for a man living in the US in 1940 was 1.69 m, compared to 1.76 m today. According to a study in collaboration with SizeGermany, Melliand (2005) asserts that the average physical stature of men in that country has increased by 16 cm since 1875. The research generally points to massive recent changes in stature everywhere, and the male clothing industry has certainly been slow in adapting.

10.2.4 Weight and density variation

The industry faces another significant issue: human body mass. According to our research, weight and density are the most variable factors among individuals, in terms of anthropometric data such as body part dimensions and circumferences. Ruff (2000b) has found that weight is determined by the interaction between genetic and environmental factors. The average weight of a North American male older than 20 is 87 kg, which is heavier, compared to Europeans (older than 20): French men at 77.1 kg, and Germans, who have larger frames, at 82.4 kg. Table 10.1 shows frame size measurements for North America.

Muscle tissue comprises 40% of total body weight. According to our research, variation in muscle density ranges from 1.6 to 4.8% for the average North American male older than 39 years, taller than 170 cm, and heavier than 72 kg. This variation has a direct impact on the choice of clothing for certain men, in terms of adjustment.

Whether people are athletic, plump or stocky, Robinet (2012) has found that overall, they are currently bigger, especially in Western nations, and the effects of this can be seen in the increasing number of unsold items returned to manufacturers by retailers unfamiliar with the changing proportions of their clients. The population has certainly gained weight, and the average

Table 10.1 Frame size based on height and weight

Height (cm)	Small bone frame (kg)	Medium bone frame (kg)	Large bone frame (kg)
153	55–60	57–62	61–66
156	56–61	58–63	62–67
158	57–62	59–64	63–68
160	58–63	60–65	64–69
163	59–64	61–66	65–71
165	60–65	62–67	65–73
168	61–66	63–68	66–74
170	62–67	64–70	68–76
173	63–68	66–71	69–78
175	64–69	67–73	70–80
178	65–70	68–74	72–82
180	66–71	69–75	73–83
183	68–73	71–77	74–85
186	69–74	72–79	76–87
188	70–76	74–81	78–89
191	72–78	76–83	80–92

Source: National Institutes of Health (NIH), 2011.

frame size has also grown; what has changed most significantly, however, is the fat mass index among men.

10.2.5 Fat mass index

Jackson and Pollock (1978) explain that the fat percentage index is crucial to better understand how adipose tissue (fat) composes body density, and they offer six levels to categorize fat percentage:

- ➢< 5% - minimum required level of fat (extremely thin);
- ➢6–13% - athletic – very low level of fat (very thin);
- ➢14–17% - sportive – low level of fat;
- ➢18–25% - average level of fat;
- ➢26–29% - overweight – above average level of fat;
- ➢> 30% - obese – extremely high level of fat.

10.2.6 Overweight

According to Robinet (2012), more than a quarter (26%) of the adult population in France is overweight, and 8.3% are obese. Similarly,

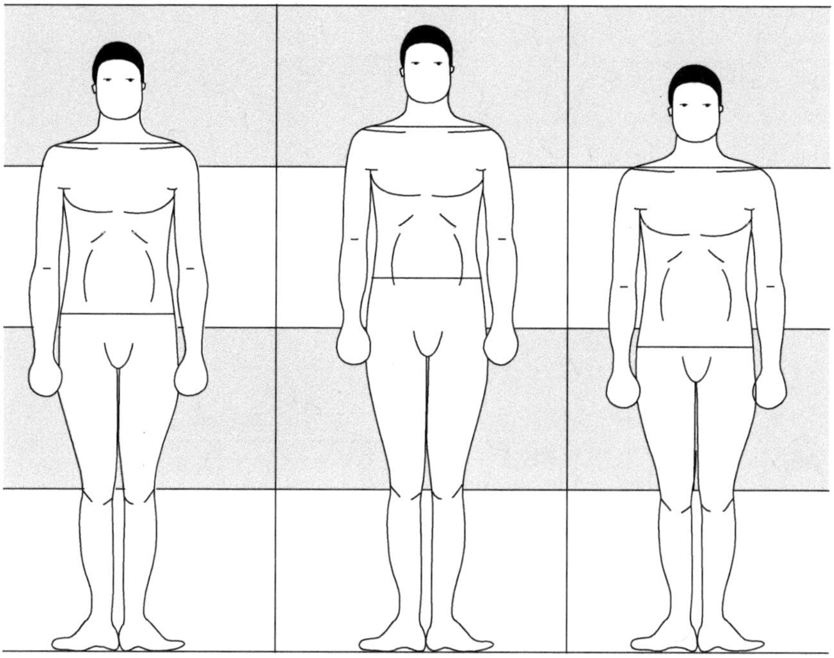

10.1 Torso proportions.

Faust (2012) has found that just under one in four Canadians (24.1%) by Statistics Canada (2011) are overweight, compared to one in three Americans (34.4%) by USA Centers for Disease Control (2011). A Statistics Canada study notes that the Canadian waistline is expanding, particularly among men aged 60–74. Among adolescents, the waistline of 15 years old is 6 cm greater than it was 37 years ago, and they weigh on average 2 kg more than their elders did at their age. This extra abdominal weight has naturally brought with it significant disproportions to several areas of the body.

10.2.7 Proportional body areas

In measuring the body, the proportions of certain areas are often used to relate two or more measurements. The most common proportion utilized by the apparel industry is that between the legs and torso, in three general categories: proportioned, short chest/long legs, long chest/short legs, as shown in the following Fig. 10.1.

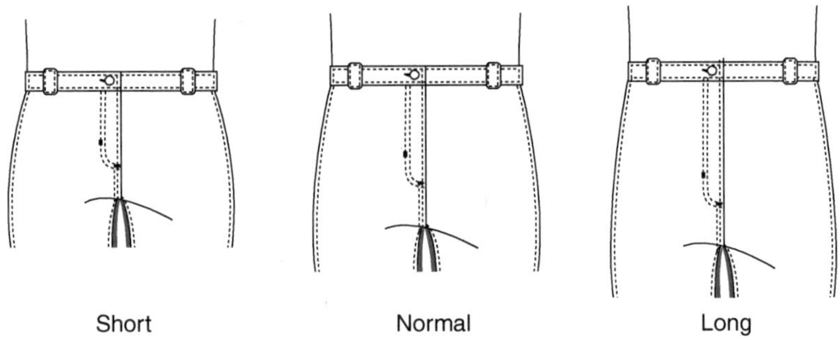

10.2 Crotch length.

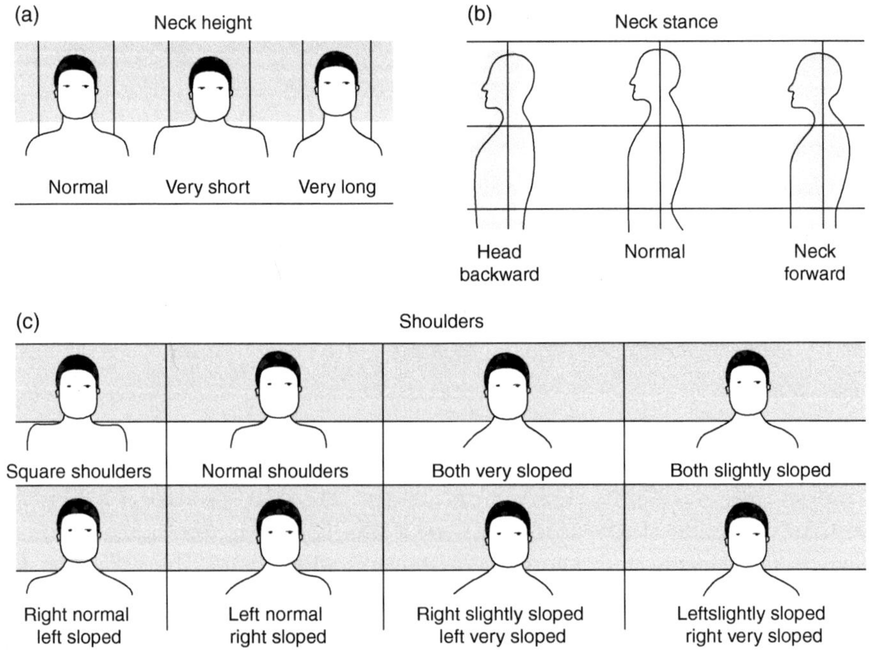

10.3 (a–c) Neck and shoulders.

The crotch height in pants is determined by body proportion. This is important because if there is not enough crotch in front, pleats will form on the inner leg. Manufacturers classify crotch length in the ways shown in Fig. 10.2.

With regards to upper body proportions, the neck height, neck stance, and shoulder types shown in Fig. 10.3 have a direct impact on shape and thus how garments fall on the body.

In addition, other factors, such as the proportional imbalance of front chest and arm silhouette, directly influence how the torso and shoulders

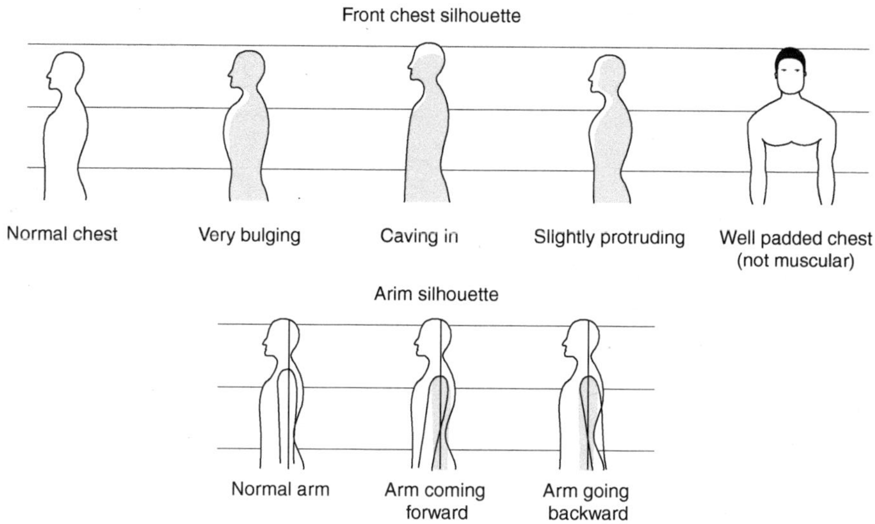

10.4 Chest and arms silhouette.

rest; it is important for manufacturers to understand the variations shown in Fig. 10.4.

10.2.8 Proportional body areas and personality types

In the 1940s William Herbert Sheldon (1961) developed an influential theory, according to which personality and somatic experience is defined by our anatomy, with particular application to the domain of sport. This theory is controversial because it is so subjective, but our research indicates that it is sometimes still applied today in the athletic world as an objective approach; Sheldon's three general somatotypes can be seen in Fig. 10.5.

Sheldon described the three somatotypes based on data that found three different types of embryological tissue: mesoderm, ectoderm and endoderm. From this research, he defined corresponding personality traits and suitable activities for each of the three basic morphologies: the mesomorph, the ectomorph and the endomorph.

The mesomorph (The warrior: bodybuilding, judo, squash)

Possessing more pronounced musculature, the mesomorph has a tendency to easily acquire muscle mass. Sheldon supposed that this type was not motivated by introspection but rather by a constant need for physical exertion and a fighting instinct.

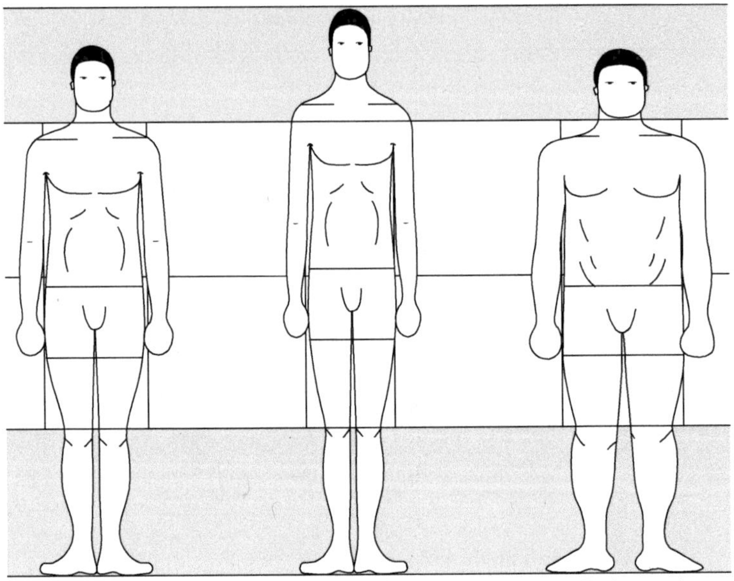

10.5 Three basic male body shapes from Sheldon's Somatotyping Theory (1961).

The ectomorph (The strategist: sprinting, biking, tennis, skiing)

Possessing a finer bone structure, ectomorphs are characterised by a slim, lithe body and restrained power over themselves, as well as a strong cerebral and emotional ability to better manage their impulses.

The endomorph (The pacifist: walking, swimming, golf)

Because of their relative girth and corpulence, endomorphs are more sensitized to their physical well-being and social relationships; they are thus very loyal to group membership and generally stay within their comfort zone. They are said to be characterised by a large natural appetite and easily express their emotions. Sheldon remarked on the rise among this group in problems related to abdominal fat and consequent lumbar problems.

Sheldon *et al.* (1954) went further in his research, explaining that these three types were frequently combined, resulting in many different variations. Size (as well as girth) plays a significant role in the development of individual personality traits; this connection can be extrapolated and refined according to diverse proportions and body measurements such as contour and weight distribution.

10.2.9 Contour and weight distribution

Contour refers to bodily curves, specifically their shape and placement, depending on bone size and structure, proportional areas, weight, muscle tone, pattern of weight distribution, and posture (Rasband and Leichty, 2006). There are five key postures among men in terms of torso contour, shown in Fig. 10.6: normal back, curved back, hump back, upper back curve/ short forward neck and upper back curve.

The abdominal or stomach silhouette has six common variations, as seen in Fig. 10.7: normal, wash board, slight, bulging, pot belly and beer belly.

Leg stance, thigh stance and seat stance are also necessary for defining contours and proportional curves, as they are important for determining body equilibrium, as shown in Fig. 10.8–10.10.

While contours offer a more comprehensive means for body measurement, it is currently only useful for adjustments and alterations. They have not yet been incorporated into size identification or labelling, which would be an interesting area of further inquiry.

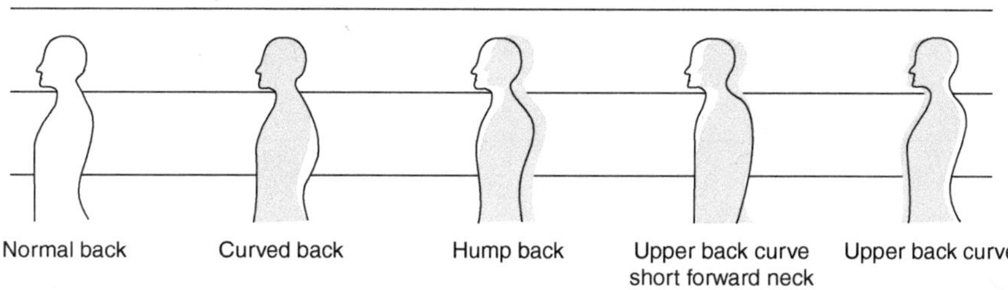

10.6 Back silhouette.

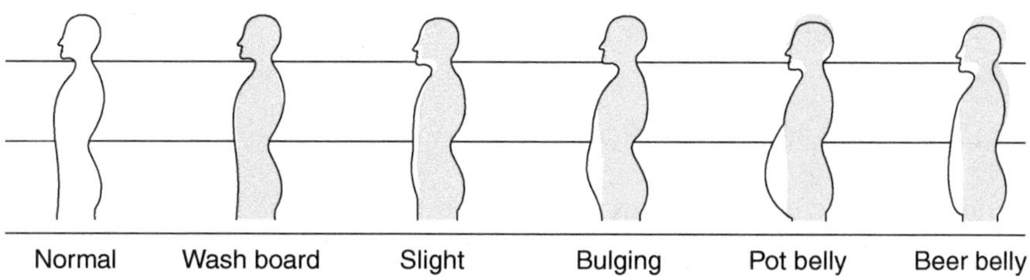

10.7 Stomach silhouette.

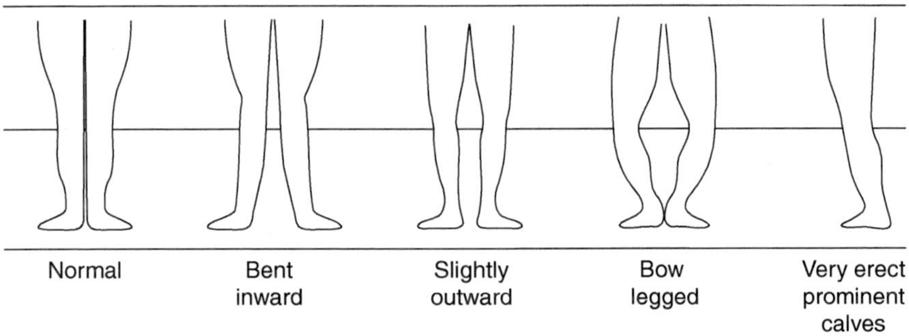

10.8 Leg stance.

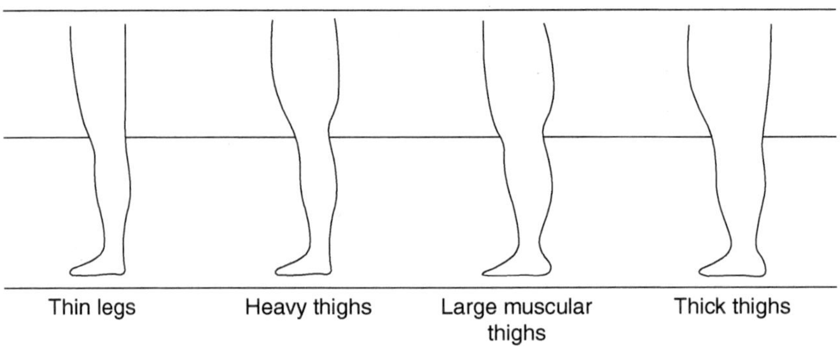

10.9 Thigh stance.

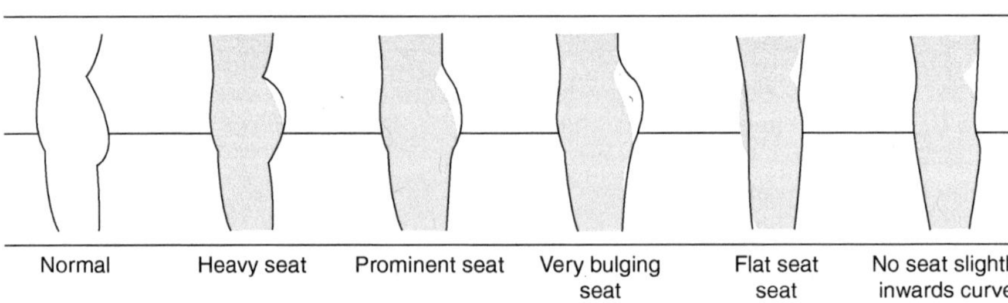

10.10 Seat stance.

10.2.10 Size identification

Research conducted in various countries between 1992 and 2010 with a male sample population between 20 and 30 years old demonstrates a large average height differentiation among various countries: Vietnam (1.62 m), France (1.77 m), Sweden (1.82 m) and the Netherlands (1.83 m). With the rise of globalisation, one of the key new challenges is that of providing well-fitting garments for a wider variety of male shapes and sizes. Different ethnic groups have different body proportions and shapes, and therefore apparel fit problems are likely to be aggravated by the diversity of figure shapes (Bougourd, 2007).

10.2.11 Appropriate size chart

Men generally know their jacket, pant and collar size, but standard measurements are never quite properly fitted for each body, which should encourage them to know their measurements with more precision. The global apparel fit experts known as the Alvanon company suggest that retailers arrange measured clothing on the shelf according to the actual measurements of their clients so they will feel more comfortable, or even generally sort them by size rather than by brand. This is a solution to the problem created by the fact that most clothes are created from broad general standards.

10.2.12 Standard sizing problems

Problems with standard sizing are becoming increasingly widespread, according to Faust and Carrier (2009a), as apparel markets become more globalised. For example, a number of retailers no longer adhere to standards, but are rather developing their own size specifications (Ulrich, Anderson-Connell and Wu, 2003) to better respond to target markets – all while still using generally accepted numeric size codes. But body proportions, as mentioned above, have changed considerably since the 1940s and 1950s, even since the 1980s. According to Norman Marks, a men's tailor Canadian family business founded in 1910, 'client measurements have changed, and garment sizes based on 1980s measurements no longer fit.' For Faust and Carrier (2009b), the problem of fit has become the main challenge for retailers because it is no longer possible for customers to select an item of clothing from the shelf, confident they have the proper size. As weight gain among men occurs mainly in the belly and thighs, and 'standard sizes' for products in the store vary considerably, finding the right fit increasingly requires determination.

10.2.13 Size Inflation

As a result, today's table of standard measurements for (North American) sizes 36 to 46 is no longer properly adjusted. Because of this, major manufacturers and retailers (within the buyer-retail supply chain) do not respect standards, creating their own size specifications either corresponding to their target market or consistent with their vanity sizing strategy. However, they still continue to use the generally accepted size labels. Without clearly established norms or standards, consumers have difficulty comparing among brands, which generates frustration and confusion when trying on clothes.

Clothes of the same size have become steadily larger over time due to the practice of 'size inflation.' Older consumers will be familiar with this, because they can still squeeze into the same size they have always bought, despite weight gain or changes in their bodies. Our research confirms that an average size-34 pair of pants for men in Canada today is 3.3–5 cm larger than the same size pair from the 1970s, equivalent to a size 36 from that era. The same 'downsizing' is also widespread in the US, where confusingly, a size 32 can be equivalent to a size 33 or 34, depending on the brand. According to some manufacturers, as consumers' weight has increased over the years, many clothing contractors have increased their measurements of their products and accessories to keep the same sizes, believing that people will feel better and thus be more likely to consume.

10.3 Understanding different male body shapes

There is nothing simple about designing apparel for consumers, nor is there a simple organisational strategy; at the most basic level, it is difficult even to comprehend the full impact of body shape and size on design.

10.3.1 Theory of body shapes

In 1968, Silvio Quattrocolo, founder of the CAD Modelling Ergonomics company, conducted research into the Theory of Anthropometric Structures. He identified six main male body shapes and supposed that every man could be classified into a morphological family with a set of common anthropometric measures (Fig. 10.11). Significantly, Quattrocolo abandoned the concepts of size, cut, extra long, small, etc., and instead started to think of clothes

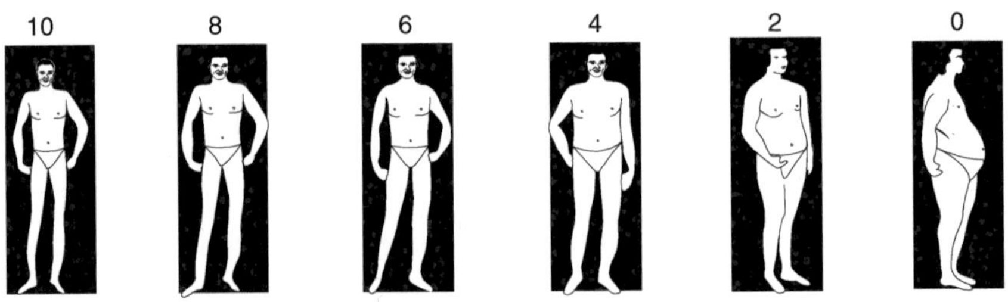

10.11 Male body shapes from *Theory of Body Shapes* (Copyright© S.I.A.E. n°9401846) by Silvio Quattrocolo.

through 'shapes'. Quattrocolo (1978) monitored and published body shape ranges and anatomical averages of product sizes and came up with the following anthropometric scale.

Each morphotype (which he labelled with the numbers 0, 2, 4, 6, 8 and 10) has a range of possible sizes (which he called 'developments'), so that a single size can be found suitable for different types. This approach is interesting because it allows for a better understanding of the difference between body shape and body size. But the question still remains: how does body shape influence the selection of clothing in terms of style and fit?

10.3.2 Different approaches

At first, we noticed little in the literature on the subject of what influences the style and fit of male apparel, which can be attributed to the common perception that finding properly fitting clothes is easy (or unimportant) for men. This has changed somewhat in recent years with the arrival of social types such as 'metrosexual', 'übersexual' and 'retrosexual', and others that have altered the perception of male attitudes towards clothing and fashion. Also, consumers in search of their identity are more frequently adopting certain styles, or even changing their shape in order to conform to a particular look.

Next, fit refers to more than just simple measurements and adjustments, and rather necessitates an understanding of the needs and (sometimes hidden) desires of consumers. The evaluation of fit is thus a complex process in which the relationship between garment and body can be considered from several different perspectives. Melliand (2005) has suggested that people generally tend to be highly subjective in this regard. It should be noted that there exists a difference between a garment that fits and one that makes someone feel good and encourages them to value themselves. Just because a pair of jeans fits well doesn't mean they are automatically the right pair. Indeed, as Ashdown (2007) found in her research, two individuals with the same measurements wearing the exact same clothes can nevertheless perceive fit in radically different ways, according to the subjective perceptions of their own bodies.

Finally, we noticed that male consumers are becoming more discerning and particular, no longer interested only in what the large mass retailers are offering. The web has played a significant role in changing consumption habits, according to Pine and Korn (2011), allowing consumers to be more aware of what's available and also what they want. In this context, the once faithful male consumer is growing tired of mass-market products and is searching instead for more authenticity from apparel manufacturers. More seasoned as a shopper, he also understands the mechanisms at work by mass retailers trying to distract the consumer; he is certainly looking for an experience, but not at any price.

10.3.3 Changes in body shape among young males

We have moved from the product era to the experience era, currently guided by a strong individualism. From this, clothing manufacturers must now respond to the demands of consumers, not just in making high quality products, but garments that satisfy the growing appetite for unique, individual products, and in such a way that personal expression is associated with the fit of the clothes.

Physical appearance

Men are acutely sensitive to their physical appearance. According to our research, the average young man feels a bit intimidated by the images surrounding him, which can in turn develop into a sense of inferiority or make them feel uncomfortable about their bodies. The same is increasingly true for men over 40, as evidenced by the intense proliferation of superficial consumer culture and male-centric advertising for products such as underwear and aftershave. But young males have adapted more readily to the recent cultural discourse, creating radical shifts in consumption habits, and they are subsequently more interested in modifying their shape.

This body revolution goes even further, as liposuction, pectoral implants and jaw enhancements are becoming more commonplace. Dissatisfaction with body shape is one of the main reasons that men vary their lifestyle. Men are increasingly obliged to be active, exercise and enhance their muscles in order to keep pace with standards imposed by society. The impact can be stark, often creating problems through the frequent gain and loss of weight, and exacerbated by fat-free dietary regimes, which the body adapts to by increasingly conserving fat reserves, thus leading to massive weight gain or bodily deformation. As a result of all this, we see more fit and muscled men, and we also see more obese men.

With regard to social, psychological, physiological and sartorial factors, the fit of a garment is a subject of great debate. For example, the trend towards lower sizes is heightened because of the extreme social pressure towards thinness and hypersexualisation, which results in confusion and conflict between consumers' tastes and what supposedly constitutes a good fit. It raises the question: is it the clothes that accentuate the body, or the body that accentuates the clothes? It's becoming more and more difficult to answer this question, which is a problem for the clothing industry.

Virtuality vs. reality

In addition to the revolution of the body, there is the recent spread of augmented or virtual reality (VR). The gulf between real figures and those considered perfect is not only a question of perception, but it has become a game for some clients. Indeed, VR (online communities, etc.) currently offers individuals 3D avatars with ideal bodily proportions that can stimulate consumption habits. And this trend towards augmented reality can easily have the negative effect of disenchanting consumers and increasing their dissatisfaction. It is thus necessary to understand the how the consumer's desire is influenced by the global saturation of the apparel market. One of the major problems is that customer dissatisfaction can occur more quickly than ever before, primarily because of the accessibility to such a wide array of online products.

There is a common perception among consumers that the apparel for sale is the same everywhere and that the shopping experience is not improved by buying in store. Thus, why bother wasting time there since everything can be found on the web? Retail stores sometimes offer products with up to 70% markdown, but even that is not incentive enough to change the matter. Consumers increasingly prefer to shop in real-time on the web unless the in-store experience is exceptional. As this trend continues, the new generation will redefine the industry with the advent of collaborative online technologies, such as co-creation and coproduction. Furthering this trend is another technological development: interactive phone applications using augmented reality in real-time; GPS data or computer-generated sensory input has led to such advances as intelligent catalogue recognition and the sorting of clothing by silhouette, history and preference. By enhancing one's current perception of reality, adding objects, shape recognition and computer assistance, information about the real world surrounding the user becomes more interactive and digitally manipulatable.

Problems with manufacturing standard sizes

Many problems occur at the manufacturing level, with profound effects on the industry. For example, our research reveals that young men (under 21) are increasingly looking for shirts altered for more breadth in the shoulders and arms, and manufacturers do not quite know how to adapt their sizing specifications to address growing upper body size. Also, as we become more focused on performance values – comfort, quality and durability of materials, and of course fit – ergonomics has become increasingly important for product development. In this new context, where it's essential to understand the rapid and profound changes occurring in male body types, pattern makers must stop using single-size, 'basic block' templates. Instead, they could develop two or three templates based on different shapes (geared to new or changing body types) for each style, labelling each according to the different block pattern used.

According to Workman (1991), manufacturer clothing sizes are highly inconsistent and create significant losses in terms of potential sales. Moreover, consumers have access to a wider variety of products, and simultaneously, less time for shopping. Thus, more attention must be paid to attracting consumers to products, and keeping them loyal. To do this, companies must focus on three areas: 1) developing products based on appropriate fit, 2) creating a particular style DNA (Distinctiveness, Novelty, Attributes) and 3) developing a comprehensive branding strategy.

Manufacturers and designers must ask the following question: how can we be truly different? The answer is relatively basic: they must adopt a development strategy based on clear and simple brand identification, attached to a specific style DNA, produced for precisely defined niche markets. To do this, a complete and updated understanding of the concept of fit is essential. Specifically, companies should adopt new measurement charts organized by shape, as this can serve as the key defining feature that really distinguishes the brand.

10.3.4 Canadian case study

In the American industry, clothing size standards were created following anthropometric studies conducted in the 1940s. The aim was to satisfy customers by allowing manufacturers to fine-tune their production in terms of sizing. Today in Canada, the government, through the Canadian General Standards Board (CGSB), provides a code for each size that corresponds to specific body and clothing measurements. For men's apparel, the CGSB uses the size chart introduced in the US in the 1950s. However, as mentioned above, certain problems have arisen recently in terms of fit as body proportions have changed radically since the time these standards were put in place. The 2009 interview with Norman Marks confirms: 'In the past, the best-selling two-piece suit was made of a size 40 jacket and size 33/34 trousers. Nowadays, the same suit with size 33/34 trousers will be offered to the new generation of customers with a size 42, or even 44, jacket.' The resulting increase in purchase returns for clothes both in stores and on the web has created headaches for retailers and has great consequences for brand image. According to a Canadian market research customer experience survey conducted by Synovate Business Consulting in 2008, 58% of Canadians claim that it is very difficult to find clothes that fit them perfectly, and 77% find that sizes vary from store to store. As a result, it is important for stores to really know their clientele and offer clothes that fit customers adequately in order to increase their sales volume per customer.

The following issues must be addressed by the garment industry:

- the templates (blocks) used to create basic patterns are inadequate;
- the sizing standards and measurement charts have become obsolete;
- the sizing per territory/population rapidly changes; and
- some information is hidden from customers, and must be decoded by manufacturers.

We conducted a case study that considers tailored men's clothing in terms of recent developments in mass production. After several years of field observation and drawing empirical conclusions, it turns out that in order to succeed in the business, there are two necessary and basic aspects to identify: 1) necessary measurement data such as length, circumference and figure (body shape) and 2) fundamental variables concerning materials, such as density and behaviour.

10.3.5 Research plan

We studied the consequences of industrialisation from a diverse array of perspectives, such as the problems of standard sizing and the problems of fit seen in terms of both traditional production methods, as well as digital advances in the business. We conducted interviews with experts in the Canadian industry, including three master tailors, two pattern makers and ten manufacturers, and collected complete measurement data for 60 male subjects, aged 18 to 69. We ensured that all of them were relatively active, engaging in physical activity (of various types) at least once per week. This showed that individuals' weight and shape varied over the course of a year, and enriched the data by showing variations in density and figure. Such a large data set allowed us to identify interesting, even unexpected, avenues of inquiry, especially in terms of the analysis of body density, the way that measurements are taken, as well as the differences that appeared between manual measurements and body scans.

Our results confirmed that it is often difficult to wear a standard-sized garment and that the evaluation processes, more subjective than objective, had a fundamental impact on the configuration of variables that determine what constitutes the most appropriate fit. Through the comparison and evaluation of alteration results for a tailored garment, we were able to understand how an individual's weight affects the gathering of virtual data. With this in mind, we created a database for a supervised training model intended to create a comprehensive classification of different shapes and types of fit. This in turn allowed us to introduce a cleaner and more comprehensive model for mass customisation of clothing, one which defines general guidelines, yet uses appropriate calculations specific to each client. Moreover, using neural networks, we were able to define a classification through the training model and a set of general rules intended to define 1) body shapes and 2) the fits necessary for the identification of our strategic variables.

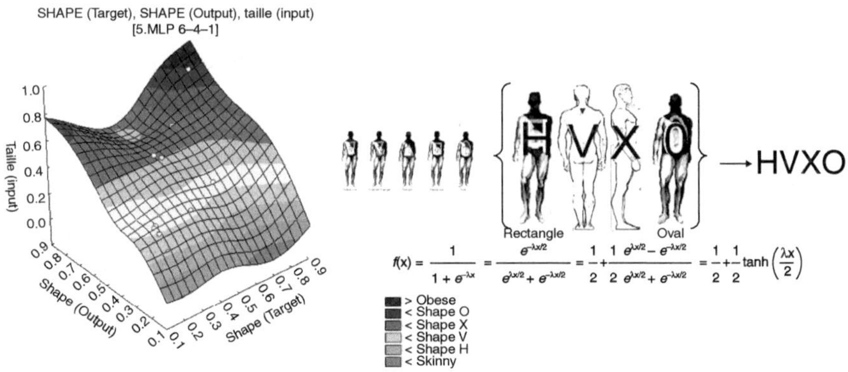

10.12 Our four male figures.

The results of this study allowed us to identify the essential guidelines for making mass-produced custom-measured clothes. Following the validation by neural networks, the results confirmed the importance of body density on the other variables. From this we drew other conclusions, according to algorithms that allow us to facilitate product development and see how fit should be configured to respond to certain predictive categories in terms of ideal shape type.

Body shape

The results generated four general figures, associated with fit, using a cluster-type approach based on our six initial shapes (Fig. 10.12).

Body Type

These four body shapes are assembled to allow us to better understand how each group influences the selection of garment style and fit.

H-type figure

H

Figure H represents a slimmer, ectomorphic man, who often has difficulty finding clothes that fit. We often noticed over the course of our research that individuals in this general category were more picky and even critical towards certain products and brands. Even though they are thin, they're still not satisfied with most products on the market in terms of fit.

V or Y-type figure

V

Males with a V or Y shape tend to have proportionally smaller buttocks, bigger chests and wider shoulders, which creates a V/Y-shape in the torso. These mesomorphic males are generally considered attractive in our society. Males with this shape sometimes have great difficulty finding shirts, suits and jackets that fit and are in stock.

X-type figure

x

X-shaped males are characterised by prominent buttocks and shoulders, naturally muscular arms and thighs, but with a slimmer waist. They often have many problems finding a good fit. For example, without a large enough crotch in the front, trousers will pull at the buttocks and pleats will form at the thighs.

O-type figure

o

Although it seems simple, this is a highly complex shape as the O could be more or less squeezed in the middle and a wide spectrum of sizes exists. Men generally deposit fat around their waists and abdomens due to a lack of estrogen, producing rounder, so-called 'apple-shaped,' or endomorphic bodies. Unlike women, the stomach region is generally wider than the hip region in men. We noticed that O-shaped subjects were less likely to complain because many products are already adapted to this shape. What they are looking for is ease and comfort.

Posture and figure variation

In our analysis of men's figures, there are obviously certain limits. For example, we only had 60 subjects, and some do not fit into the standards governing men's tailoring such as the classification of men into eleven or just different figures: stooping, short neck, square, normal, long neck, sloping, head forward, erect, corpulent, tall and thin, and large shoulders. Thus, we must take posture into account, excluding figure abnormalities that would cause fitting problems.

Body shape and aging

Of course, bodies do not stay the same, they change as we age. As men get older their metabolism slows down. Fat tends to gather around the belly, and while it accumulates rather easily, it can be incredibly difficult to get rid of.

10.4 Sizing requirements for male apparel

This section discusses the sizing requirements for male apparel by providing examples from the sizing systems of North America, Europe and Asia, presenting basic body measurements and size equivalents among different countries, according to Table 10.5 and 10.6 (Section 10.5.1 and 10.5.2).

10.4.1 Sizing system

To create a sizing system, the population is first divided into different body types, based on dimensions such as height and ratios between body measurements (Yu,

2004). In terms of menswear, body length and drop value (the difference between chest and waist circumference) are the two main measurements determining the definition of figure type. As Yu (2004) explains, the International Organization for Standardization (ISO) developed a revised sizing system in 1991 classifying body type according to drop value. Also, Maier (2000) has defined drop value as the difference between the chest and the waist measurement, and conducted a study comparing existing sizing systems throughout the world.

In the UK, the British Standards Institution (BSI) has identified four average body heights: short (166 cm), regular (174 cm), tall (182 cm) and extra tall (190 cm), and each has two figure types: mature (drop value of ± 14 cm) and athletic (drop value of ± 18 cm). The German sizing system places men's figure types into nine different categories: slim, normal, stocky, large, short stocky, large waist, short large waist, tall and athletic (with drop value of ± 20 cm). In North America, four body heights – short, regular, tall and extra tall – are normally identified with four types of figure: slim, regular, large and big. The French identify six main body heights in cm: 162, 168, 174, 180, 186 and 192, and seven types of figure: athletic, slim, normal, stocky, large, stout and large-waisted, as well as seven representative human body morphotypes: normal, athletic, light overweight, squat, potbellied and high potbellied (Robinet, 2009). In Argentina, the Czech Republic, Denmark, the Netherlands and Finland, men's figures are all categorised according to drop values (Yu, 2004). All sizing systems have both similarities and differences, although size intervals for men in various countries are usually classified according to height and drop value. According to Maier (2000), in the US, the gap between sizes is ± 5 cm. In Europe, size changes every 4 or 6 cm. As Yu (2004) explains, the drop value of each size remains constant in some countries (e.g. 12 cm in Japan, 14 cm in China and 15 cm in the United States). In

Table 10.2 Body size identification of men's figure type

Sizes identification	Designation	Body height	Type of figure	Drop
XSS-XLXS	Extra Short sizes	152–164 cm	Kid	–12 cm to –8 cm
XS-XLXS	Short sizes	167–174 cm	Short	–12 cm to –8 cm
S-XXL	Slim sizes	177–190 cm	Slim	–14 cm to –8 cm
S-XXL	Regular sizes	175–186 cm	Normal	–14 cm to –10 cm
MT-XXLT	Stocky sizes	183–192 cm	Solidly built	–12 cm to –6 cm
MXT-3XLXT	Large sizes	193–204 cm	Strong	–10 cm to –4 cm
XS-XXL	Athletic sizes	168–190 cm	Athletic	–20 cm to –12 cm
MT-XXLT	Tall sizes	183–192 cm	Elegant	–16 cm to –10 cm
MXT-3XLXT	Extra Tall sizes	193–204 cm	Elongate	–16 cm to –10 cm
2XL-5XL	Big Regular sizes	175–182 cm	Stout and large waist	+6cm/+8cm/+10 cm
2XLT-5XLT	Big and Tall sizes	183–192 cm	Tall and large waist	+4cm/+6cm/+8 cm

Table 10.3 Global categorisation of four male figure types

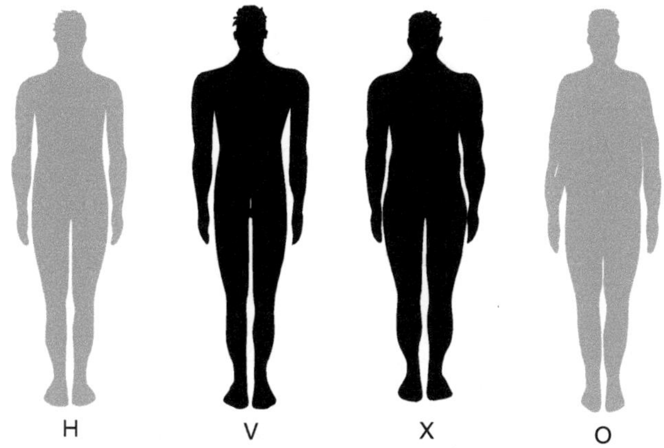

	H-type figure	V-type figure	X-type figure	O-type figure
American size	L44	L44	L44	L44
Body height (cm)	180	180	180	180
Chest measurement (cm)	112	112	112	112
Waist measurement (cm)	102	92	96	120
Drop (cm)	–10	–20	–16	8
	116	110	118	134

other countries, such as Switzerland and Scandinavia, the drop value of the smallest size is 15 cm, but it decreases by 1 cm for each larger size.

According to our research, a normal figure has a drop value of 10–14 cm, while heavier men have a drop value of 6–12 cm and men with an athletic figure have a drop value of 12–20 cm, as shown in Table 10.2. Key body dimensions divide the population into size groups, and the selection of global key dimensions must satisfy the requirements of size identification, designation, body height, type of figure and drop value.

After summarizing various theories of representative human body morphotypes, we arrived at the following categories: normal, athletic, strong and overweight or potbellied, as shown in Table 10.3.

- **Normal – H-type figure:** Male population with a normal drop value of 10–14 cm, normal weight and body fat.
- **Athletic – Y-type figure:** Male population with an athletic drop value of 12–20 cm, and a lower weight.

- **Strong – X-type figure:** Male population with a drop value of 12–16 cm, with a larger shape. Usually, need to have hips drop value (difference between the chest measurement and the hips measurement) for long jacket, suit and adjusted shirt.
- **Overweight – O-type figure:** Slightly overweight male population with a corpulent shape and a drop value of ± 6–8–10 cm, and a mean chest girth measurement less than the waist.

The theory of representative human body morphotypes can help pattern makers develop 'basic block' templates based on different shapes (geared to new or changing body types) for each style, labelling each according to the different block pattern used. According to Wargnier (2010), the drop posture is not dependent on other variables, because though a man may change his size during his life (especially if he gains or loses weight), he will not necessarily change the way he holds his body.

10.4.2 Body measurement chart

This body measurement chart (Table 10.4) developed by Wargnier (2010) is developed from a drop of –12 for someone 180 cm tall, meaning that we consider a normal posture as having a 12 cm difference between the chest measurement and the waistband measurement. To work with drop –8 or –10, simply shift the measurement taken. For example, for drop –15 use a chest measurement of 100 with a waistband measurement of size 85, for a difference of 15 cm. For size 46-40, 46 = blazer size based on a 92 cm (46 x 2) chest measurement, and 40 = pants size based on a 80 cm (40 x 2) waistband measurement.

A common perception exists that it is difficult for men to express exactly what they want in regards to fashion and clothing; a thorough measurement program will go a long way in solving this problem.

10.4.3 Size adaptation and 'sticky information'

Many authors have done research on size adaptation; however, few of them have sought to identify the problems related to sizing and to the so-called hidden data that customers don't always reveal (regarding, for example, ease allowance, fullness, etc). There are mass markets for some customized products, as demonstrated by the emergence of mass-produced though customized apparel (Zipkin, 2001). Agrawal, Kumaresh and Mercer (2001) indicate, however, that the main problem of mass customisation is related to the preparation of products according to customers' requirements.

Von Hippel (1998) states that because of their lack of knowledge and experience, consumers sometimes do not know what they really want. It is thus important to simplify their requests by offering them some guidance. Doing

Table 10.4 Body measurement chart with drop –12 cm for a 180 cm male

Measurements		Evolution Size	Sizes fit					
			44- 38*	46- 40*	48- 42*	50- 44*	52- 46*	54- 48*
1	Chest	4	88	92	96	100	104	108
2	Waistband	4	76	80	84	88	92	96
3	Full hip	4	90	94	98	102	106	110
4	Back waist length	0.3	44.1	44.4	44.7	45	45.3	45.8
5	Front waist length	0.3	39.9	40.2	40.5	40.8	41.1	41.4
6	Front waist length from base of back neck	1	55.1	56.1	57.1	58.1	59.1	60.1
7	Neckline	1	38	39	40	41	42	43
8	Back shoulder	1	37	38	39	40	41	42
9	Front shoulder	1	32	33	34	35	36	37
10	Shoulder length	0.3	13.7	14	14.3	14.6	14.9	15.2
11	Arm length	0	64	64	64	64	64	64
12	Wrist	0.2	17.2	17.4	17.6	17.8	18	18.2
13	Body length	0.2	154.4	154.6	154.8	155	155.2	155.4
14	Waist to floor	(–0.2)	112.1	111.9	111.7	111.5	111.3	111.1
15	Waist to hip length	0	20	20	20	20	20	20
16	Crotch depth	0.2	27.1	27.3	27.5	27.7	27.9	28.1
17	Crotch	1.6	67.9	69.5	71.1	72.7	74.3	75.9
18	Waist to knee	0	62	62	62	62	62	62
19	Leg inner	(–0.4)	85	84.6	84.2	83.8	83.4	83
20	Thigh	1.8	52.7	54.5	56.3	58.1	59.9	61.7
21	Calf	0.9	35.1	36	36.9	37.8	38.7	39.6
22	Back-of-knee measurement	0.7	32.6	33.3	34	34.7	35.4	36.1

Source: Wargnier, 2010.

so not only requires knowing a customer's measurements and style, but also obtaining information that he usually doesn't reveal. Information that Von Hippel (1994) defines as 'sticky' is that it is hidden by a customer; information that, in certain cases, provides companies with a key competitive advantage and significant opportunities for innovation. It is generally difficult for manufacturers to obtain information that is either confidential or usually perceived to be so irrelevant that consumers wouldn't think to reveal it. This unknown data, regarding ease allowance, fit, proportion and the like, are essential to the production of custom-made garments, and according to Ashdown (2007), these are the most common reasons for purchase returns at clothing stores.

10.5 Examples of sizing systems for male apparel

We suggest you take a quick look at the example of sizing systems for male apparel used by the apparel industry. This section presents the results of our data collection from our manufacturers which we divided into five sections: men's shirts, men's suits, men's belts, men's shoes and hats, gloves and mitts.

10.5.1 Men's shirts

Table 10.5 Men's sports shirts, T-shirts and dress shirts

Men's sports shirts, T-shirts							
European	36	37/38	39/40	41/42	43/44	45/46	
American	XS	S	M	L	XL	XXL	
	86 cm	94 cm	102 cm	110 cm	118 cm	129 cm	
Dress shirts							
American sizes	XXXS	XXS	XS	S	M	L	XL
	14	14.5	15	15.5	15.8	16	17
Italian sizes	XS	S	M	L	XL	2XL	3XL
	36	37	38	39	40	41	42

10.5.2 Men's suits

Men's suit size conversions can vary widely depending on style and brand. Sizes commonly offered are regular, short and tall, with a drop value of –6 inches, as well as big, and big and tall, with a drop between –4 and –6 inches, depending on the manufacturer. In North America, suit market is ± 7 inches drop and sizes 46 and 48 have ± 6 inches drop.

Table 10.6 Specialty sizes, suits and pants

Height									
Regular	5′8″ to 5′11″								
Tall	6′ to 6′3″								
Short	5′4″ to 5′7″								
Big	5′8″ to 5′11″								
Big & tall	6′ to 6′3″								
Regular, short & tall									
Chest	34	36	38	40	42	44	46	48	50
Waist	28	30	32	34	36	38	40	42	44
Drop	*–6*	*–6*	*–6*	*–6*	*–6*	*–6*	*–6*	*–6*	*–6*
Big, big & tall									
Chest	52	54	56	58					
Waist	48	50	52	54					
Drop	*–4*	*–4*	*–4*	*–4*					
Suits									
American sizes	XXXS	XXS	XS	S	M	L	XL		
	34	36	38	40	42	44	46		
Italian sizes	XS	S	M	L	XL	2XL	3XL		
	46	48	50	52	54	56	58		
Men's Jeans, blacks, pants, trousers: Length									
European	34	36	38	40	42	44	46		
American sizes	25/26	27/28	29/30	31	32	33	34		

Table 10.7 Men's belt sizes

	XS		S		M		L			XL		2XL	3XL
Inches	31	33	33.5	35.4	35.5	37	37.4	38.25	39.4	41.3	43.3	45.28	47.2
Centimetres	80	83.8	85	90	90.17	94	95	97.15	100	105	110	115	120
Waist (inches)	28	29	30	32	32	33	34	35	36	38	40	42	44

Table 10.8 Men's footwear sizes

American	5	5.5	6	6.5	7	7.5	8	8.5	9	9.5	10	10.5	11	11.5	12	13	14	15	16
European	37	37–38	38	39	39–40	40	40–41	41	42	42–43	43	44	44–45	45	46	47	48	49	50
UK / Australian	4	4.5	5	5.5	6	6.5	7	7.5	8	8.5	9	9.5	10	10.5	11	12	13	14	15
Japanese (cm)	23.5	24	24.5	25	25.5	25.5	26	26.5	27	27.5	28	28.5	29	29.5	30				
Chinese/Taiwanese	37	38	39	40	41	42	43	44	45	46	47	48	49	50	51				

Table 10.9 Hat, glove and mitt sizes

Hat sizes:					
	S	M	L	XL	
	(6 7/8 - 7)	(7 1/8 - 7 1/4)	(7 3/8 - 7 1/2)	(7 5/8 - 7 3/4)	
Head circumference	21 1/2″ - 22″	22″ - 22 3/4″	22 3/4″ - 23 1/2″	23 1/2″ - 24 1/4″	
Glove and Mitt Sizes:					
	XS	S	M	L	XL
Knuckle circumference	6 1/2 - 7″	7 1/2 - 8″	8 1/2 - 9″	9 1/2 - 10″	10 1/2 - 11″

10.5.3 Men's belts

Store-bought dress belts are usually measured with a range of pant sizes (Table 10.7). You must choose your belts two or three sizes larger than your pants to get a good fit. Men's belt length conversion are standard but If you wear a 32 inches trouser waist, a belt labeled 34-36 inches will probably be in the right neighborhood.

10.5.4 Men's shoes

Footwear size conversions can vary depending on the manufacturer, and sizing may vary accordingly (Table 10.8). Because of the many discrepancies between sizing systems and differences between brands, it is never a good idea to buy shoes based solely on simple conversions. The foot is a three-dimensional object, and it's impossible to determine the best fit with two-dimensional measuring tools. Thus for footwear, the inside of the shoe toe to heel should always be measured and listed in both centimetres and inches.

10.5.5 Hats, gloves and mitts

Hats should be measured around the largest part of the head, while gloves and mitts sizes should be based on the circumference of a flat hand at the knuckles, excluding the thumb (Table 10.9).

10.6 Future trends

In an age where standardisation is ubiquitous, the current catchword in the fashion world is 'unique'. Thus a popular trend has emerged where consumers are encouraged to creatively alter products so they will be one-of-a-kind. This is why, as Joseph Pine (1993) suggests, even in mass manufacturing markets, companies should include more customisation and personalisation of details in all aspects of the production process, from design to final touches.

10.6.1 Mass customisation

This concept is not exactly new. In the 1990s, mass customisation was the new frontier being explored in the fashion and business worlds. But today, it's become an imperative; as Pine (2011) states, it's a must for companies that want to remain competitive. In retrospect, Pine admits that this trend has become more popular than previously expected, speculating that the fashion and clothing industries have profited more quickly due to the simple fact that everybody is unique.

10.6.2 Affordable well-fitting clothing

With the recent surge in the use of new telecommunications media, consumers are more demanding and better informed. They are no longer satisfied with standardised products that force them to make compromises. The internet has influenced customers' buying habits by creating needs that can (and must) be satisfied instantaneously. In the clothing industry, these expectations not only imply having to constantly provide consumers with new options in terms of style and colour, but also allowing them to find affordable, well-fitting clothing and making it available to them almost as rapidly as if it were sized in the standard way. In order to meet these expectations, clothing companies must now propose custom-made products. Indeed, brands that offer personalised products (i.e. mass customisation) are taking over both traditional and online stores. This is now possible through the identification of key body measurements necessary to produce well-adjusted and well-fitting garments. However, being able to take these measurements effectively and efficiently is, of course, crucial. The future will hopefully bring with it better measurement tools, as well as more comprehensive sizing charts based on body shape and size.

10.6.3 Traveling tailors and e-tailoring sizing requirements

Mass customisation offers the opportunity to meet, or even surpass, consumers' expectations, whether one aims for pure customisation (one garment for one individual) or merely a highly precise market segmentation. Yet to reach this objective, one needs to acquire and manage an extensive knowledge of consumers' needs and preferences. The fashion and clothing industries are well-placed to embrace this future trend, as they are currently improving their measurement and pattern adaptation techniques, as well as making their manufacturing processes more flexible. A good example of this is the traveling tailor trend of the last few years, in which tailors (primarily from Hong Kong) travel to the US, Europe, Australia and Japan. They provide a more personal service to their customers and give them an opportunity to see fabric samples and meet the tailor in person. Traveling tailors move from city to city and set up in a local luxury hotel for a short period of time to meet clients and ship the final products within 3–4 weeks.

The e-tailor is a person who makes, repairs or alters clothing professionally, especially suits and men's clothing, via online cameras and body-scan tailoring techniques. The goal here is to meet client demands at any time, while offering a variety of mass-produced products adapted to the client's particular shape. Clients thus gain something specifically suited to them, and also benefit from the tailor's access to the wholesale market. This trend is reinforced by the evolution of global markets, as well as the return of the demand for custom-made clothing. At first, it seemed that this type of service would be applicable only to luxury products, but as it gains popularity, it's becoming increasingly affordable. Moreover, e-tailoring is giving manufacturers access to strategic information vital to creating enhanced sizing standards and requirements, allowing them to develop cluster-shape sizing models and, in certain cases, allowing for a 'co-fitting' approach.

10.6.4 Co-fitting

Social networks are particularly interesting to fitting specialists because they reorganise large amounts of detailed information from personal profiles on user consumption habits, and allow for optimal market segmentation and measureable results in terms of a better fit for particular brands. We are currently witnessing the drifting of opinion leaders towards social networks, where their influence over users is more diffused, subtle and thus enhanced. This has changed the framework of the clothing industry, because the advent of Facebook, Twitter and YouTube – combined with fashion reference systems such as mybestfit.com and truefit.com – has created a proliferation of homemade media, such as 'hauling' videos, where shoppers show off and/or critique what they just bought and how it fits using a web camera or smartphone application.

In addition, the arrival of the concept of 'co-creation' has had a direct impact on the management of online products and introduced the practice of 'co-fitting', resulting in exchanges among consumers and web users regarding fit, directly on an online platform set up by manufacturers. For both manufacturers and retailers, co-fitting offers an active, co-creative collaboration with their clients and has thus become an agent for the necessary transformation of product development processes. In this case, companies still focus on branding and the DNA of different style lines, but co-fitting extends beyond this, becoming a way to better satisfy a wide variety of customers by including them in the process. Internet users have thus started to focus less on design and more on the fit and feel of products through the process of co-creation. This trend will soon extend into the world of athletic clothing and products to create more personalised, high-performance and value-added products.

10.6.5 E-sizing

Another current trend is e-sizing, which is centred on performance and has allowed companies to position themselves as the major brands in the industry, such as Adidas, Nike and UnderArmour. E-sizing allows manufacturers to sell clothes through sizing charts on the web and, in certain cases, shows them in comparison with the competition in order to promote the fit of their own brand. This is the inverse of the concept of 'vanity sizing' because it is far more objective, allowing customers to know whether certain garments are truly appropriate for their own shape and contours. On the other hand, it has created a situation where it is more difficult for manufacturers and retailers to retain absolute control over their message.

10.6.6 Manufacturer 'Fitthinking'

The goal of 'fitthinking' is to create a context for mass production that better responds to the needs and desires of clients, another necessary step in the development of mass customisation. Manufacturers have lately realized the need to design and produce durable clothing, all while providing an excellent and unique experience for the client. This awareness is reflected among contractors by the need to identify what customers specifically desire in terms of economy and functionality, responding also to the total product life cycle and durability requirements. This approach creates a relationship between manufacturers and clients at the same time as creating and identifying product niches.

Fitthinking is a new theory based on the results of our own research that aims towards *lean* thinking, trying to do better work with less waste. Thus, consumers will buy less clothing and accessories because they are finding products that fit better and last longer. The target consumers here are those who are creative and environmentally conscious, those who don't consume merely for consumption's sake, and those who adhere to the 'slow wear' movement, a philosophy of fashion that emphasizes the connection between appearance and ethics. Far different than 'fast fashion', slow wear considers fair trade economics and ecology, as well as other values that contribute to responsible and sustainable clothing consumption. With this development, companies will be better able to understand what is feasible in terms of the customisation of mass-produced products, and thus be better able to establish a business strategy. Fitthinking includes the client in all steps of the process, from conception to fabrication of the garment, and offers companies a relatively easy way to exceed the needs of their clients. Moreover, it can provide a knowledge base of customer preferences and desires, thus creating privileged access to a highly segmented market over the long term.

10.6.7 Conclusion

In conclusion, these new trends will allow for the development of intelligent information systems, which when applied to the design and sizing requirements for male apparel, will allow for the tracking of consumers' changing shape, correlated to consequent changes in preferences, choices and habits. At a time when innovation is of the utmost importance, current technology alone – such as three-dimensional body scanners and simulation techniques – will not allow for the full realisation of mass customisation, because the key rather lies in simplifying business practices, as well as acquiring and managing information. Manufacturers must develop a strategy of fitthinking, which should be seen from the beginning not only as an innovation but as a structured investment that allows for unexpected solutions in terms of fit, feel and product design. This strategy should be considered an investment, as there must be a final result, a real output; indeed, if it doesn't affect the market, then it's not truly an innovation.

10.7 Sources of further information and advice

In order to understand consumers' increasingly technology-infused experiences and also how to create new value for customers with strategies that fuse the real and the virtual, Joseph Pine and Jim Gilmore's *Infinite Possibility: Creating Customer Value on the Digital Frontier* is highly recommended. They explain the various dimensions shaping the experience economy and identify the seismic shifts that digital technologies have brought to the business world. It will also be interesting to follow the development of Microsoft's Xbox Kinect technology. A video called 'Kinectshop: The Next Generation of Affordable Shopping' outlines several promising implications for buying customised, well-fitted clothes online.

Having clothes custom-made is an exceptional experience. In order to get a feel for it, we recommend visiting a traveling tailor. There are also a number of companies such as www.indochino.com/traveling-tailor that exist on the web. Also, with regard to the development of new sizing systems based on body shape and size, interesting work is being done by the CAD Modelling company, found at www.cadmodelling.it/english/, which offers a more accurate look at different male body shapes.

The world of mass customisation has never been as vibrant or accessible. To experience the wide variety of services available, visit the following sites:

- mass-customization.de
- nikeid.nike.com
- fits.me

- me-ality.com
- trueFit.com.

Finally, regarding my research, including the concepts fitthinking, co-fitting and eSizing, please don't hesitate to send an email to bellemare.jocelyn@uqam.ca.

10.8 References

Agrawal, M., Kumaresh, T. V., and Mercer, G. A. (2001). 'The false promise of mass customization', *The McKinsey Quarterly* **3**(1), pp. 62–71.

Ashdown, S.P. (2007). *Sizing in clothing: Developing effective sizing systems for ready-to-wear clothing*, Woodhead Publishing Limited, Cambridge.

Bougourd, J. (2007). 'Sizing systems, fit models and target markets'. In *Sizing in Clothing*, Woodhead Publishing Limited, Cambridge, pp. 108–51.

Chen, C.M. (2006). 'Fit evaluation within the made-to-measure process', *International Journal of Clothing Science and Technology* **19**(2), pp. 131–4.

Faust, M.E. (2012). 'Are your clothes properly sized for your market?' Conference of Fit and Size, Vestechpro, Montreal, April 4 2012.

Faust, M.E., and Carrier, S. (2009a). '3D body scanning's contribution to the use of apparel as an identity construction tool'. In *Digital Human Modeling*, Computer Science, Springer Berlin / Heidelberg, pp. 19–28.

Faust, M.E., and Carrier, S. (2009b). 'A proposal for a new size label to assist consumers in finding well-fitting women's clothing, especially pants: An analysis of size USA female data and women's ready-to-wear pants for North American companies', *Textile Research Journal* **79**(16), pp. 1446–58.

Gangong, W. F. (2001). *Review of Medical Physiology*, Lange Medical, pp. 392–7.

Jackson, A. S., and Pollock, M. L. (1978). 'Generalized equations for predicting body density of men', *British Journal of Nutrition* **40**, pp. 497–504.

Maier, A. (2000). 'International size chart comparison for men'. *Knitting Technology* **22**(1) 32–6.

Melliand, T. (2005). 'Current problems associated with body sizes of German men'. *Deutscher Fachverlag* **86**(5), pp. 372–3.

Otieno, R., Harrow, C., and Greenwood, G. (2005). 'The unhappy shopper, a retail experience: Fit and affordability', *International Journal of Retail and Management* **33**(4), pp. 298–309.

Pine, J., and Korn, K.C. (2011). *Infinite Possibility: Creating Customer Value on the Digital Frontier*. Berrett-Koehler Publishers, San Francisco, CA.

Pine, B.J. (1993). *Mass Customization: The New Frontier in Business Competition*. Harvard Business School Press, Boston, MA.

Quattrocolo, S. (1978). *Teoria delle Conformazioni*. Firenze, Italia.

Rasband, J. and Liechty, E. (2006). *Fabulous Fit: Speed Fitting and Alteration*, 2nd ed. Fairchild Publications, New York, NY.

Robinet, P. (2012). 'Are your clothes properly sized for your market?' Conference of Fit and Size, Vestechpro Montreal, April 4 2012.

Robinet, P. (2009). 'Virtual prototyping: From concept to 3D design and prototyping in hours', In *Transforming Clothing Production into a Demand-driven,*

Knowledge-based, High-tech Industry: The Leapfrog Paradigm. Eds. Lutz, Walter, George-Alexander Kartsounis and Stefano Carosio. Springer Verlag, London.

Ruff, C. B. (2000a). 'Body mass prediction from skeletal frame size in elite athletes', *Journal of Physical Anthropology* **113**, pp. 507–17.

Ruff, C. B. (2000b). 'Body size, body shape, and long bone strength in modern humans', *Journal of Human Evolution* **38**, pp. 269–90.

Serwatka, A. (2008). Contribution à la modélisation et à la simulation de vêtements sur mannequin adaptatif. Thesis, Université des Sciences et technologies de Lille I.

Sheldon, W.H. (1961). 'New developments in somatotyping technique'. Lecture delivered at Children's Hospital, March 13, Boston.

Sheldon, W.H., Dupertuis, C.W. and McDermott, E. (1954). *Atlas of Men: A Guide for Somatotyping the Adult Male at All Ages*. Harper and Brothers, New York.

Sun, H., and Su, Z.-Q. (2009). 'Establishing a fit model of clothing brand supply and demand factors based on consumers' cognitive perspective'. Management and Service Science proceedings, MASS 09 International Conference. Wuhan, China, pp. 1–4.

Ulrich, P.V., Anderson-Connell, L.J., and Wu, W. (2003). 'Consumer co-design of apparel for mass customization', *Journal of Fashion Marketing and Management* **74**(4), pp. 398–412.

Von Hippel, E. (1994). ''Sticky information' and the locus of problem solving: Implications for innovation', *Management Science* **40**(4), pp. 429–39.

Von Hippel, E. (1998). 'Economics of product development by users: The impact of 'sticky' local information', *Management Science* **44**(5), pp. 629–44.

Wargnier, C. (2010). *Devenir Modéliste, le vêtement masculin*, ESMOD éditions, Paris.

Workman, J.E. (1991). 'Body measurement specifications for fit models as a factor in clothing size variation', *Clothing and Textiles Research Journal* **10**(1), pp. 31–6.

Yu, W. (2004). 'Objective evaluation of clothing fit'. In *Clothing Appearance and Fit: Science and Technology* Eds. Jintu Fan, Winnie Wing-man Yu, Winnie Yu, L. Hunter. Woodhead Publishing Limited, Cambridge.

Zipkin, P. (2001). 'The limits of mass customization', *Sloan Management Review* **42**, pp. 81–7.

11

Male and female consumers: segmenting consumers in the apparel market by body shape and other factors

P. ROBINET, IFTH, France and S. CARRIER, Université du Québec à Montréal, Canada

DOI: 10.1533/9781782422150.2.221

Abstract: The apparel industry has, for years, used a geometric representation or a letter-based system to identify fit and body shapes. Although useful, this approach is limited when one wishes to adapt garments to a specific population subgroup. This chapter presents the main findings of the French Sizing Campaign. It describes how principal component analysis (PCA) was used to identify both male and female population morphologically different subgroups. It presents the reader with a proven methodology to arrive at population segmentation for apparel purposes.

Key words: sizing, fit, apparel, morphology.

11.1 Introduction

Even though industries such as fashion, fitness and apparel retail define body types for different purposes, their body-type classifications are similar. They use the alphabet (H, O, A, X, R, I, S), geometric shapes (rectangular, oval, triangle), names of fruits (apple, pear) and other distinctly shaped things (hourglass, bean, heart) to classify body types (Simmons and Istook, 2003). Although very useful for classification purposes, these groupings provide only limited information when it comes to adapting apparel to various population subgroups. This chapter focuses on male and female clustering in a manner conducive to targeted marketing. It describes the procedure used in the French body scanning campaign, as well as its most important results.

11.2 Techniques to cluster body shapes

One of the most challenging tasks in finding clusters (or groups) within the French anthropometric database has to do with the sheer quantity of data.

The French Sizing Campaign allowed us to obtain the body measurements and 3D Scans from a statistical sample of 11 500 people (male, female and children) from 5 to 79 years of age. Clustering (or cluster analysis) is a data mining technique that deals with the extraction of the implicit knowledge, data relationships or other patterns not explicitly identified in the databases, by grouping related records. A cluster is a collection of objects that are similar to one another and are dissimilar to the objects in other clusters. The goal of clustering is to find intra-cluster similarities and inter-cluster dissimilarities through the discovery of hidden patterns that lead to the formation of meaningful groups (clusters) of objects (in our case, the objects are the people in the sample). The advantage of using the clustering approach is that one can detect the natural groups, in the form of clusters, based on different body types; along with other information such as anthropometric ratio characterisation (categories, etc.).

Hierarchical clustering methods build a cluster hierarchy, i.e. a tree of clusters also known as a dendrogram. Hierarchical clustering methods are categorized into agglomerative (bottom-up) and divisive (top-down). In our case, the agglomerative method was used. An agglomerative clustering starts with one-point clusters and recursively merges two or more neighbouring clusters. Hierarchical methods facilitate the handling of any form of similarity or distance. They may be applied to various attribute types. However, most hierarchical algorithms do not revisit and improve intermediate clusters after their construction. The algorithm we used in our analysis takes into account the consolidation of the partitioning and improves the quality of the clusters.

The French Sizing Campaign database has a huge attribute set that needed to be considered when finding clusters. There exists a maximum number of attributes above which the performance of the clustering algorithm will degrade rather than improve. The first step is to perform a list of body measurements driven by the product and the body modelling. To create good clusters, our initial task is to reduce the dimensions of the data set and find the clusters in new dimensional space. There are different techniques available to accomplish this objective, among which is the Principal Component Analysis (PCA).

In apparel related research, seven to nine body shape categories (or clusters) are generally used as a starting point: hourglass, bottom hourglass, top hourglass, spoon, rectangle, oval, triangle, inverted triangle, etc. Figure 11.1 shows the statistical methodology used to identify clusters of body shapes (Walter *et al.* 2009).

We used the same methodology to analyse the different body shapes found in the male and female populations. We first present our methodology and results for the female population, as this is the object of most discussions in the literature, hence the best 'known'. We then continue to the male population study.

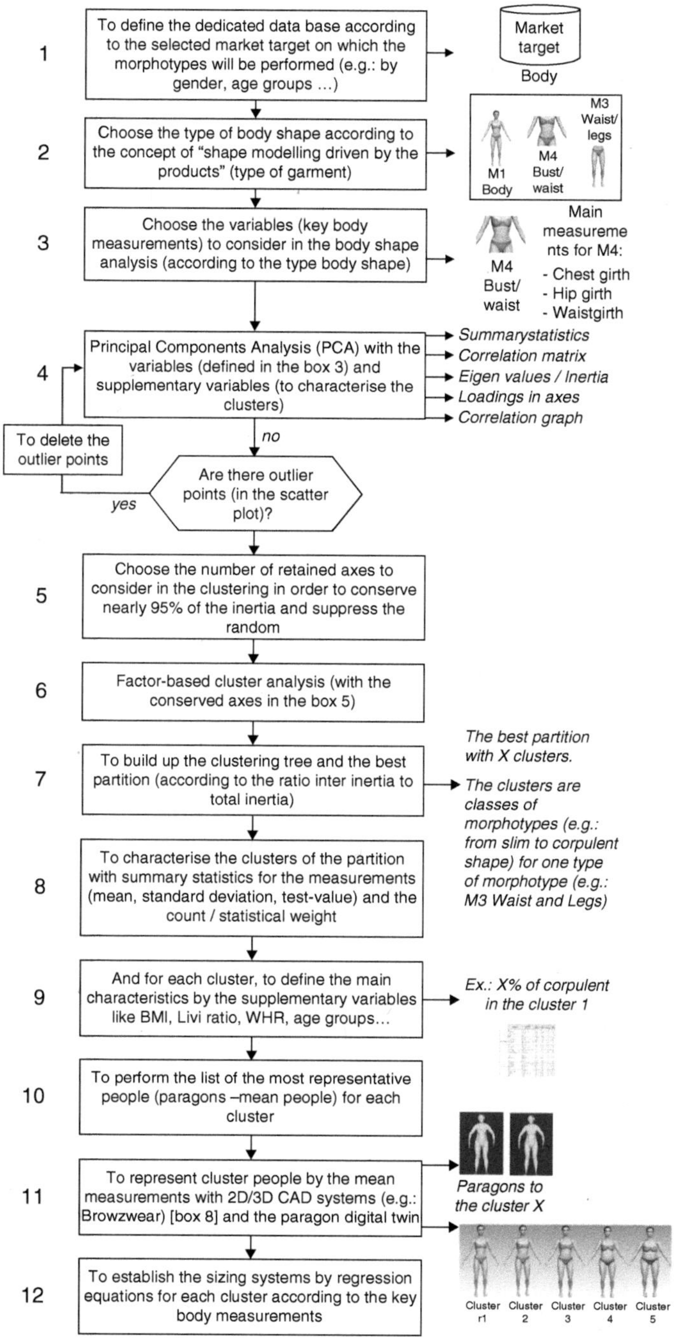

11.1 Methodology to identify body shape clusters.

11.3 Body shapes amongst the female French population

For our analysis, we worked with six 'active' variables (variables taken into account in the statistical calculations) and nine 'illustrative' variables (variables which illustrate the statistical results and which are not taken into account in the calculations). The active variables (see Fig. 11.2 for body location) are:

- bust girth;
- under-bust girth;
- waist girth;
- abdomen girth (iliac crest);
- high hip girth (EIAS);
- hip girth.

The variables considered as illustrative are:

- height;
- weight;
- body mass index (BMI);
- age;
- waist-to-hip ratio (WHR);
- bust-to-waist ratio (BWR);
- waist-hip conformation;
- bust-waist conformation;
- chest-hip conformation.

Once the initial clustering, and verification of results, was performed, the clusters were characterised. Characterisation of the clusters is an important stage of the process. For each cluster identified, we looked to find similarities in our sample groupings to the nine shapes identified by the American study conducted by $[TC]^2$ in the early 2000s, using the mean measurements along with a visual analysis of the body scan images.

Based on those visualisations, we identified seven body shapes (associated with the seven clusters). Figure 11.3 shows the 'most representative' individuals in each cluster. We then compared these with paragons in order to confirm the shapes. Our clusters could the characterised as (Fig. 11.3):

- cluster 1: bottom hourglass;
- cluster 2: trapezoid;
- cluster 3: small spoon;
- cluster 4: hourglass;

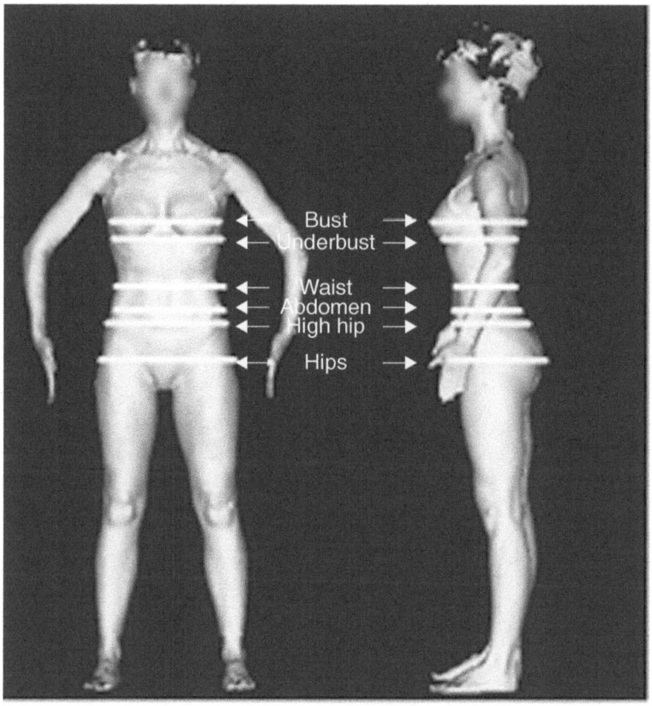

11.2 Correct location of measurements used in the study.

- cluster 5: oval;
- cluster 6: large spoon;
- cluster 7: rectangle.

Of the 5226 female subjects, over 25.58% were classified in the bottom hourglass category, followed by the small spoon group (23.88%), trapezoid (18.35%), hourglass (12.67%), large spoon (12.17%), rectangle (5.32%), and oval (2.03%) (Fig. 11.4).

Table 11.1 presents a comparison of importance of the various groups in the American and French populations. In both populations, one can see the bottom hourglass shape is the most frequently found (25.6% for the French and 40% for the American study). Yet the distribution is different for the next shapes: whereas the spoon represents a share of 17.1% for the US, 36.1% of French women fall in this category (divided into two parts: the small spoon with 12.2% and the large spoon with 23.9%). The hourglass shape is more frequent in the US population with 21.6% against 12.7% for the French. The rectangle describes 5.3% of the French female population against 15.8% for the US.

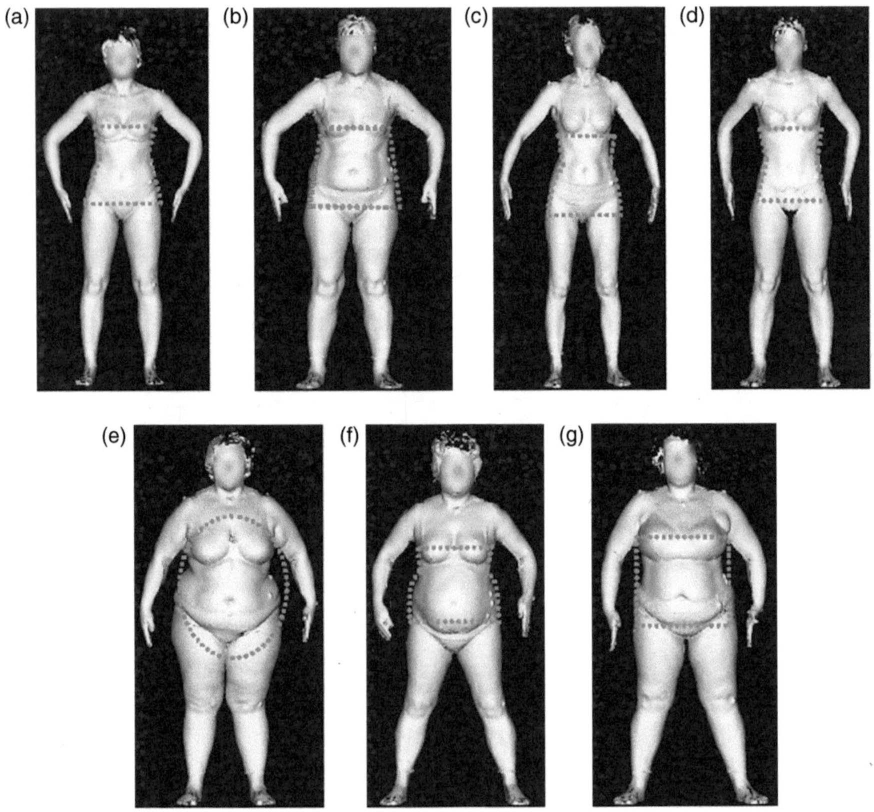

11.3 Representative individuals for each cluster. (a) (cluster1) Bottom hourglass; (b) (cluster 2) trapezoid; (c) (cluster 3) small spoon; (d) (cluster 4) hourglass; (e) (cluster 5) oval; (f) (cluster 6) large spoon; (g) (cluster 7) rectangle.

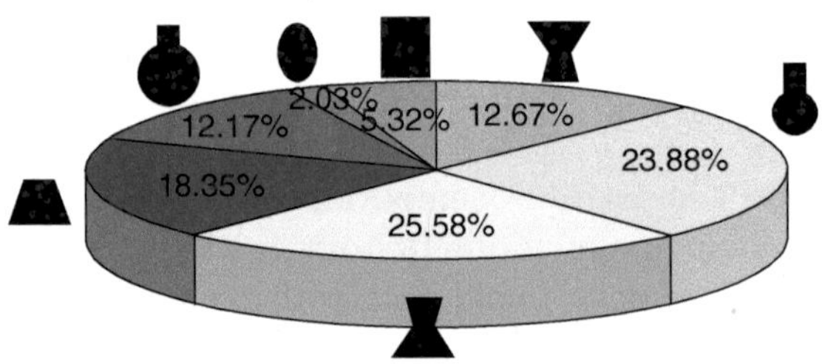

11.4 Distribution of body shapes (for female subjects).

The explanation of the causes behind these differences clearly exceeds the scope of our mandate. One may surmise that they are related to ethnic origins, average age of the population, eating and exercising habits, etc. The

Table 11.1 Comparison between the French and the American study

	French study (%)	American study (%)
Bottom hourglass	25.58	40
Hourglass	12.67	21.6
Spoon	23.88 (small) 12.17 (large)	17.1
Rectangle	5.32	15.8
Oval	2.03	3.6
Trapeze	18.35	1.8
Inverted trapeze		
Diamond		
Top hourglass		

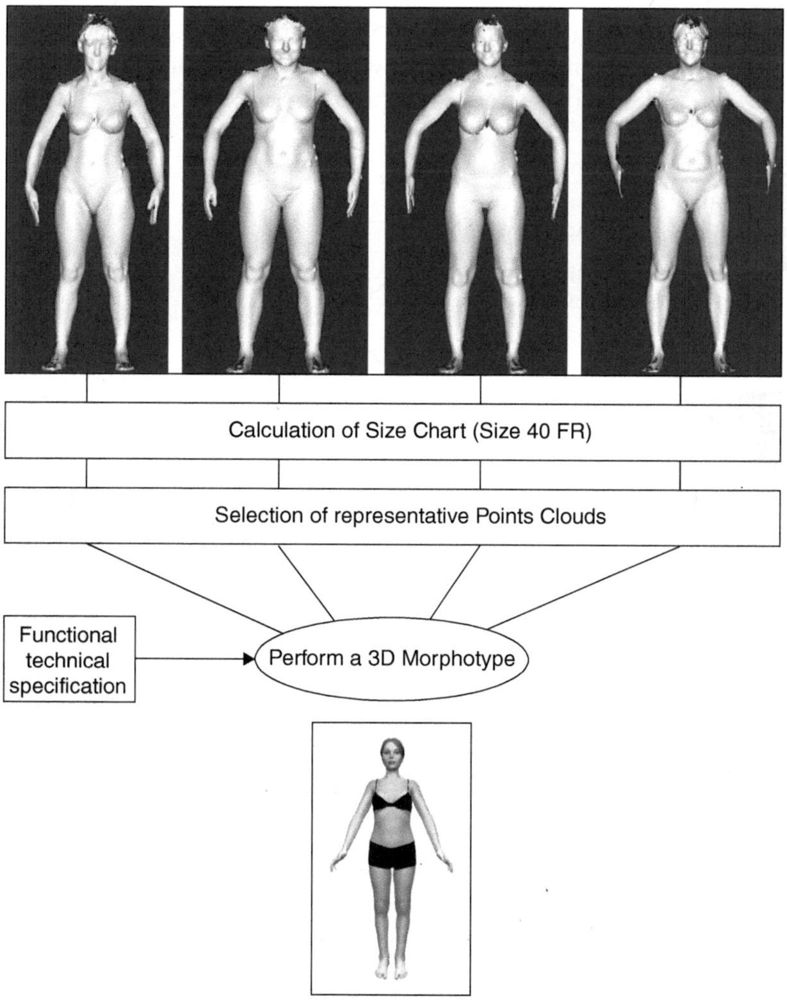

11.5 'Sizing' a size 40 garment.

obvious conclusion though is that (1) the determination of a sizing system and (2) the selection of target markets or even (3) grading templates cannot simply be transferred from one country to another. Apparel sellers have to adapt to regional (national) differences. Figure 11.5 summarises the steps one would need take in order to 'size' a (French) size 40 garment to its targeted population.

11.4 Body shapes amongst the male French population

Once we had completed the female population analysis, we focused on the male population using the same approach as described above. Yet, since this segment of the population has been much less studied than the female segment, we had much less existing information and prior research to work from.

As with the female population, we identified both active and illustrative variables. The active variables were:

- chest measurement;
- waist measurement;
- abdomen measurement;
- hip measurement;
- basin measurement.

Our illustrative variables were:

- height;
- weight;
- Body Mass Index (BMI);
- age;
- waist to basin ratio.

Basically, we had to create and name our clusters, as nothing existed in the literature. In the following text, we present our general findings. As one may see from Fig. 11.6, we identified six clear clusters in the population.

11.4.1 Cluster 1: the athletic type

Our cluster # 1 groups a set of 717 people, that is to say 17.76% of the male sample population of 3986 individuals. The main characteristics within this group are:

- 92.74% of these males present a normal weight (normal BMI), with a mean BMI of 20.75 kg/m^2;

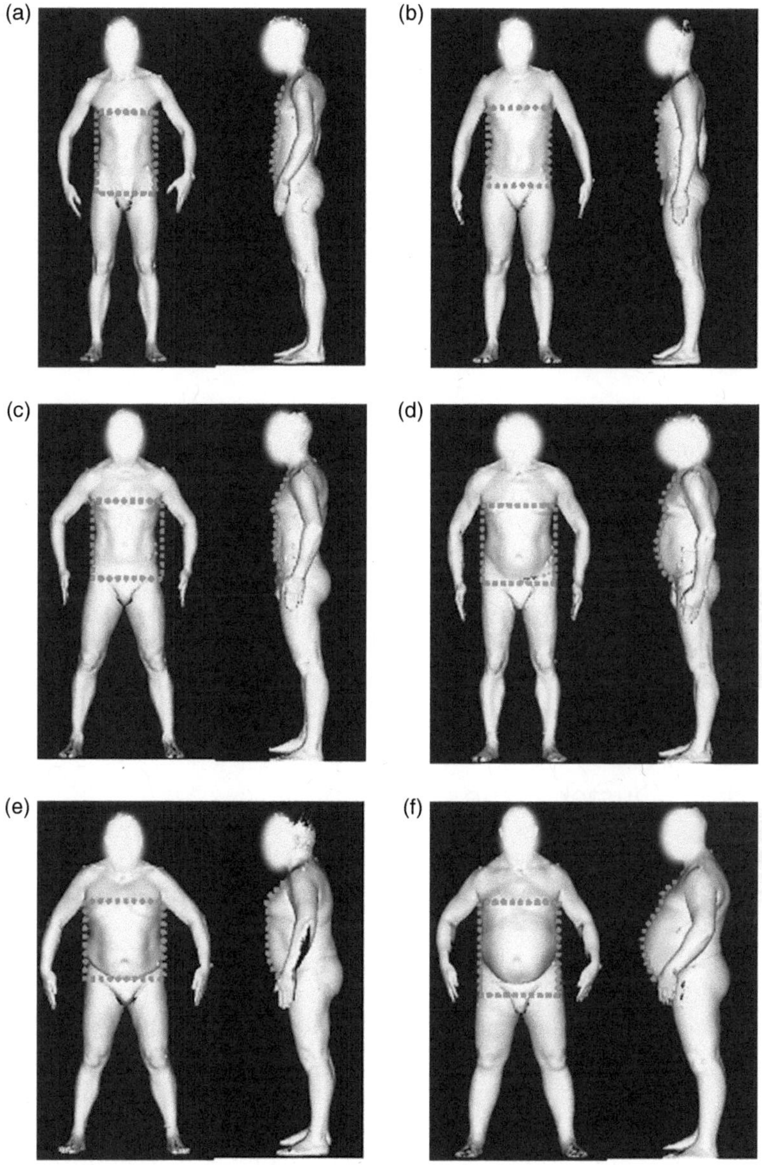

11.6 Male population clusters: (a) (cluster 1) normal; (b) (cluster 2) athletic; (c) (cluster 3) light overweight; (d) (cluster 4) squat; (e) (cluster 5) potbellied; (f) (cluster 6) high potbellied.

- the majority are not yet 35 years old, representing 70.28% of the population;
- the athletic conformations are quite prevalent with the high proportion of drop 8, drop 10 and drop 12, with respectively 26.95%, 22.41% and

8.95% of this class (the 'drop' is the difference in measurement between the chest and the waist – the higher the 'drop' the more athletic the body is considered to be);
- the medium and low heights are the most prevalent with 22.76% people from the 'short' group and 45.54% for the 'medium' height group; the mean height within the group is 173.55 cm;
- pants sizes 36, 38 and 40 are the most represented (that is to say 31.01%, 38.85% and 16.53% of the population respectively); for the jackets, sizes 44 and 46 formed a majority (28.22% and 37.89%).
- τhe android shape is the most characteristic of this cluster with a proportion equal to 94.77%.

High BMIs are little or not represented in this group, nor are extreme jacket and trousers sizes. The very tall constitue only 4.23% of this group. Drops not presented previously (drop −4, drop −2 and drop 0) represent a tiny fraction of this group. In terms of specific observed measurements within this cluster, let us mention that the mean chest girth is 90.89 cm against a waist girth of 74.62 cm and a hip girth of 91.73 cm. We notice that the chest girth and the hip girth are relatively close, with a defined waistline. This observation is strengthened by significant conformations (especially between the waist and the hip with a mean equal to 8.55 cm).

11.4.2 Cluster 2: the normal type

Our second cluster was the largest with 29.68% of our male population belonging to this group, that is to say 1198 people. The main characteristics are:

- 89.50% are of a normal weight, with a mean BMI equal to 23.30 kg/m^2 (higher than in the previous cluster);
- 'athletic' drops of 8 and 10 are the most frequent, respectively counting for 28.93% and 16.56% of the group.
- the highest percentages of our population were in the 21–25 year-old group (at 11.97%) and the 26–35 year-old group at 29.35%; the 36–45 year-olds represent 22.7%, yet this is misleading as they also account for a larger percentage of the sample population than the 21–25 year-old subset;
- 76.81% of the males found in this cluster are of an android shape;
- sizes 40, 42 and 44 are the most frequent for the trousers (with respectively 29.01%, 36.28% and 22.28% of the sample) while, for the jacket, sizes 48 and 50 are the most prevalent;
- males in this group come mostly from the Southeast region of France (26.29%).

The squat conformations and extreme sizes are little represented in this cluster. The older population (56–70 year-old) make up 29.61% or a third of the cluster.

11.4.3 Cluster 3: the portly type

Cluster 3 represents 28.22% of our sample, that is to say 1139 individuals. The main characteristics observed in this group are:

- 65.75% have a BMI that classes them as overweight, whereas 33.95% have a BMI in the normal weight category;
- the mean age within this cluster is 45.33, with the 'over 36' being a majority with 24.84% in the 36–45 range, 23.82% in the 46–55 group and 25.92% in the 56–70 category;
- drops of 6 and 4 are most characteristic, respectively counting for 30.20% and 22.69% of this sub-group;
- the sizes **44**, **46** and **48** hold the majority for trousers (at the level of 24.31%, 37.90% and 23.69%) and for jackets, the sizes **50**, **52** and **54** are the most present.

The extreme sizes are little represented as well as the obesity. Under 35 year-olds are present at the level of 25.42%. For this class, a main characteristic emerges from the analysis of the mean measurements: the chest girth is 103.96 cm against a hip girth of 100.82 cm. The waist girth is less defined than previously (mean of 91.61 cm), making reference to a more concentric shape. The conformations are smaller than the mean, confirming the previous shape.

11.4.4 Cluster 4: the squat type

Cluster 4 holds16.96% of the males. It represents a set of 685 people. The main characteristics of this cluster are:

- based on BMI, 81.08% of them are overweight and 16.60% fall in the obese category: indeed, the mean BMI is 28.30 kg/m^2;
- 88.44% are of a ginoid shape, confirming the previous point;
- the majority fall in the 'stout' conformation: drops of 4, 2 and 0 are present respectively at the level of 30.73%, 17.23% and 6.83% (this last group being characteristic of the cluster as they represent 44% of the people with a drop 0 in the whole male population);
- this cluster is mainly composed of 'over 46 year-olds': the 46–55 category counts for 29.51% with another 33.72% falling in the 'over 56 year-old' group;

- trouser sizes 48, 50, 52 and 54 are almost the norm (93.81% of the cluster), with associated jacket sizes of 54, 56 and 58 (for a total of 83.40% of the cluster);
- although people in the 'tall' category only account for 13.19% of this group, one may say that it is a characteristic.

Athletic conformations are little represented (drops of 8, 10 and 12) which is also the case for the 'young' groups. Small (apparel) sizes account for minimal proportions of the cluster. In this cluster, we note that mean measurements are fairly similar to one another, identifying a rectangular shape: indeed, the mean chest girth is 110.55 cm, the waist girth has a mean of 101.06 cm and the mean hip girth is 105.36 cm. The waist-to-hip ratio is relatively important (0.96 in mean and near to 1), again emphasizing the equality of the various girths.

11.4.5 Cluster 5: the potbellied type

Cluster 5 makes up 6.31% of the male population, that is to say 255 persons. The main characteristics observed in this class are:

- 67.47% of the representatives of the cluster are considered obese (based on BMI) whereas the overweight make up for 22.58% of the population of this cluster; one must remember that that only 3.86% of our surveyed population falls in the obese category, so a percentage of 22.58% in this cluster is something to underline; the mean BMI is 31.98 kg/m^2;
- 97.24% present a gynoid shape and a majority (59.8%) is of a stout conformation (drop of 4, 2 and 0);
- again we find that this cluster is mostly made up of older people, with 66.11% of its population coming from the 'over 56 year-old' group;
- the tall and the very tall make up 55.59% of the cluster: the mean height is 178.6 cm;
- the most prevalent jacket sizes are 58 and 60 (56.53% of the cluster) while for the trousers, the most characteristic size is 54 (34.92%).

As with the previous cluster, few representatives of cluster 5 fall in the 'small' category. The same observation can be made for the athletic conformations (which represent only 14.68% of these persons, drops of 8, 10 and 12). We note that the formula on which the drop is based is little adapted to stout conformations. Insofar as the observed measurements are concerned, we notice that the various girths (chest, waist, hip) have a mean close to 112 cm, except for the chest girth which has a mean of 119.71 cm. The breast of these people is more developed than the belly and the hips yet with a certain equality of the measurements.

11.4.6 Cluster 6: the seriously potbellied type

This cluster is definitely our smallest, with only 1.07% of the population, or 43 individuals. The main characteristics of this population are:

- 100% present a gynoid shape, a shape characteristic of the obese;
- people considered 'tall' are most characteristic (54.15%);
- the stout conformation is very present (drops of 0 and 2 add up to a percentage of 51.81%);
- 41.57% of the people in this group come from the Northeast region of France.

A majority of the people in this cluster are obese (based on BMI index); we can therefore say that obesity is a characteristic of this cluster. We did not find anyone with a trouser size lower than 50; likewise for sizes lower than 54 for the jacket. The normal and the overweight are not represented in this cluster. The normal conformations (drop 6, drop 8, drop 10) are little or not present. The measurements within this cluster are very important, confirming the obesity of its representatives: the mean of the main measurements are: 134 cm for the chest, 133.13 cm for the waist, and 131.74 cm for the hip.

11.5 Conclusion

Although most of the literature on body shapes and sizing has looked at the female population, one cannot say that this sex has the monopoly on variability in body conformations. Our survey of the French population shows that males also come in different sizes and shapes. We have identified six clusters:

- **Normal:** Males with an athletic drop, a normal weight, and a bust girth mean equal to 90.89 cm.
- **Athletic:** Males with a drop of 8 or 10, a normal weight, and a bust girth mean equal to 97.71 cm.
- **Portly:** Males with a drop equal to 4 or 6, overweight, and a bust girth mean equal to 103.96 cm.
- **Squat:** Males with a corpulent conformation, overweight or obese, and a bust girth mean equal to 110.55 cm.
- **Potbellied:** Males with a corpulent conformation, obese, and a bust girth mean equal to 119.71 cm.
- **Seriously potbellied:** Males with a drop of 0 or 2, obese, and a bust girth mean equal to 134 cm.

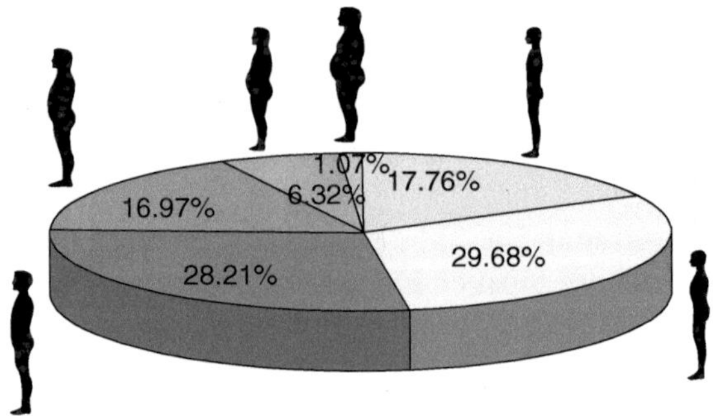

11.7 Clusters of males in the French population.

The two most important clusters (in terms of population percentages) are the athletic and portly clusters (see Fig. 11.7). The discussion on the necessity to adapt female apparel fit to the different body shapes (H, X, A, etc.) has been going on for decades. Little attention has been afforded to the male population, which has, to this date, either dressed with off-the-rack apparel, focusing on one specific body conformation, or in tailor-made. Our survey proves that adapting apparel to different shapes and sizes may be just as important for the firm targeting the male population as it is for the business focusing on the female population.

11.6 References

Simmons, K.P. and Istook, C. (August 2003), Body measurement techniques-Comparing 3D Body scanning and anthropometric methods for apparel applications, *Journal of Fashion Marketing and Management*, **7**(3), 306–332.

Walter, L., Kartsounis, G-A., and Carosio, S. (2009), Transforming Clothing Production into a Demand-driven, Knowledge-based, High-tech Industry, pp. 106–16 (Patrick Robinet)

12
Pregnant women: understanding pregnant women's shape, sizing and apparel style preferences

M.-E. FAUST, Philadelphia University, USA

DOI: 10.1533/9781782422150.2.235

Abstract: Manufacturers/retailers recognize the benefit of focusing on specific target markets, and the importance of working with the right information such as accurate body measurements and idiosyncrasies and varying tastes related to the one target market. The concept of specific, appealing and well-fitting garments is now being applied in this chapter to a dynamic target market: the pregnant woman.

Key words: maternity, maternity wear, target market, fit and sizing, apparel, style.

12.1 Introduction

> Life could change on a dime… One minute you're standing in your office discussing … the next you're standing in your bathroom staring at a stick that has somehow managed to register two skinny pink lines…
>
> Lisa Kogan, *O, The Oprah Magazine*, 'O's guide to Life', p.42, 2007

'It doesn't take that much "people watching" to discover that figures are different from one another' (Rasband and Lietchy, 2006, p. 19). Height and weight, as well as women's body morphology, vary widely between individuals. According to the legendary William Herbert Sheldon (1898–1977), individuals can be grouped into three major *somatotypes* (Alic, 2001). Nowadays, many brands within the apparel industry recognize this phenomenon and consequently produce brand names (lines or collections) for a specific body shape (Faust *et al.*, 2006). Consumers then, consciously or unconsciously, identify brands that fit their body shape and size (Faust and Carrier, 2009a). The question that arises is how women manage the acquisition of garments when their body shape is changing radically, such as when they become pregnant.

The literature review presented in this chapter first establishes the identifying characteristics of a target market, and establishes mothers-to-be as a viable target market. Next it recognizes the body transformation during pregnancy by trimester to inform changes in fit as well as changes in activities (i.e. exercise) to identify the demand for particular garments. Garment tastes and styles are also viewed through a brief history of maternity wear. Thus we identify the perceived changes in size, hence garment needs, which are critical issues for maternity wear in the apparel industry and women in this target market. In addition, since previous studies support the fact that women are judged by their appearance, and show that when women who are satisfied with their body/ silhouette tend to wear clothes that are closer to their body, whereas when they are unsatisfied they tend to wear clothes with a looser fit (Lennon, 1992), fit becomes an even more important parameter in the appearance of, and satisfaction with, maternity wear/ apparel and body cathexis. This chapter focuses on the dynamic market of mothers-to-be and our results identify the types of garments needed for expectant mothers as their body changes during the three trimesters of pregnancy. It should be beneficial to both manufacturers, retailers and consumers.

Lastly, since this research focuses on the body's physical transformation, women are asked if they are willing to be measured either manually or with the use of 3D body scanner. In this case, results are not that exciting.

12.2 Identifying target markets

The legendary Dr William H. Sheldon (1898–1977) introduced, last century, the theory of *somatypes*. His theory described three basic human body types, meaning that individuals can physically be grouped into three major

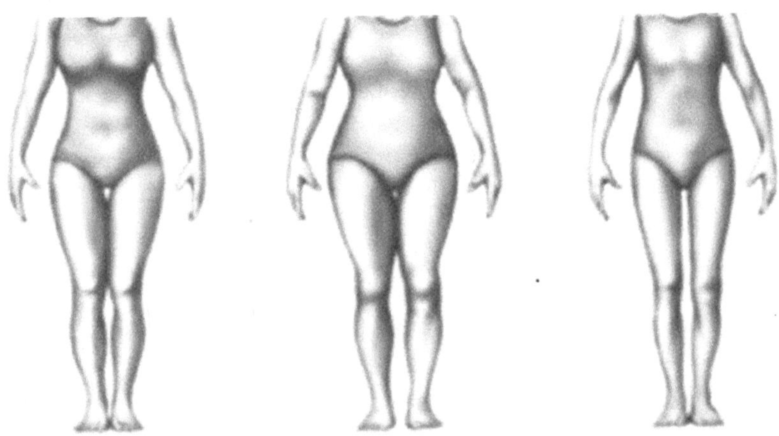

12.1 Mesomorphy, endomorphy, and ectomorphy body shapes.

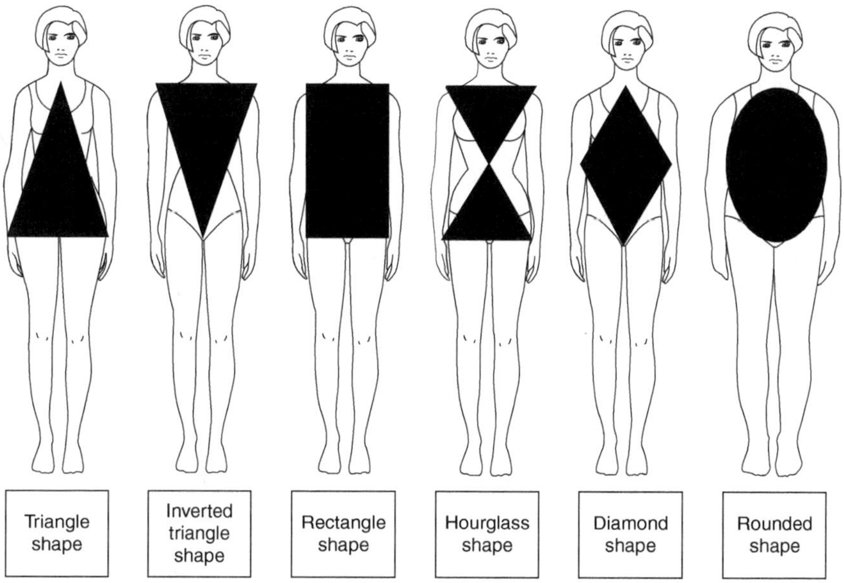

12.2 Different body figures. (*Source*: Rasband and Lietchy, 2006, p. 19 *Fabulous fit*, Fairchild Publications (US), an imprint of Bloomsbury Publishing Plc..)

somatotypes: (1) *endomorphy*, (2) *mesomorphy*, and (3) *ectomorphy,* as shown in Fig. 12.1 (Sheldon, 1940).

Yet, according to Rasband and Lietchy (2006, p.19) 'It doesn't take that much "people watching" to discover that figures are different from one another'. Height and weight, as well as women's body morphology, vary widely between individuals (Fig. 12.2).

Attention to specific target markets has been growing more and more over the last few decades in the apparel industry (Roebuck, 1995). For example, a decade or two ago, there was a lack of appealing, well-fitting garments for older women (55 or 65 years old and over), plus sizes women, petites and tall, etc. (Goldsberry *et al.,* 1996a, 1996b; Horne *et al.*, 1999; Yoo *et al.*, 1999; Oldmand Kind and Hathote, 2000; Otieno *et al.,* 2005). According to the literature, more research and awareness were paid to these markets and as a result more manufacturers/retailers recognize their needs and focus on them. They became more effective and improved consumer satisfaction. Because more and more manufacturers/retailers recognize the benefit of focusing on specific target markets, the importance of working with the right information and accurate body measurements becomes a must, along with the proper garment selection for each target market.

According to Day (1990), identification and justification of a target market is based on the definition of segmentation, i.e., identify a set or groups of homogenous people. To fulfil this definition the group must be:

- identifiable so that they can be reached;
- durable so that profit can be realized before the characteristics of the segments disappear;
- measurable in terms of sales volume and rate of growth;
- substantial to justify the allocation of resources to serve the segments;
- sufficiently distinctive in their behaviour in the marketplace.

Goldsberry *et al.* (1996a, b) and Horne *et al.* (1999) successfully identified older females as a target market, whereas Otieno *et al.* (2000) and Oldham Kind and Hathcote (2000) identified plus sizes as a viable target market, in the US as well as in the college females segment. Using the same definition, we investigated if mother-to-be is a target market, i.e. if it is clearly identifiable and distinctive. Mothers-to-be may wear maternity clothes for a few months, but they seem to be a very homogenous 'never ending' group of women and thus a durable group. As the garments are specific to the physical condition, it is a measurable group and can be substantial. This segment can easily be measured with national censuses in terms of volume (reported births) and is relatively easy to estimate in term of sales. To the extent that mothers-to-be may work, money will be spent for a work wardrobe and/or to fill the need for other specific types of garments. It appears, *a priori*, unquestionably to justify the allocation of resources to serve this segment. Our results confirmed that pregnant women are sufficiently distinctive in their behaviour in the marketplace and constitute a target market; as such, maternity wear is an important part of the women's apparel industry with distinctive needs.

12.3 Key stages in pregnancy

Pregnancy, the condition of a woman's offspring developing in the uterus, can occur between the ages of 10 and 60 (*Medicine.net.* 2008). However, the majority of pregnancies occur between the ages of 20 and 35 (Stoppard and Younger-Lewis, 1995). During this period of pregnancy, the body changes its physiological and homeostatic mechanisms to ensure the provision of necessities for the foetus. Weight gain results from increased appetite, fat deposition, and growth of the reproductive organs and foetus, as well as increased blood volume and water retention. Women can gain anywhere from 2.5 to over 45.3 kg (5 to over 100 pounds) during pregnancy (Jimenez and Bacallao, 1995).

In America, during the past 60 years, a significant number of changes in the recommendations concerning maternal weight gain have occurred (Murkoff, 1984; David and Wardler, 1994). It went from the 1940s to the 1960s trend to be relatively low, to a shift in the 1970s to encourage women to eat according to their appetite; whereas today's women are encouraged to gain only 11–15.8 kg (25–35 pounds), less if the woman is already overweight (Claesson *et al.*, 2008) and more, up to 18 kg (up to 40 pounds) if the woman is underweight (Davies and Wardle, 1994). However, according to Jimenez and Bacallao (1995) reality may be different. Jimenez and Bacallao (1995) looked at the weight gained and compared it to studies of previous decades. The results of their study, done between 1980 and 1983, showed an average weight gain of 61.9 pounds with a standard deviation of 10.1. When compared with Hanes' study of the 1970s, where the average weight gain was 64.3 pounds with a SD of 15.4 and HES' study of the early 1960s, with an average weight gain of 63.7 pounds and a SD of 13.8, they concluded that variation in terms of weight gain is statistically similar over the past few decades.

Pregnancy duration is typically 38 weeks after conception (9 months and 2 weeks), i.e., approximately 40 weeks from the last normal menstrual period (Stoppard and Younger-Lewis, 1995). Duration is somewhat arbitrarily divided into three unequal trimester periods as a means to simplify reference to the different stages of prenatal development. Each trimester has its own predictable developments for both the foetus and the mother-to-be. Results presented in Rutter *et al.* (1984) illustrate a series of highly predictable growth patterns for pregnant women, the effect of weight gain on posture, and relationships of fetal carriage. Their pilot study included eight subjects: four *endomorphic*, two *mesomorphic,* and two *ectomorphic*. The pilot study's objective was to evaluate the ease with which anatomical landmarks could be identified on subjects at term and to assess measurement session protocol, measuring instruments, and the survey recording form. Their final sample was done on one hundred and five subjects randomly selected from a specific maternity clinic, where each subject was measured twice. The first time they were measured was during the first trimester i.e., at the time of their first visit between week 4 and 16 (in this case) and again before any physical development and at term during pre-labour conferences, i.e., the end of the third trimester.

12.3.1 First trimester

The first trimester, week one to week twelve, is a period during which the outer layers of the embryo grow and form a placenta for the delivery of essential nutrients through the uterine wall, or *endometrium*. During this period, not only are the muscular fibres of the uterus getting thicker and longer but the uterus itself is increasing, putting pressure on the bladder. At this point not much can be perceived of the belly, however changes in the upper part of the female body are more visible (Fig. 12.3). Since the cardio rhythm is increasing, the '*galactophores canals*' are multiplying and the size and weight of the breast can increase considerably (Stoppard and Younger-Lewis, 1995). Some women become minutely conscious of any changes in their body, and the baby may not seem real, thus women may focus on themselves (Martell and Mitchell, 1984). Seventy percent of pregnant women who

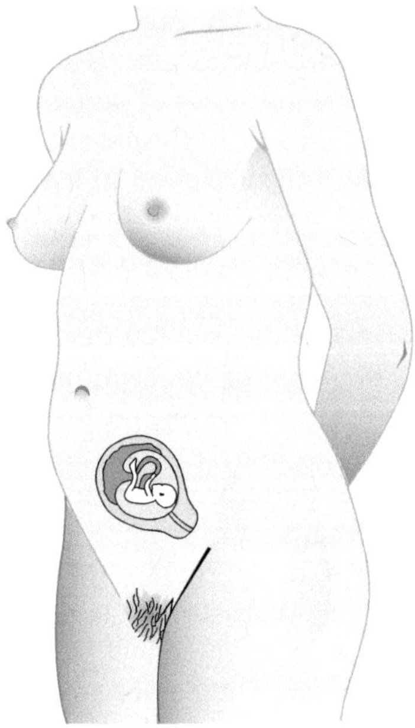

12.3 Body profile for the first trimester of pregnancy. (*Source*: Stoppard and Younger-Lewis, 1995, p. 170.)

experience morning sickness see it end before the end of this first trimester. Lastly, most miscarriages also occur during this period.

12.3.2 Second trimester

During the second trimester, from week 13 to week 28, the development of the foetus can be more easily monitored and most women feel more energized. The movement of the foetus may be felt, and the reproductive organs can be recognized, making it possible to distinguish the foetus as male or female. It is during this second trimester that women begin to put on weight, especially if it is their second pregnancy (Rutter *et al.*, 1984). The uterus expands in order to adapt itself to the size of the foetus (Stoppard and Younger-Lewis, 1995); and the belly (Fig. 12.4) becomes very upright. At the beginning of the second trimester, even if there is little evidence of pregnancy it is not unusual for an enthusiastic mother-to-be to wear maternity clothes as a 'validation' of her pregnant state (Olds *et al.*, 1984).

12.3.3 Third trimester

Lastly, the third trimester, from the beginning of the 29th week, often approximates the point of viability, or the ability of the foetus to survive, with or

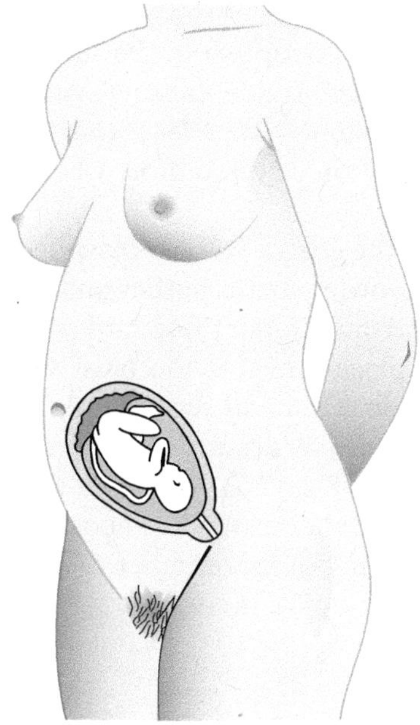

12.4 Body profile for the second trimester of pregnancy. (*Source*: Stoppard and Younger-Lewis, 1995, p. 170.)

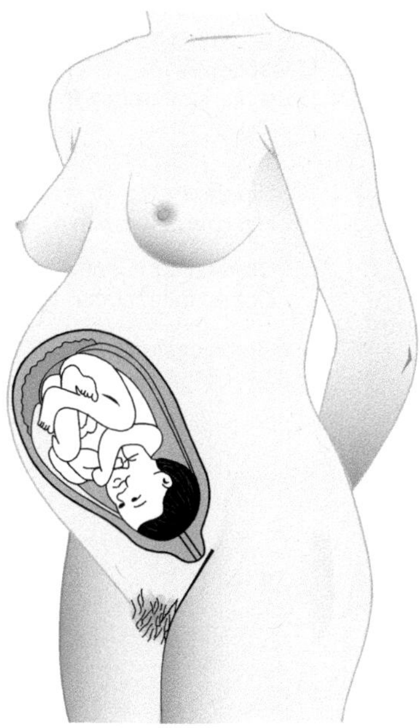

12.5 Body profile for the third trimester of pregnancy. (*Source*: Stoppard and Younger-Lewis, 1995, p. 170.)

without medical help, outside of the uterus. The foetus begins to move regularly and can become quite strong and be disruptive to the future mother. This period of her pregnancy can be uncomfortable, causing symptoms like weak bladder control and back-ache (Rutter *et al.*, 1984). The woman can feel the foetus *rolling* and it may cause pain or discomfort when it is near the woman's ribs and spine.

It is during this third trimester that the foetus will be growing the most rapidly, gaining up to 28 g per day. The growing foetus pushes on the digestive apparel and compresses the stomach and the bladder (Stoppard and Younger-Lewis, 1995). Final weight gain takes place, which is the most weight gain throughout the pregnancy. Breasts become full and sensitive. At this point, the woman's belly may change shape as the belly drops, due to the foetus turning in a downward position ready for birth (Fig. 12.5). Moreover the woman's navel will sometimes become convex due to her expanding abdomen.

Interestingly, results may vary depending of the number of pregnancies. According to Rutter *et al.* (1984) whose sample contained women who were

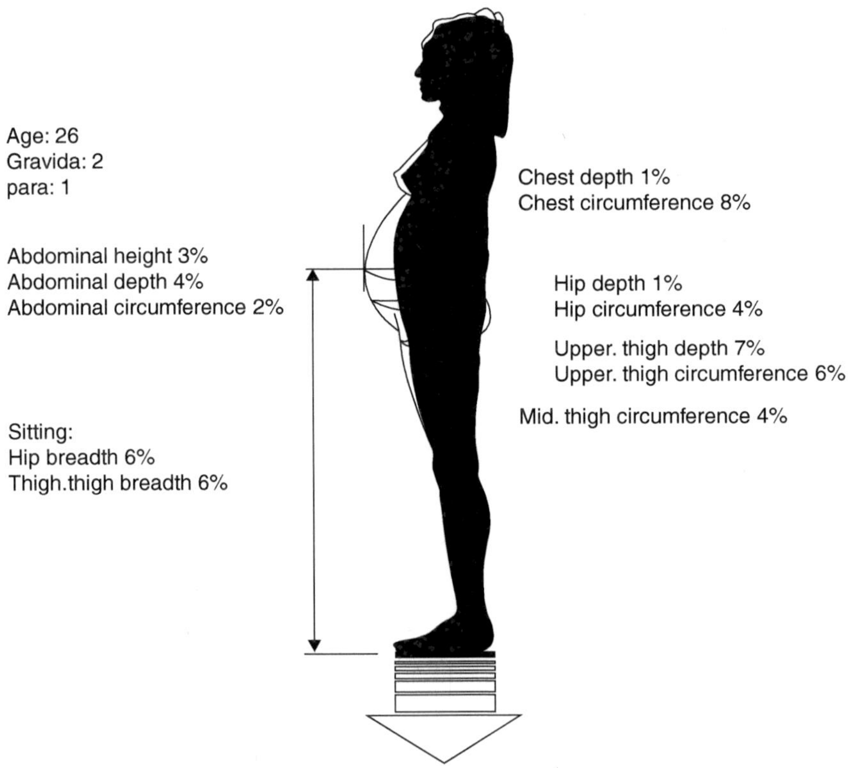

12.6 Abdominal extension during pregnancy. (*Source*: From Rutter *et al.*, 1984, p. 42.)

almost 26 years old (Fig. 12.6), had been pregnant twice (including current pregnancy) and had produced one viable birth, second time pregnant women gained 10.5 kg (17% increase over pre-pregnant weight). In addition, state the authors, they had increases of 2.8 cm in abdominal extension circumference and 8.2 cm in abdominal extension depth.

12.4 Pregnancy and apparel choice

During pregnancy, the foetus obliges the future mother to curve her back in order to counterbalance the weight of the abdomen. The posture of the vertebral cervical is not as straight as it would otherwise be, and may cause pressure on the articulations, especially during the second part of the pregnancy (Stoppard and Younger-Lewis, 1995). According to Chan (2000), physiological changes that appeared during the pregnancy, such as slightly higher body temperature and posture changes, require pregnant women to alter their purchasing patterns for clothes. Many choose more natural fibres such as cotton and wool blends along with a bit of elastane for comfort. Surprisingly, as of today only fourteen maternity patents were approved in the US for skirts for pregnant women. Today, patents are registered and classified in Classes 219 and 221 of the Patent and Trademark Office.

According to the literature, during the time a woman is pregnant, some activities are recommended, some should be practiced under supervision, and some should be avoided (Stoppard and Younger-Lewis, 1995). Swimming, yoga and relaxation are most often recommended as well as walking, stationary cycling and low-impact aerobics (Klusmann *et al.*, 2005). For those with prior expertise, certain equilibrium or balance activities such as horseback riding, skating and skiing could be practiced under supervision until the belly is voluminous (Stoppard and Younger-Lewis, 1995). All contact sports and weight lifting should be avoided, since they increase the risk of abdominal trauma. Discouraged activities include boxing, hockey, wrestling, football, soccer, gymnastics, hang gliding, vigorous racquet sports, and scuba diving (Stoppard and Younger-Lewis, 1995). This may not involve lifestyle changes for many women during their pregnancies.

There is little evidence that the maternity wear market makes use of this detailed information on physical changes and activities. Popular magazines such as the North American 'FitPregnancy' or those from ACP magazines in Asia for mothers-to-be may devote only a small portion of their magazine to promote a 'comfortable' suit for yoga, nursing bra, a supporting belt, a few fashionable garments and a comfortable pair of walking shoes. Very little is published about the needs for specific maternity wear. These types of magazines mentioned before mainly focus on the new member of the family. The following brief history of maternity wear reveals changing tastes in addition to changing sizes and activities.

12.5 Maternity wear: a brief history

According to Chan (2000), until the beginning of the new century, no research had been undertaken regarding functionality and comfort to fulfil the requirements of pregnant women. Mothers-to-be often ended up buying clothes that were not suitable for their needs. The first dress made specifically for maternity was a cheap cotton wrap-around housedress created in the early 1900s (Bucknum Brinley, 1985). Before that time, women wore their usual clothes for as long as they could. When they could no longer 'squeeze' their ripe bodies into a single dress, it seems that they just hid themselves from the light of day until after they delivered. Traditional *wrapping* clothing such as the sari or caftan obviously accommodated pregnancy more easily (Rubenstein, 1995 in Manlow, 2007).

Moreover, according to Peteu and Gray (2009), it is only after 1890 that garments begin to meet special feminine needs. These authors examined 1126 patents from the US Patent and Trademark Office to determine the primary purpose of garments/skirts (fashion, wearability, use and manufacturing) and how garments evolved to accommodate sports and professional activities as well as women's unique physiology, such as maternity and sanitary protection. Peteu and Gray (2009) state that it was in 1907 that the first maternity skirt was patented in the US (classified in Class 211). The waistband of that skirt was adjustable to accommodate growth.

After the Second World War, styles evolved to emphasize a women's procreative role. Garments called attention to the abdomen, accentuating the hips and breasts (Rubinstein 1995). Even then, maternity wear itself was discreet. According to Haugland (2010), it is in 1956, when Princess Grace's pregnancy was confirmed, that maternity wear finally became fashionable. Grace Kelly, an icon of fashion, who was known for classic simplicity, got public attention when showing in public. People were fascinated by the Princess' maternity wardrobe (Haugland, 2010). She was photographed in November 1956 wearing a Dior 'tailored suit' with a gently flared jacket, which obviously showed she was pregnant. Again she was on the front page of the *Paris Match* in January 1957 '...wearing a pastel chiffon maternity evening gown by Dior. At that time, when expecting, women had little fashion choice beyond the unbecoming yet ubiquitous maternity smock. Princess Grace of Monaco attracted international attention with her fashion-wise maternity wardrobe' (Haugland, 2010, p.76). The Princess was named on the Best-Dressed List for 1956, during her pregnancy, making 'maternity fashion' no longer an oxymoron.

Recently, maternity fashion has gained prominence, especially amongst today's celebrities. One may remember Ms. Kardashian or Ms. Klum.

In addition to this, the low rise pants, the halter tops, and even rings and jewels for the navel focus attention and make pregnancy more obvious.

In the last few years, a major shift has occurred in maternity wear; a style Rubinstein (1995) called the 'pregnant look'. The days when new moms had to feel like they were running around in their nightgowns are long past. The pregnant look may be desired because it is for a short period of time (Garrett, 2010) and the industry needs to keep abreast of the changing tastes.

12.6 Meeting the needs of a target market: body measurement and fit satisfaction

Understanding a target market, including body measurements and garment preferences, is very important, but it is also challenging for apparel companies; thus most retailers focus on one specific niche (Kerin *et al.*, 2007). In this manner, they hope to satisfy their consumers and in return get consumer loyalty. Chan (2000) defines satisfaction as 'an outcome of purchase and use from the buyer's evaluation of the rewards and costs of the purchase in relation to the anticipated consequences' or 'normative expectations the consumer thinks the product ought to be in terms of specific elements such as size'.

In order to satisfy their target market, manufacturers adapt the sizes and shapes of garments to specific groups. As mentioned, William Herbert Sheldon looked at thousands of photographs, studying physical traits and physical build, and grouped individuals into three major *somatotypes*: (1) *endomorphy*, (2) *mesomorphy*, and (3) *ectomorphy* (Sheldon, 1940). Women from different origins can also be distinctive or singular in their body shapes and could be defined as a target market. Indeed, women's anthropometric data analysis establishes that origin or race can make a difference in body shapes and that body shapes can be clustered geographically (Faust and Carrier, 2009b; [TC]^2, 2004). If one thinks about it, when comparing Asian

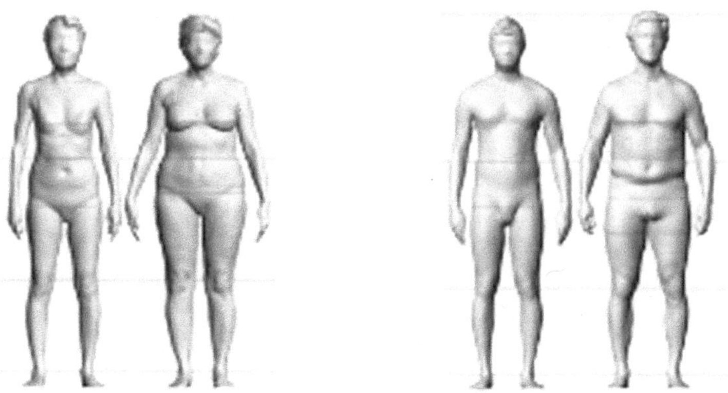

12.7 Business wire, 12 August 2008, Alvanon releases most extensive Chinese body measurement study.

women to western women it seems obvious that sizes and body shapes are not alike (Chow, 2012).

The apparel industry recognizes this phenomenon and consequently produces brands for specific body shapes. Consumers then come to identify brands that fit their body shape and size. Indeed, companies able to adapt their production to the Chinese market are leaders in satisfying Asian demand. According to Alvanon, a Hong Kong based company, the Chinese are similar in height to westerners on average. The major difference is in the girth (Fig. 12.7). Their analysis reveals significant insights into Chinese size and shape for the fashion industries. Companies need to apply different size scales as well as discern varying style preferences for their products to be successful abroad.

When it comes to maternity clothes, body sizes differ between groups, and within groups over time. Since fit is an important characteristic and since it seems that manufacturers measure and label their garment sizes according to past practice, habitually using what was done by predecessors, it appears necessary to have a full understanding of perceived physical changes and fit satisfaction from one target market, here being pregnant women.

12.7 Case study: pregnant women in Hong Kong

The questions we address in this chapter are: when and how body changes are perceived by mothers-to-be, and how do women manage the acquisition of garments when their body shape and garment needs change radically during pregnancy? This research focuses on a specific target market: pregnant Asian (Hong Kong) women; their perception of their body transformation during the three trimesters of their pregnancy and their need to change their clothes and shoes. We investigate pregnant women's satisfaction with the sizes of clothes available to them on the market for the three trimesters. Lastly we ask if women would like to serve for a 3D body measurement study.

For the purposes of this study, we first estimate pregnant women as being a viable target market. According to the national census, the volume of birth was 662 000 between mid-2006 and mid-2007 (Hong Kong Census and Statistics Department, 2007). This research showed that a high percentage, 96%, of those mothers-to-be are working, which translates into more money spent for a work wardrobe, justifying the allocation of resources to serve the market. It is sufficiently distinctive in its behaviour, viable, and therefore becomes an important target market in the women's apparel industry.

This study was conducted in Hong Kong in 2008–9. The sample size was just over two hundred women. Out of this sample, 26% were pregnant during the study, 30% has been pregnant within the last 2 years, 23% had been pregnant 2 to 5 years ago and of the rest, 23% had been pregnant within the

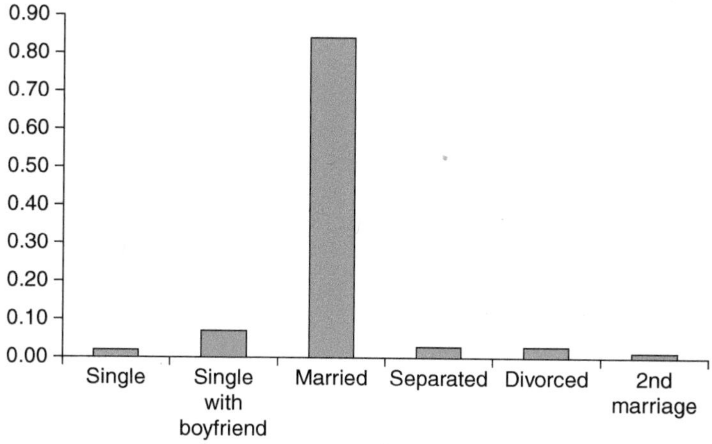

12.8 Matrimonial status distribution of our sample.

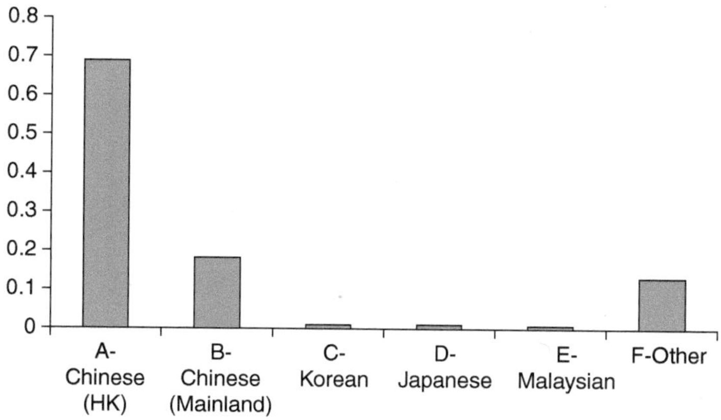

12.9 Nationality distribution of our sample.

last decade. For the majority of them (61%), it was their first pregnancy. Of our sample, 84% were married and the count went up to more than 90% if we included women with a significant other (Fig. 12.8).

Since this survey was done in Hong Kong the majority, 68%, were Hong Kong Chinese, 17% were Chinese from mainland China and the final 15% were from other Asian nationalities (Fig. 12.9).

The age distribution is consistent with that discussed in the literature review; the majority of the women (close to 70%) were between 20 and 35 years old, and of these, 14% were in the age range of 20 to 25 years old, 26% were between 26 and 30 years old; and close to 30% were between 31 and 35 years old (Fig. 12.10).

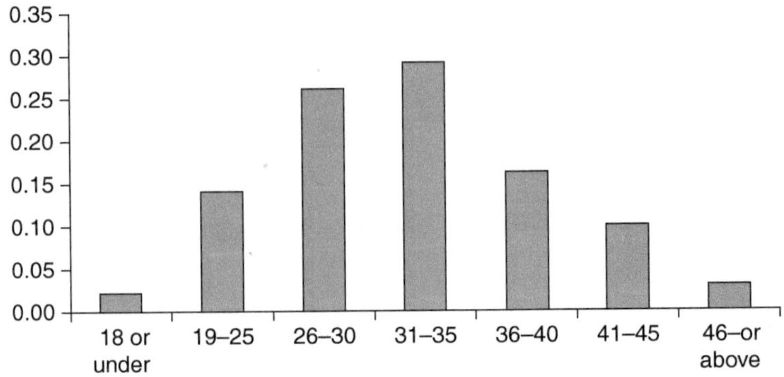

12.10 Age range distribution in percentage for our sample.

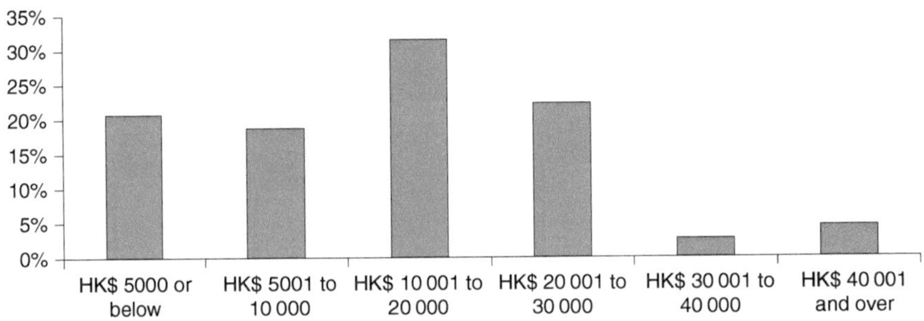

12.11 Revenue range distribution in percentage for our sample.

To complete the demographic data, the average monthly personal income varied from below HK$5000 to over HK$40 000 (Fig. 12.11).

12.7.1 Physical changes

In the first trimester, consistent with the literature, 46% of our sample reported changes in their bust. Yet contrary to the literature, 40% stated that their waistline was modified as early as in the first trimester. As expected, it is during the second trimester that most respondents perceive physical changes (Fig. 12.12). During the last trimester, while the percentage of respondents reporting physical changes declined, more than 30% of our sample identified changes in leg and feet sizes. A total of 64% of our sample perceived differences in their foot size during their pregnancy.

The percentage of women that perceived body changes for different parts of their body per trimester are as follow:

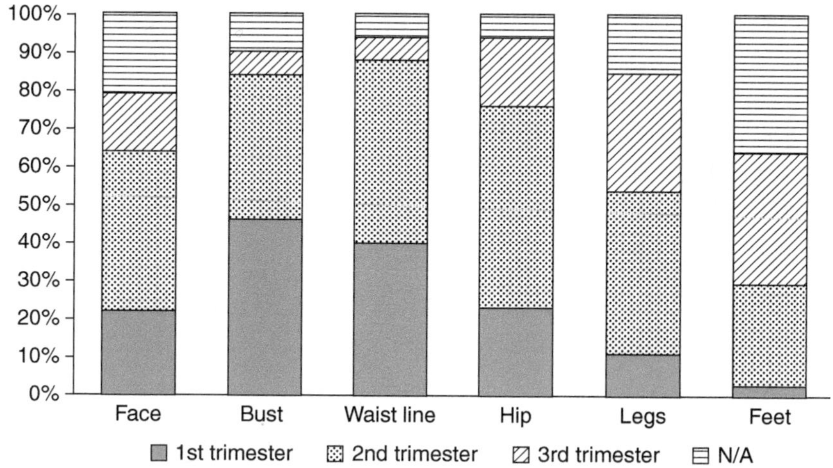

12.12 Perceived physical changes for different body parts per trimester.

- First trimester: face, 22%; bust, 46%; waist line, 40%; hip, 23%; legs, 11% and feet, 3%.
- Second trimester; face, 42%; bust, 38%; waist line, 48%; hip, 53%; legs, 43% and feet, 26%.
- Third trimester; face, 15%; bust, 6%; waist line, 6%; hip, 18%; legs, 31% and feet, 35%.

A small percentage didn't perceive any changes at all. The bar chart (Fig. 12.12) clearly indicates the second trimester as being the one with the most perceived changes, with the exception again of the bust and waist line which also occur during the first trimester.

12.7.2 Need for new garments

While size changes will necessitate new garments, so will a change in exercise/activities or a preference to emphasize the pregnant state, resulting in the acquisition of different garments during this period of life. The need for new clothes is most pronounced during the second trimester (see Fig. 12.13) with the exception of the need for a new bra (40%) which appeared earlier, which also coincides with our first results. Surprisingly, a high percentage mention they needed new panties (40%) and pants (42%) during their first trimester, although this correlates to the fact that 40% of our sample stated they perceived a change in their waistline as early as during the first trimester. Our results also highlight that close to 60% felt they needed to change their shoe size during pregnancy.

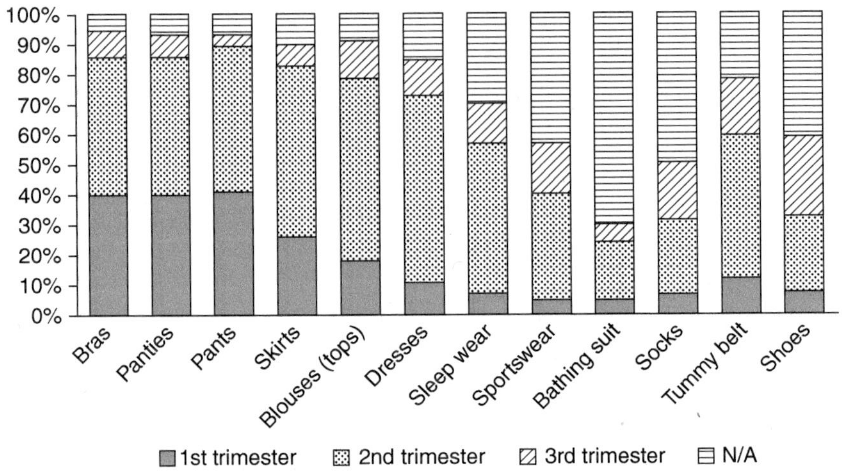

12.13 Percentage of respondents with specific clothing need by trimester.

The percentage of women that stated they needed different garments, per trimester, is as follows:

- First trimester: pants are the number one item with 42%, closely followed by bras, 40%, and panties, 40%. Skirts (26%) and blouses (18%) are the two next in terms of priority. The others such as sleepwear, sportswear, bathing suit, socks, tummy belt and shoes each counts, respectively, for less than 10%.
- During the second trimester, dresses counted for 62% and blouses 61%. These are closely followed by skirts (57%). The two next items in terms of priority are pants (48%) which are again closely followed by bras 46% and panties 46%. Contrary to the first trimester, during the second trimester women also perceive a need for other garments such as sleepwear (50%), sportswear (36%), bathing suit (20%), socks (27%), tummy belt (48%) and shoes (26%).
- Lastly, during the third trimester most of the women seem to have had less need for new clothes. Only shoes count for more than twenty percent (26%). All the other items listed before count for less than 20% in terms of new needs.

Results, illustrated in Fig. 12.13, make it easier to see that the second is the most important one.

12.7.3 Active wear/sportswear

Our survey results show that many of the pregnant women, 32%, did not do any activities/sports at all during this period. However, 36% answered

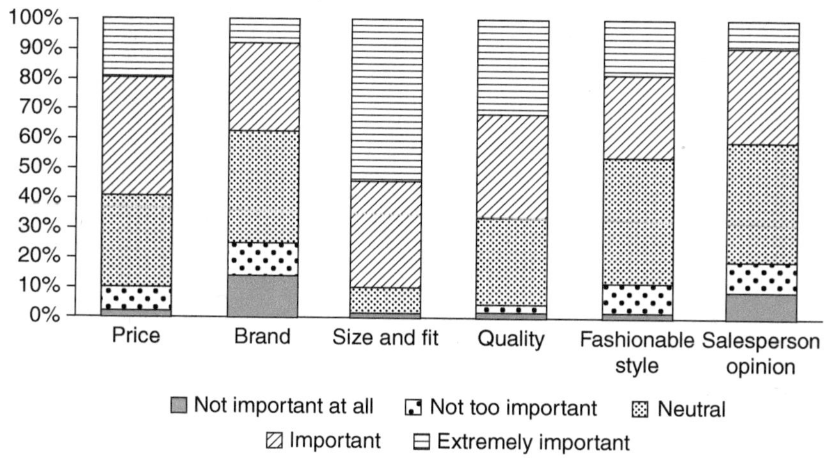

12.14 Importance of various criteria for purchase of maternity wear.

they did took part in activities such as yoga, Tai Chi or swimming. Only 2% answered they played racquet sports. When asked if they needed new clothes to continue to practice their activities/sports, 54% responded yes (either new garments, second hand ones or borrowed) whereas 46% answered they continued with the garments they had.

12.7.4 Criteria for maternity wear

To identify the importance of various criteria in the purchasing decisions of our target market, women in our sample were asked to rank each of the following: price, brand, sizes and fit, quality, fashionable styles and the salesperson's opinion, from 1 to 5. On this Likert scale, (1) denotes not important at all while (5) indicates being extremely important criteria in the decision to purchase maternity clothes. Again, surprisingly, the most important criterion is the size and fit; 55% of women stated that it was an extremely important criterion and 35% said that it was an important criterion (90% in total). This is followed by the quality and the price. According to our results, women of our sample are either neutral or pay little attention to criteria such as the brand, the trend and the salesperson's opinion. These results are shown in Fig. 12.14.

12.7.5 Garment fit and satisfaction

Figure 12.15 reports the level of satisfaction with the garment fit. Garment fit appreciation is ranked on a Likert scale of 1 to 5 (1 being not satisfied at all, 5 is very satisfied). Contrary to our expectation, a high percentage of women were

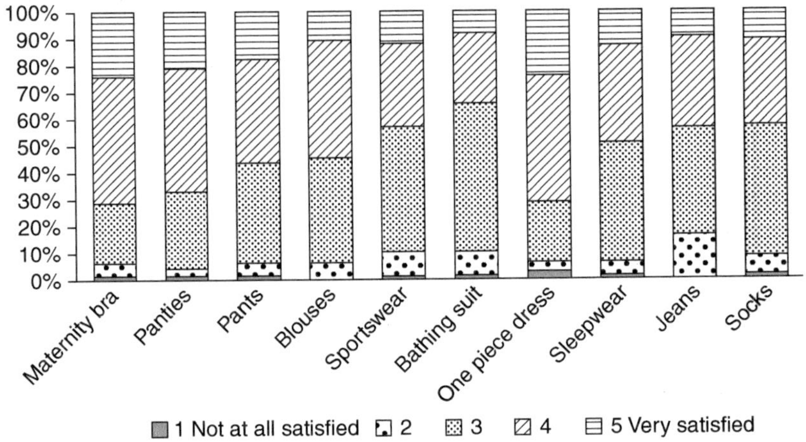

12.15 Percentage of satisfaction of our respondents with clothing size and fit.

'neutral' (wrote 3 on a Likert scale of 1 to 5) or satisfied (wrote 4) with the sizes and fit offered for maternity clothes. One could expect such results for sizes and fit of socks, for which 49% wrote 3 on a Likert scale of 1 to 5. But to our surprise, as many as 56% wrote 3 on a Likert scale of 1 to 5 for garments like bathing suit, 47% were satisfied with their maternity bra, and 40% were 'neutral' and 37% satisfied with jeans; clearly these results are unexpected, to say the least.

Although a high percentage state that they prefer a one piece dress (69%), this garment is also the one that received the lowest score for satisfaction. As many as 44% of our sample answered they were either not satisfied at all or were dissatisfied with the fit of this type of garment.

12.7.6 Serving as a model to be measured

Size in the apparel industry has always been challenging for manufacturers and retailers, and a decisive factor for consumers' clothing selection. As our results show, size and fit seems to be a critical issue for mothers-to-be. In order to improve the maternity wear size system and garment fit, pregnant women need to be measured.

We asked women of our sample if they would serve as models to be measured during their pregnancy. If so, would they prefer to be scanned or manually measured? Lastly, would they need to consult with anybody to help make the decision to take part in a study which will require body measurement such as a 3D body scanner, and, if so, who would it be?

The majority, 73%, answered they would prefer not to be measured, whereas 33% added that if they had to be measured they would prefer to be body scanned and the rest, 67%, would prefer to be manually measured.

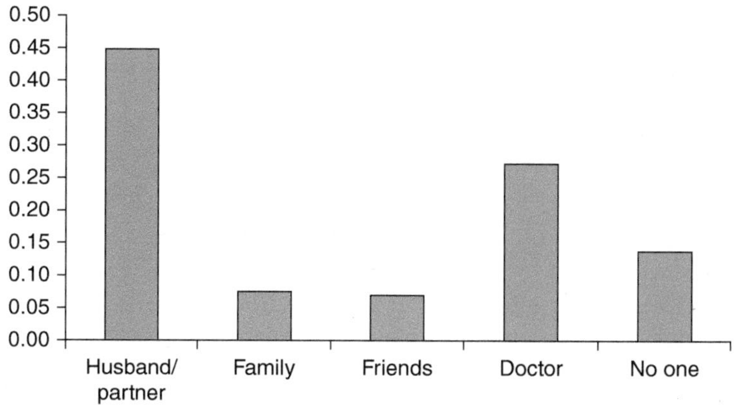

12.16 Percentage of our respondents who need to discuss (get approval) either with their husband, doctor, friends or family to serve as model to be measured.

A total of 45% state that they would need to discuss with their husband (lover) first, whereas 27% state they would need to discuss with their doctor first. Only 14% were comfortable to take this decision on their own. These results are shown in Fig. 12.16.

12.8 Conclusion

Aesthetics, size and fit all affect the capacity of consumers (in this case pregnant women) to truly evaluate how they will look and feel in a garment. This research serves as a starting point to investigate women's perception of their body shape during pregnancy and their appreciation for apparel offered on the market for this small and well identified target market. Since the result comes from a convenient sample of 203 pregnant or previously pregnant women in one specific area, Hong Kong, it provides an idea, but could easily be replicated on a larger scale or in another geographic area. Further statistical details can be found in 'An Expanding Market: Asian Mothers-to-be' by Faust *et al.* (2011). Nevertheless, this study investigated a new field of research – maternity clothes for different geographic markets and ethnicities – and thereby informs manufacturers and retailers, as well as marketers and designers, with the objective of satisfying this specific target market of pregnant women.

12.9 References

Alic, M. (6 April 2001). 'Sheldon, William Hebert (1898–1977)'. *Encyclopedia of Psychology*, in http://en.wikipedia.org/wiki/William_Hebert_Sheldon#cite_ref-eop_3-0

Alvanon (2008) Alvanon Releases Most Extensive Chinese Body Measurement Study: Analysis Reveals Significant Insight on Chinese Size and Shape for the Fashion Industries, New York, August 12, 2008, http://www.alvanon.com/news/CHINASCAN.pdf.

Bucknum Brinley. M. (1985). Maternity Style: How to look your best when you're at your biggest, St-Martin's Press, 1st Edition, New York, 174p.

Chan, C. P. J. (2000). The product development of the maternity clothes for the Hong Kong pregnant women in the period from third trimester to postpartum, Ph.D. Thesis, Institute of textiles and clothing, The Hong Kong polytechnic University, 130p.

China Demographic, Hong Kong Census and Statistics Department (2007), https://www.cia.gov/library/publications/the-world-factbook/geos/ch.html.

Chow, V. (2012). The LoDown; Here's a tall story: study finds Chinese are similar in height to Westerners, South China Morning Post.

Claesson, I.M., Josefsson, A., Cedergren, M., Brynhildsen, J., Jeppsson, A., Nystrom, F., Sydsjo, A. and Sydsjo, G. (2008). 'Consumer satisfaction with a weight-gain intervention programme for obese pregnant women', *Midwifery*, Vol. **24** (2), June, pp. 163–7.

David, K. and Wardle, J. (1994). 'Body image and dieting in pregnancy', *Journal of Psychosomatic Research*, Vol. **38** (8), pp. 787–99.

Day, G.S. (1990). *Market driven strategy*. The Free Press, New York.

Faust, M-E. and Carrier, S. (2006). 'Variations in Canadian women's ready-to-wear standard sizes', *Journal of Fashion Marketing and Management*, Vol. **10** (1), pp. 71–83.

Faust, M-E. and Carrier, S. (2009a). 'Women's wear sizing: a new labeling system', *Journal of Fashion Marketing and Management*, Vol. **14** (1), pp. 88–126.

Faust, M-E. and Carrier, S. (2009b). 'A Proposal for a New Size Label to Assist Consumers in Finding Well-fitting Women's Clothing, Especially Pants: An Analysis of Size USA Female Data and Women's Ready-to-wear Pants for North American Companies', *Textile Research Journal*, Vol. **79** (16), pp. 1446–58.

Faust, M-E., Christoffersen, S., Tan, J. and Shin, K. (2011), 'An Expanding Market: Asian Mothers-to-be', *Journal of Global Fashion Management*, 23p.

Garrett, J. (2010). http://www.sooperarticles.com/health-fitness-articles/pregnancy-articles/maternity-clothes-advice-expectant-mothers-35928.html, January 18th, 2010.

Goldsberry, E., Shim, S. and Reich, N. (1996a). 'Women 55 years and older: Part I. Current body measurements as contrasted to the PS 42–70 data', *Clothing and Textiles Research Journa*l, Vol. **14** (2), pp. 108–20.

Goldsberry, E., Shim, S. and Reich, N. (1996b). 'Women 55 years and older: Part II. Overall satisfaction and dissatisfaction with the fit of ready-to-wear', *Clothing and Textiles Research Journal*, Vol. **14** (2), pp. 121–32.

Haugland, K. H. (2010). Grace Kelly Style, V & A Publishing, 110p.

Horne, L., Campbell, L. and Scholz, C. (1999). 'Older females as a market segment for well-fitting clothing', *Journal of Fashion Marketing and Management*, Vol. **3** (3), pp. 236–44.

Jimenez, R. and Bacallao, J. (1995). 'Prognostic performance of several anthropometric indicators for predicting low and insufficient birth weight', *American Journal of Human Biology*, Vol. **7** (3), pp. 302–11.

Kerin, R.A., Hartley, S.W. and Rudelius, W. (2007). *Marketing the Core*, 2nd edition, McGrawHill International Edition.

Klusmann A, Heinrich B, Stöpler H, Gärtner J, Mayatepek E, Von Kries R (2005). 'A decreasing rate of neural tube defects following the recommendations for periconceptional folic acid supplementation'. *Acta Paediatr.* Vol. **94** (11), pp. 1538–42, doi:10.1080/08035250500340396, PMID 16303691, http://www.blackwell-synergy.com/openurl?genre=article&sid=nlm:pubmed&issn=0803–5253&date=2005&volume=94&issue=11&spage=1538.

Kogan, L. (2007). O the Oprah Magazine, O's guide to Life, p.42.

Lennon, S.J. (1992). 'Categorization as a function of body type', *Clothing and Textiles Research Journal*, Vol. **10** (2), Winter, 18–23.

Manlow, V. B. (2007). Designing Clothes: Culture and Organization of the Fashion Industry. Transaction Press, New Brunswick, N.J.

Medicine.net. http://www.medterms.com/script/main/art.asp?articlekey=11895. Retrieved 16 Jan 2008.

Murkoff, H. (1984). *What to Expect When You're Expecting,* Workman Publishing, New York, 616p.

Oldham Kind, K. and Hathcote, J.M. (2000). 'Speciality-size college females: satisfaction with retail outlets and apparel fit', *Journal of Fashion Marketing and Management*, Vol. **4** (4), pp. 315–24.

Olds, S.B., London, M.L., Ladewing, PA. and Wesley, A. (1984). *Maternal-Newborn Nursing: A Family-Centered Approach.* 2nd Edition, London, England, Addison-Wesley Pub. Co., nursing Division.

Otieno, R., (2000). 'The Role of Garment Sizing in Creation of Customer Satisfaction: Indications from Focus Group Responses', *J. Fashion Mark. Management*, Vol. **4** (4), pp. 325–35.

Otieno R., Harrow, C. and Lea-Greenwood, G. (2005). 'The Unhappy Shopper, a Retail Experience: Exploring Fashion, Fit and Affordability', *International Journal of Retail & Distribution Management*, Vol. **33** (4), pp. 298–309.

Peteu, M. C. and Gray, S. H. (2009). 'Clothing invention, improving the functionality of women's skirts', *Clothing and Textiles Research Journal*, Vol. **27** (1), pp. 45–61.

Rasband J. and Lietchy, E. L. (2006). *Fabulous Fit: Speed Fitting and Alteration*, 2nd Edition, Fairchild Publications, New York, 432P.

Roebuck, J. A. (1995). Anthropometric methods: designing to fit the human body, Human Factors and Ergonomics Society, Ottawa, Canada, 194P.

Rutter, B., Hagger, J. Daigle, G., Smith, S., McFarland, N., and Kelsey, N. (1984). 'Dimensional Changes Throughout Pregnancy: A Preliminary Report', *Carle Select Papers*, Fall 1984, Vol. **36** (2), pp. 38–46.

Sheldon, W. H. (1899–1977) (1940). *The varieties of human physique; an introduction to constitutional psychology*, 4th edition. New York, Harper and Brothers, 347P.

Stoppard, M. and Younger-Lewis, C. (1995). *Etre Femme (French),* Sélection du Reader's Digest; 1st edition, Montreal, Canada, 226P.

[TC]^2 (2004). *Size USA, Let's Size Up America…* The National Sizing Survey: Body Measurement and Data Analysis Reports on the US Population Report, Cary, North Carolina, USA.

Yoo S., Khan, S. and Rutherford-Black, C. (1999), 'Petite and Tall-sized consumer segmentation: Comparison of fashion involvement, satisfaction and clothing needs', *Journal of Fashion Marketing and Management*, Vol. **3** (3), pp. 219–35.

13

Plus size Black and Latino women: the implications of body shape and size for apparel design

C. J. PATTERSON, Northwestern University, USA

DOI: 10.1533/9781782422150.2.256

Abstract: While a particular representation of plus size has been accepted as 'fashionable', the same representation has rendered many Black women and Latinas invisible within the fashion industry. This chapter will explore the links between fashion designing, aesthetics, and silhouettes within Black and Latina groups to offer ways in which to view, value, and (re)consider plus size Black female and Latina bodies.

Key words: plus size fashion, plus size Black women and Latinas, body size discrimination, plus size entrepreneurs, Black and Latino women's fashion.

13.1 Introduction

The fashion world is a vast world made up of its own culture, aesthetics and economic opportunities. It is also a microcosm of social, political and cultural structures that replicate inequalities for marginalized persons, causing some to work harder for access into this world, while barring others from it altogether (Ash and Wilson, 1993). Plus size Black women and Latinas appear to be two of these marginalized groups, making the fashion world a fertile site in which to interrogate inequality, agency and power with regard to how it has considered these women over time and what these women have done in response to such considerations. This chapter will explore issues of access and agency between the fashion world as an industry and Black and Latino women (both Black and Latino are used here as umbrella terms to refer to people of Black African and/or non-European descent from Spanish-speaking Latin American countries, including, but not limited to: African Americans, Caribbean Americans and African immigrants; Puerto Ricans, Dominicans, Mexicans, Cubans, and people in the Americas, respectively, with ties to the US via birth, residency and European colonial formations. It is important to note that there are Latinas of African descent who are racialized and identify as Black in the United States). Exploring these issues will address the implications of their absence in fashion, particularly high fashion, and the ways in which they negotiate and maneuver around industry limitations. In 'Plus size models in the fashion world' I briefly outline the emergence of plus size model popularity in fashion in the US and investigate the current state of plus size fashion with regard to the visibility of Latinas and Black women. With particular interest in body size, race and silhouette, Black and Latina models are juxtaposed with conventional aesthetics. The next section, 'Causality and reality: body size among Black and Latino women', explores some of the social and cultural processes that these women experience that may help to form their identity. This section also includes voices from notable bloggers in the 'Plus Size Blogosphere' (a network of plus size models, designers and consumers, bloggers and commentators that are representative of the plus size market). In 'Plus size clothing designers who hire Black and Latina models', I interrogate and analyze some popular plus size clothing designers, focusing on those who have featured Black and Latina models in their advertisements. Lastly, 'Conclusion and future trends' discusses the direction in which plus size fashion appears to be headed, centering on contributions from and representations of plus size Black and Latino women. This section also demonstrates how the plus size blogosphere seemingly functions as a database for plus size Black and Latina models.

13.2 Plus size models in the fashion world

As the 'obesity epidemic' spreads worldwide, many local, federal and international agencies have become more active in monitoring and regulating body size. The fashion industry has not gone untouched by this phenomena, evidenced in 2006 when Madrid, under pressure from its local government, banned models with a Body Mass Index (BMI) below eighteen, as well as by clothing designer Elena Miro opening Milan's fashion week with her plus size creations that same year. This careful attention to body size, and the narratives surrounding its linkages to health, have risen in the fashion world, particularly among designers, models and consumers, to address how the obesity epidemic, along with western beauty ideals, have undoubtedly influenced fashion culture. However, noticeably absent from this dialogue, and influential fashion events like Elena Miro's runway show, is the inclusion of plus size Black and Latina models. While more designers and models of color are now entering into the fashion industry, Black and Latina plus size models continue to struggle to find a space that fits their distinctive aesthetic, and are calling for more conversation between stakeholders who share power over what is labeled, defined and upheld as fashionable.

13.2.1 The fashion industry as an art world

The fashion industry and what is fashionable has never been a monolith. The 'high fashion' print, mainstream/commercial, and runway sectors that are upheld, refined and redesigned by designers, students, consumers and entrepreneurs are some of the components of its ever-changing *art world*. Howard Becker (1982) defines art world as a network of people, acting cooperatively under the auspices of a shared understanding of 'conventional means of doing things' to produce the kind of art that its particular art world is known to produce. This means that people in the fashion world (and arguably fashion industry) have a sense of what is acceptable and may be excluded based on their participation in (or outside of) the art world. Fashion empires like Gianni Versace and Christian Dior only acquire status based on fashion industry people subscribing to, and thus validating, their respective concepts of fashion. The same can be said about their models.

Using 'straight-size' models, referring to women sizes double zero through four in the US, is a practice that has not always been the standard. In both print work and design, women (a vast majority of them white) had 'hourglass' shape figures that were considered to be 'rubenesque', voluptuous, or curvy. 'Stick-thin' models were not made popular until the 1960s when English models Jean Shrimpton and Twiggy (Lesley Lawson) debuted their thin builds in Vogue magazines. This uprooting of the voluptuous figure set a trend for designers to use extremely thin models for both print and runway. This shift occurred based on what Becker would describe as 'non-standard' works, works that are able to surface because there often exists a subsidiary distribution channel and/or 'adventurous entrepreneurs and audiences' that make a space for these non-standard works in the art world. Members of the art world thus began to subscribe (or Becker would say adapt) to 'what existing institutions can handle' (Becker, 1982, p. 28). The designers, editors and photographers who employed models such as Shrimpton and Twiggy would shift the boundaries of 'conventions' and thus open a lucrative gateway for models with thinner bodies.

In a majority-white fashion world, where social and institutional factors such as racism and sexism do have an effect, fashion world members make conventional choices when it comes to models for commercial, editorial or runway campaigns. These conventional choices, based on European standards of beauty, place non-thin models and women of color outside of the fashion world. Although Black straight-size models, like Iman and Naomi Campbell, and white plus size models, such as Kate Dillon and Tara Lynn, have undoubtedly worked hard to gain worldwide acclaim, they still represent a portion of people very similar to conventions of fashion, albeit not completely so. Black and Latina plus size models, and those who hire them for their various fashion campaigns, have made notable attempts to fold plus size fashion into the conventional arena. Eunice Walker Johnson of Ebony Fashion Fair fame, responding to her constituents' request to diversify fashions on the runway, incorporated more athletic and plus size models in her shows to represent the variety of Black women's bodies. Under Kenneth Owen's direction, the plus size fashion segment of the traveling fashion show became the most popular, and perhaps the most impactful, portion of Fashion Fair. However, Black and Latina plus size models still face an uphill challenge to become as conventional as straight-size and white models.

13.2.2 The difference between plus size women and plus size models

When American plus size model Melissa 'Emme' Aronson gained nationwide fame in the mid-to-late 1990s, she became the face of plus size fashion. Arguably one of the most beautiful women in the world, Emme made plus size modeling a reputable business, but her success created more challenges for Black and Latina models. Nearly a decade after Emme was named one of *People's* '50 Most Beautiful People' for the second time, *Glamour* published a photograph of plus size models thought to represent a more realistic view of the American woman (see Fig. 13.1). The magazine touted its November 2009 issue as supportive of the 'curvy' size acceptance movement prevalent throughout a number of fashion circles. While thankful that such a prestigious magazine included a picture of more 'average'

13.1 From left to right: Crystal Renn, Amy Lemons, Lizzie Miller, Ashley Graham, Kate Dillon, Anansa Sims and Jennie Runk, featured in the November 2009 edition of *Glamour* Magazine. (*Source*: Photo credit: Matthias Vriens-McGrath/Trunk Archive.)

looking women, many readers still believed that these women were not true depictions of plus size women. Instead, some readers asserted that these women were 'thin' in comparison to what they considered the average American woman.

This consideration of plus size model body size unearthed an issue of silhouette type that could constitute the difference between plus size women and plus size models. Silhouettes with similar bust and hip measurements and smaller waist measurements appear to be the preferred silhouette for plus size models. Plus size women, both overweight and obese, come in varied shapes and sizes, some arguably larger than a commonly known American size 14, and some of their silhouettes would be considered completely outside of the fashion world. For example, a woman with bust, waist, and hip measurements of 42--53--39 would less likely be chosen for a high fashion plus size model campaign than a plus size model on the size 8 to 14 continuum. Thus, the models pictured in *Glamour* (Crystal Renn, Amy Lemons, Lizzie Miller, Ashley Graham, Kate Dillon, Anansa Sims, and Jennie Runk) were strikingly close to straight model proportions, and much like Emme, still seemed to represent the conventions present in the fashion world.

13.2.3 The effects of fame

Noticeable in the *Glamour* photograph is not only the absence of a Latina model, but also the lone Black woman featured. Anansa Sims, the talented daughter of supermodel Beverly Johnson (the first Black model to appear on the cover of *Vogue*) appeared in the issue to seemingly diversify the group of women. However, her presence connotes how celebrity is yet another factor when casting Black or Latina plus size models. This is not to say that there are no plus size Latina celebrities, but Sims, who holds an MBA, has been connected to both the fashion and entertainment worlds since birth. Her parents' connections, combined with her appearance and talent, afforded her opportunities to build relationships with the fashion and entertainment elite, evidenced by her beginning her modeling career at the age of 13 with the Wilhelmina agency (Leavitt, 1992; Swope, 2010). The idea of fame, debatably a form of exceptionalism, has been discussed as a source of success for many Black and Latino people (Hajnal, 2007). Plus size women in particular, such as Queen Latifah, Liv Tyler, Preta Gil, and Oprah, have been able to negotiate and maneuver their way through the fashion world not only because of their skill sets, but also because of the success and (socioeconomic) status bestowed upon them by their celebrity. Plus size fashion figures, such as Toccara Jones, Rosie Mercado, Jordan Tesfay, and Nicole Esperanza Zepeda, entered into televised or publicized competitions to gain some facet of celebrity that in turn catapulted their careers. Some plus size models entered into the fashion industry as straight-size models, and after either health concerns and/or lifestyle choices, chose to re-enter as plus size models, tapping into an already-established list of connections. Those without fame or celebrity status have decreased chances of entering into the elite world, and this has been especially true for Black and Latina plus size models.

Anansa Sims' inclusion in the *Glamour* photo may have symbolized progress on the surface, but underneath the glamour of the photo remains the idea that she is indeed an 'anomaly.' Her presence in the photograph exhibits even more severely that being a plus size model equates to not only being 'not plus size,' but also white and exceptional. Those models who are non-white, have non-conventional silhouettes, and little to no ties to the celebrity elite have to create their own networks to gain opportunities into (or close to) the fashion world. These are the types of networks that will be discussed in a later section.

13.3 Causality and reality: body size among Black and Latino women

Black and Latino women, like the fashion industry and its participants, are anything but a monolith. Just as they live all over the world and have a myriad of social and cultural practices, they define body size in a plethora of ways. However, some of their approaches to body size are similar because they appear to define them in a cultural manner rather than a medical manner. In some Black and Latino communities, a woman with a larger body size is normative and accepted, where in mainstream communities she can be labeled overweight or obese, thus not accepted (Zikalala, 1994; Beauboeuf-Lafontant, 2003; Parker *et al.*, 1995). In contrast, a woman that fits within the normal BMI range might be considered healthy in some mainstream communities, but can be labeled otherwise in other communities. Therefore, people define body size relative to the communities they live in, enter and navigate, and each community has its own set of structures and institutional mechanisms that helps it to define, value or reject particular body sizes.

13.3.1 A brief sociology of Black and Latino plus size women in the US

Many of those body size-regulating structures stem from socioeconomic stratification, environmental racism, and institutionalized racism that target females' bodies more prevalently than those of males (Yancey *et al.*, 2006). Sociologists have found that Black female and Latina adolescents have a more positive body image than their white counterparts, but once they move from their indigenous environments, they begin to take on practices of the dominant group, including anorexia, bulimia, and body image ideals (Thompson, 1992; Parker *et al.*, 1995; Lovejoy, 2001; Hrabosky and Grilo, 2007; Bates *et al.*, 2008). Black and Latino women face a combination of racial and gender discrimination, single-parent households, poverty, and a shortage of employment opportunities that not only have financial implications, but also psychological and physiological implications (Turner and Avison, 2003; Cohen *et al.*, 2006).

Cortisol, a hormone that increases with levels of stress, has been found to increase appetite and body mass, particularly along the abdomen and thighs (Epel *et al.*, 2001). These factors make it no coincidence that Black woman and Latinas lead the US in obesity rate, based on BMI measurement. According to the US Department of Health and Human Services' Office of Minority Health 2012 report, 80% of African American women aged 20 and older are overweight or obese, while Mexican-American (Latina) women 20 and older follow closely behind at 78%. The percentages of both overweight and obese African American and Mexican-American women are much higher than their white counterparts, who measure in at 60.3%. However, some Black and Latino women, along with scholars in medical and sociological fields, believe that their larger body sizes are normative, and that the BMI scale is biased and discriminatory along categories of race, socioeconomic status, and gender (the formula is the same for men and women, but women tend to be considered overweight or obese at lower BMIs than men), and is not an indicator of health, thus making 'overweight' and 'obese' just as relative as any other body size categories (Kwan, 2009; Saguy and Riley, 2005).

13.3.2 What plus size women in the fashion world are really saying about body size

The dynamic between the social, cultural, and medical processes of determining body size leads many women to ask the question: if there are more overweight and obese (or curvy, thick, fat, or voluptuous) Black and Latino women in the US than any other group, why would they not be depicted as 'models' for plus size clothing? Considering the fashion world's aesthetic (de)valuing of overweight and obese Black women and Latinas, it appears that both groups are marginalized from the very arena in which they would seem to have the most contributions. However, their absence from the mainstream media does not exclude their opinions on the matter.

Acting as a subsidiary distribution channel or what Becker would label as 'adventurous entrepreneurs and audiences', plus size designers, models, show promoters, photographers, and consumers have gathered together (some very purposefully) to comprise what has become known as the plus size blogosphere. In the plus size blogosphere, 'fatshionistas' collaborate on and exchange ideas about a bevy of topics including, but not limited to fashion, body size discrimination, health, and modeling opportunities. Many bloggers have inside connections to plus size designers, who may showcase items on their websites, while others simply provide commentary on the day-to-day musings of what it means to be a plus size woman. Many of these bloggers are Black women, but the plus size blogosphere continues to grow exponentially and covers topics directed to women from different racial, ethnic and socioeconomic backgrounds.

Gabi Gregg, founder of *GABIFRESH*, stated that she started the blog as a 'premiere fashion blog for trendy plus size women', that was 'dedicated to girls like me; the ones who are so in love with an industry that is, well, so not in love with us'. She is an ambassador for the Eloquii brand and has debuted many company products on her website. Although she labels herself a stylist, the former New Yorker and current Chicagoan has been featured in *Plus Model Magazine*, founded by former Latina model Madeline Figueroa-Jones. On *GABIFRESH*, Gregg unequivocally thanked the editors of *Plus Model Magazine* for honoring 'true' plus size women like herself (she is currently a size 18) rather than choosing a slimmer industry standard size 12. Clearly aware of plus size discrimination from the fashion world, Gregg's 'just like me' aligns with what many plus size bloggers have to say about the state of fashion: plus size women simply are not being represented by models who look like them.

Across the blogosphere, many models, bloggers, and photographers comment on what the 'ideal' plus size model should look like:

> There's a difference between being plus-sized and being a plus-sized model. From what I can tell it's the same height requirements as a regular model, height and weight proportoniate [*sic*] and usually a size 12. I get e-mails from plus-sized 'models' that are 5'2" amd [*sic*] weight [*sic*] 250#. Not exactly model material. (Photographer, Modelmayhem.com)

The photographer's statement is one of the central reasons why the plus size blogosphere was established. Recognizing the struggle for the fashion world to accept silhouettes other than the 'conventional' plus size aesthetic, many bloggers use their websites to create opportunities for plus size models and women alike. Jessica Kane (2011) of *Fatshion Chic* utilizes her business and 'fatshion' acumen to (re)build confidence in plus size women as they unapologetically love who they are. In 2006, she started *SKORCH Magazine*, which focuses on 'real women with real style.' She has completed the nationwide Full Body Photo Project (with corporate sponsorship from Lane Bryant, IGIGI, Monif C. and others) that gave away over $2300 in prizes to a vast majority of Black women.

Another advocate for fat fashionistas, or fatshionistas, is the highly opinionated Bella Styles. Based in Virginia, she is a fashion stylist, event planner, plus size model, fashion editor for Kane's *SKORCH* magazine, and Director of the Plus Academy VIRGINIA (a training and development program exclusively for plus size women). She publishes a blog similar to *Fatshion Chic*, but she does not hesitate to engage with audience members about size discrimination. Karl Lagerfeld's recent comments about Grammy-recording artist Adele being 'a bit fat' put the blogger up in arms:

> Having a 'smaller' frame isn't for everyone. That's what pisses me off with society. We feel that you have to be below a size 6 to be 'accepted' in the world. That's what's been brain washed into our heads for years. Many times producers and industry people tell entertainers that they 'must' lose weight or be a certain size in order for records to sell. Ummm, I think last night Adele proved them wrong. (Styles, 2012)

Kane's *SKORCH* and Bella's assertion that societal expectations have deleterious effects on people's social and psychological development position the two 'adventurous entrepreneurs' outside of the fashion world, but situate them as creators of 'subsidiary distribution channels' that one day may shift the aesthetic values associated with plus size models by redefining plus size silhouettes and offering products that legitimize this newer negotiation of beauty.

Gabi Gregg is also considered a key player in the plus size blogosphere, but the recent controversy surrounding her depiction in a 'fatkini' on her website sparked a difference of opinions (Gregg, 2012a,b; Huffington, 2012; Sieczkowski, 2012). On *GABIFRESH*, Gregg received hundreds of comments from supporters who commended her for her courage to pose in a two-piece, and encouraged her to continue in the crusade for curvy acceptance. Her photograph gained momentum and inspired thirty other women, ranging from size 12 to 32, to share their pictures in a gallery featured on the feminist website, *XOJane*. There, the feature and Gregg faced much more scrutiny from both plus size and non-plus size people. Some plus size female commentators mentioned that Gregg, at a size 18, had no stretch marks and cellulite, was well proportioned and could 'get away' with wearing a bikini. Other non-plus size people reprimanded her for celebrating obesity and ignoring one of the US' most prevalent health concerns. There were also some comments that alluded to Gregg's 'smooth' skin and skin color. Gregg appeared on the *Today* show and answered concerns raised from the website. She professed that she is actually a supporter of health and was not promoting obesity. On the contrary, she wanted to promote a healthy self-esteem for women and girls who should be happy with the body that they 'are in right now'. She openly admitted to having cellulite and stretch marks, and that they simply were not visible in the photographs (Huffington, 2012). Although Gregg was able to address some of the concerns raised on the website, there still remains some truth to the observations. Gregg is proportionate when compared to other bloggers in the blogosphere, and she also has fair skin and curly hair, two phenotypical characteristics that the fashion world seem to favor over darker skin and more tightly coiled hair (Edmonds, 2010). However, a 'preference' for lighter skin and straight(er) hair is surprisingly under-discussed on most of the blogs that promote size acceptance and reprimand fat discrimination. This absence shows that some blogosphere members hold different opinions with regard to the intersections of race, racial discrimination and body size.

One online writer, Tasha Fierce (2010), strongly lays a foundation for why there is such a scarcity of plus size women of color in the fashion world. She states that while many people conflate fatness and blackness, she is confused as to why mainstream publications that publish plus size work continue to exclude Black women and other women of color. Despite the existence of many Black plus size models, an enormous base of Black plus size consumers, and networks within the fat fashion blogosphere being spearheaded by Black women, Fierce says that the face of 'acceptable curvy is white'. Her assertion clearly displays the boundaries between what is and is not acceptable, or conventional, in the fashion world, but also how these boundaries have been reproduced and modified in the plus size blogosphere world as well.

Fierce proceeds by explaining that there is 'definitely a line between acceptable fat and unacceptable fat' (Fierce, 2010). She indicates that those fat women who are 'fortunate enough' to be considered thick (or proportionate), especially with fat in specific places such as the posterior, breasts and hips, may be afforded more opportunities to work in the fashion (and music) industry, but could succumb to hypersexualization in both industries. Fierce is speaking specifically about their potential objectification as 'Sapphire' or 'Jezebel', two *controlling images* that Patricia Hill Collins (2000) would assert permeate the global imagination of who Black women and Latinas are.

Fierce goes on to say that Black 'thick fat' is different from Black 'just fat' and can lead to the asexual stereotype of 'mammy', another controlling image who Collins (2000, p. 84) describes as dark, asexual, and corpulent, all characteristics marked 'unattractive' to European men. If those in the fashion world continue to prefer and praise physical characteristics closest to European beauty standards, then people who look like 'mammy' are relegated to a dumping ground for what is considered unattractive in the entire fashion world. Fierce's commentary raises questions about race, beauty, and sexuality in ways that other bloggers only discuss superficially or not at all. In a perceived 'post-racial' society, she unearths a possible truth about why plus size women of color are indeed still struggling to gain notoriety in the fashion world, and shows how much more social and economic work must be done by the plus size fashion blogosphere and its networks, especially with regard to race.

13.3.3 Latinas and the plus size industry

Latina bloggers, such as Monique Frausto, Olga Gonzalez Ramos, and Farrah Estrella, acknowledge the large amount of work being done, especially by Black and Latina plus size models, but make it a point to indicate that there are areas that still need improvement. Frausto, one of the first Latina bloggers in the blogosphere, set up a directory for Latina bloggers worldwide and encourages them to share their thoughts on fashion, health, parenting, politics and relationships. She describes herself as a 'fat fashion blogger' and on her personal blog, *Curves and Chaos*, openly admits, 'Size discrimination, doesn't just end with straight sizes either. The plus size fashion world is guilty of this too' (Frausto, 2011). Cognizant of a racialized history of fashion, she and other Latina bloggers disrupt the Black-white binary of fashion and motivate other ethnicities to participate as well (Belsha, 2012).

Latina plus size models are also some of the most prominent faces in plus size modeling (Anna, 2012; Black, Latina and Fabulous, 2012; Frausto, 2012; MadeinSpanglish, 2012). As discussed earlier, characteristics such as lighter skin and straight or loosely curled hair are closer to European beauty standards and are favorable in the fashion world. However, Afro-Latinas, as women of Indigenous, African and European descent (much like African American women) offer a spectrum of skin tones, facial features and hair types, and also face much of the same skin tone and hair discrimination as other darker-skinned models. Latina plus size models recognize that although they have made great strides in fashion, they still represent a small portion of the fashion world in general (Jones, 2007).

Overall, both Latinas and Black women know that there is a need for more women of color in the modeling sector of plus size fashion, and recognize the discrepancy for opportunities between women who resemble a European aesthetic and those who do not. Recurring themes of 'true' plus size silhouettes, size discrimination and lack of professional opportunities are highlighted throughout each of their blogs. Some women are more vocal about racial discrimination than others, while others avoid the topic altogether. Nonetheless, these women, working as entrepreneurs and serving as fat enthusiasts, have created a network of strong plus size voices that have in turn created an unyielding space dedicated to the lives and experiences of plus size women.

13.4 Plus size clothing designers who hire Black and Latina models

People connected to the fashion world may claim that being 'exceptional' is characteristic for anyone involved in fashion: exceptional designers, exceptional beauty, and an exceptional eye for fashion. However, as Becker has shown, these exceptions are made conventional in the fashion world, especially in high fashion where only a select few enjoy the benefits of inclusion (Ash and Wilson, 1993). If being 'exceptional' is conventional in the fashion world, then the meaning of exceptional can inevitably be redefined. Just as those in the fashion world have set and maintained parameters for boundaries in society, those outside of the world push back against those boundaries, upsetting that particular social order (Manlow, 2007).

Plus size designers and clothing companies provide opportunities for plus size models and women to appear fashionable, and thus beautiful, by wearing their products. American designers such as Michael Kors, Donna Karan, and Ralph Lauren produce highly fashionable plus size clothing, but they rarely, if ever, employ a considerable number of Black or Latina plus size models to market their clothing (as of today, Ralph Lauren only features two models on its plus size section, one white and one Black, while Michael Kors' website has straight-size models modeling all clothing). This section will focus on those companies that predominantly hire Black and Latina plus size models and those that pay careful attention to differences in silhouettes.

IGIGI by Yuliya Raquel, Lane Bryant, and Monif C. are popular companies that produce clothing that is similar to what a straight-size model would wear, but considers the forms, contours and silhouettes of plus size bodies. Currently, they are the leading companies hiring Black and Latina plus size models for both marketing campaigns and runway shows. However, a majority still fail to depict a variety of silhouettes.

IGIGI by Yuliya Raquel, based in San Francisco, features one Latina and one Black plus size model for their marketing campaign, but they have an abundance of white and 'racially ambiguous' models for marketing. Even so, IGIGI by Yuliya Raquel makes it clear to customers, through its core values and principles published on its website and Facebook page, that the company is designing clothes for 'full-figured fashionistas' who will help 'transform the world's view of beauty' by wearing 'fashion-forward, form-fitting, true-fitting, high-quality plus-size clothing'. The company also asserts that its clothing is made in the US in limited quantities. While IGIGI by Yuliya Raquel may design clothing for silhouettes outside the conventional fashion world, it continues to leave Latina and Black models outside of its own conventions.

Lane Bryant, one of the first and most popular plus size fashion clothing companies in the US, started out clothing expectant mothers and 'stocky' women. Now, the plus size powerhouse makes and sells clothing for plus size women all over the world. Moderately priced, the clothing is mass-produced for women of American sizes 14 through 26. The company has been a staple of plus size American women, and has sponsored clothing for many plus size events throughout the plus size blogosphere as well as film and television. Entertainers such as Queen Latifah, Jennifer Hudson, and Mo'Nique have worn (and applauded) Lane Bryant for providing fashionable clothing. However, now that these women have gained fame in their respective industries, it seems all of them have lost weight since their debut and tend to wear more high fashion (and not necessarily plus size) designers. Currently, the Lane Bryant website features the most variety of models of color, but their 2012 runway show during Full Figure Fashion Week featured an overwhelmingly white majority.

Ashley Stewart, founded in Brooklyn, New York by Joseph Sitt of Thor Equities, began to clothe plus size Black women in urban areas during the early 1990s. Sitt, the real estate tycoon who also owned and managed The Children's Place and Marianne Stores (a retail outlet serving Latinas), fervently believed that consumer power in urban areas was ignored, and so he took on several real estate ventures to persuade big companies to invest in his retail properties (Issenberg, 2006). Ashley Stewart, named after Laura Ashley and Martha Stewart, initially hired women from the community to work in its stores and was also a strong employer in the US Welfare to Work program (Owen, 2008). Sitt sold his share of Ashley Stewart in 2000, but the chain continues to feature a majority of Black women in both its print and runway campaigns.

Monif C., a New York-based popular design company consisting of a Barbadian mother and daughter duo, started after a trip to Europe that frustrated the daughter of the duo. Monif Clarke, like many plus size fashion designers and bloggers, was unpleased with the plus size market and began designing her own line of clothing. The models that are used on the Monif C. website vary racially and by silhouette, with a preference given to 'truer' plus size women and models of African descent. Monif C.'s 'Marilyn' dress has been worn by many celebrities, including Emme, Sherri Shepherd, Jill Scott, Jennifer Hudson, Gabourey Sidibe, Kim Coles, Raven Symone, Angie Stone, Tocarra and Amber Riley.

Many of the plus size clothing companies dedicate their designs to 'the plus size woman', but they seem to highlight proportionate silhouettes in their marketing and runway campaigns. Are they doing so to align with fashion world conventions? Or has the idea of varied silhouettes simply not yet been introduced to them? Companies such as Lane Bryant and Ashley Stewart appear to engage with consumers through community events and plus size organizations, while other companies, such as Michael Kors and Ralph Lauren, seem to only reap the profits from tapping into the plus size market share. Either way, plus size fashion companies have recognized the buying power of plus size women. What remains is the reality that the fashion world has yet to acknowledge all of the contributions that plus size women of African descent have had on fashion all over the globe.

13.5 Conclusion and future trends

There have been many women of African descent who set up their own businesses in order to meet the needs of their communities (Alexander, 1982; Gill, 2001), and plus size models, designers and bloggers of color have picked up where these women left off. Not allowing their needs to lay stagnant in the hands of big companies, these women are carving their way into the fashion world one outfit at a time. Some, like Gabi Gregg, work with large clothing companies and magazines to promote size acceptance and provide opportunities for plus size models. Others, like Madeline Figueroa-Jones, created her magazine because she felt the void for plus size Latinas was simply too large. The plus size blogosphere is expanding faster than ever, and as more 'fatshionistas' of color and a variety of sizes begin to publish their experiences living in a narrow fashion world, I believe the fashion world will begin to pay increased attention to the group.

While the fashion world may be slow to include plus size women, the market for plus size fashion is more dynamic. Some stores flourish, such as Hot Topic's Torrid, which has been offering fashionable clothing to plus size teenagers and young adult women for more than a decade. Torrid employs 'unambiguously large models' and has seen growth and profit, proven by their expansion to over 170 stores across the US (Bellafante, 2010). Also experiencing success are popular mega-retailers such as Target that offer small, but fashion-forward, plus size sections for their plus size customers. Currently, Target, like The Gap and Old Navy, are moving toward offering most, if not all, of their plus size selections solely online. On the contrary, the Limited's Eloquii brand is closing both its online and retail store doors to customers, after just one short year of business. A large majority of Eloquii's patrons, mostly women looking for more sophisticated styles they can wear to work, have been devastated by its closing, citing the company's lack of advertising and presence in the plus size blogosphere as the main reasons for its demise (Denee, 2013). This is more evidence that the plus size blogosphere has a significant influence on the fashion industry.

Aside from retailers, there are productions of plus size fashion shows emerging in cities and towns throughout the country. They feature a diverse group of women from various backgrounds and show that high fashion and plus size fashion have a lot in common. Future trends will most likely involve an increase in sales for plus size retail, but hopefully also more inclusion of plus size models with unconventional silhouettes. Many plus size fashion shows, especially those produced by Full Figure Fashion Week founder Gwen DeVoe and her associates, have made it a priority to include a diverse number of women based on race and silhouette. Both designers and bloggers attend these events, which often designate a period of time for them to network with the audience and models. If representatives from large clothing companies make it a priority to attend these fashion shows, then they can urge their companies to realistically represent the women with whom they come into contact. The plus size fashion world will inevitably become larger than it is now, and it would behoove all of the fashion world to take notice and take the opportunity to recognize the contributions of plus size women.

Because there is a limited amount of literature published on plus size fashion, I utilized social media to obtain most of my information. The plus size blogosphere is vast, so it would be wise to begin with some of the blogs mentioned in the 'Causality and reality' section to direct your interests. *GABIFRESH*, *Curves and Chaos*, and *The Curvy Fashionista* are some of the most popular blogs to visit. Currently, social media dominates the marketing and communications industry, so many plus size commentators, designers, models, and boutiques have created accounts on sites such as Facebook and Twitter. Companies such as Eddy & Bri, Curvy Closets, and Elite Plus, Inc., share information about castings, fashion shows and shopping with the entire Facebook community. International clothing companies have also taken advantage of social media by setting up an online presence through Facebook, Twitter, and their own personal websites. Searching through any of these sources will give the reader a glimpse into the plus size fashion world, particularly through the lens of Black women and Latinas.

13.6 References

Alexander, L. (1982) *Blacks in the history of fashion*. New York, Harlem Institute of Fashion.

Anna, C. (2012) Plus size fashion: the rise of the full figured Latina model, Los Angeles, Curvy Market Blog. Available from: http://curvymarket.com/blog/2012/08/22/plus-size-fashion-the-rise-of-the-full-figured-latina-model/ [Accessed 30 August 2012].

Ash, J. and Wilson, E. (1993) *Chic thrills: a fashion reader*. Berkeley, University of California Press.

Bates, L. M., Acevedo-Garcia, D., Alegría, M. and Kreiger, N. (2008) 'Immigration and generational trends in body mass index and obesity in the United States: results of the national Latino and Asian American survey, 2002–2003', *Am J Pub Health*, **98**, 70–77.

Beauboeuf-Lafontant, T. (2003) 'Strong and large Black women? exploring relationships between deviant womanhood and weight', *Gender Soc*, **17**, 111–21.

Becker, H. (1982) *Art worlds*. Berkeley, University of California Press.

Bellafante. G. (2010) Plus-size wars, New York, New York Times. Available from: http://www.nytimes.com/2010/08/01/magazine/01plussize-t.html?pagewanted=all&_r=0 [Accessed 1 May 2013].

Belsha, K. (2012) Blog by Latinas connects women around the world, Miami, Voxxi. Available from: http://www.voxxi.com/blogs-by-latinas-connects-women/ [Accessed 2 September 2012].

Black, Latina and Fabulous. (2012) Out & about; more plus-size fashion week fun! Available from: http://blacklatinafabulous.wordpress.com/2012/06/18/out-about-more-plus-size-fashion-week-fun/ [Accessed 26 September 2012].

Cohen, S., Doyle, W., and Baum, A. (2006) 'Socioeconomic status is associated with stress hormones', *Psychosomatic Med*, **68**, 414–20.

Collins, P.H. (2000) *Black feminist thought: knowledge, consciousness, and the politics of empowerment*. 2nd ed., New York, Routledge.

Denee, M. (2013) An open letter to the Limited brands and Eloquii, Los Angeles, The Curvy Fashionista. Available from: http://thecurvyfashionista.mariedenee.com/2013/03/an-open-letter-to-the-limited-brands-and-eloquii/ [Accessed 1 May 2013].

Edmonds, A. (2010) *Pretty modern: beauty, sex and plastic surgery in Brazil*, Durham, Duke University Press.

Epel, E., Lapidus, R., McEwen, B. and Brownell, K. (2001) 'Stress may add bite to appetite in women; a laboratory study of stress-induced cortisol and eating behavior', *Psychoneuroendocrinology*, **26**, 37–49.

Fierce, T. (2010) Where my sistas at? the underrepresentation of Black plus size models in mainstream fashion, Los Angeles, Tasha Fierce. Available from: http://www.sexandthefatgirl.com/2010/04/19/where-my-sistas-at-the-underrepresentation-of-black-plus-size-models-in-mainstream-fashion/ [Accessed 12 February 2012].

Frausto, M. (2011) Why size discrimination? New York, Curves and chaos. Available from: http://curvesandchaos.com/2011/07/why-size-discrimination.html [Accessed 2 September 2012].

Frausto, M. (2012) Huff post Latinovoices, Latinas emerge in the plus size modeling industry, New York, TheHuffingtonPost.com, Inc. Available from: http://www.huffingtonpost.com/monique-frausto/plus-sized-latina-model_b_1639124.html [Accessed 5 July 2012].

Gill, T. M. (2001), ' 'I had my own business…so I didn't have to worry': beauty salons, beauty culturalists, and the politics of African-American female entrepreneurship', in Scranton, P., ed. *Beauty and business: commerce, gender, and culture in modern America,* New York, Routledge.

Gregg, G. (2012a) Fatkini 2012, New York, Gabifresh. Available from: http://www.gabifresh.com/2012_04_01_archive.html [Accessed 18 December 2011].

Gregg, G. (2012b) The XOJane and Gabi Fresh fatkini gallery: 31 hot sexy fat girls in skimpy swimwear, New York, SAY Media. Available from http://www.xojane.com/fun/gallery/fatkini [Accessed 18 June 2012].

Hajnal, Z. L. (2007) 'Black class exceptionalism: insights from direct democracy on the race versus class debate', *Pub Op Qtrly*, **71**, 560–87.

Hrabosky, J. I., and Grilo, C. M. (2007) 'Body image and eating disordered behavior in a community sample of Black and Hispanic women', *Eat Behav*, **8**, 106–114.

Huffington Post. (2012) Huff post woman, Gabi Gregg, blogger who started 'fatkini' photos, talks body image on the 'Today' show, New York, TheHuffingtonPost.com, Inc. Available from: http://www.huffingtonpost.com/2012/05/30/gabi-gregg-fatkini-today-show_n_1556480.html [Accessed 18 June 2012].

Issenberg, S. (2006) How I did it: Joe Sitt, Thor Equities, New York, Inc Magazine. Available from: http://www.inc.com/magazine/20060101/qa-sitt.html [Accessed 9 May 2013].

Jones, M. F. (2007) Plus Model Magazine, An Interview With Larissa Laurel, the New PLUS Industry Editor!, Vancouver, V.I.E. Available from: http://www.plusmodelmag.com/General/plus-model-magazine-article-detail.asp?article-id=104110667 [Accessed 18 June 2012].

Kane, J. (2011) Full Body Photo Project, Vancouver, *Life & Style of Jessica*. Available from: http://fatshionchic.blogspot.com [Accessed 12 February 2012].

Kwan, S. (2009) 'Framing the fat body: contested meanings between government, activists, and Industry', *Soc Inq*, **79**, 25–50.

Leavitt, S. (1992) People Archive: Like mother, like…New York, Time, Inc. Available from: http://www.people.com/people/archive/article/0,,20113286,00.html [Accessed 18 June 2012]

Lovejoy, M. (2001) 'Disturbances in the social body: differences in body image and eating problems among African American and white women', *Gender Soc*, **15**, 239–61.

MadeInSpanglish. (2012) Shine Latina, Latinas are the top names in plus-size fashion, Sunnyvale, Yahoo! Inc. Available from: http://shine.yahoo.com/latina/latinas-top-names-plus-size-fashion-154600906.html [Accessed 10 September 2012].

Manlow, V. (2007) *Designing clothes: culture and organization of the fashion industry*, New Brunswick, Transaction.

Owen, J. (2008) An eye for opportunity, New York, SternBusiness, the alumni magazine of NYU Stern. Available from: http://w4.stern.nyu.edu/sternbusiness/spring_2008/sternInCity.html [Accessed 9 May 2013].

Parker, S., Nichter, M., Nichter, M., Vuckovic, N., Sims, C. and Ritenbaugh, C. (1995) 'Body image and weight concerns among African American and white adolescent females: differences that make a difference', *Hum Organ*, **54**, 103–14.

Saguy, A. and Riley, K. (2005) 'Weighing both sides: morality, mortality, and framing contests over obesity', *J Health Politics Policy Law*, **30**, 869–921.

Sieczkowski, C. (2012) Gabi Fresh's 'fatkini' gallery on XOJane receives overwhelming applause, New York, The International Business Times, Inc. Available from: http://www.ibtimes.com/gabi-freshs-fatkini-gallery-xojane-receives-overwhelming-applause-photos-700997 [Accessed 18 June 2012].

Styles, B. (2012) Singer Adele Proves That She's Not 'Too Fat' To Win 6 Grammy's, Now Take That Karl Lagerfield, Richmond, *Trendy Curves by Bella Styles*. Available from: http://bellastyles.com [Accessed 16 February 2012].

Swope, T. (2010) Anansa Sims and David Patterson, Columbus, Columbus Media Enterprises. Available from: http://www.columbusmonthly.com/October-2010/Wedding-album-Anansa-Sims-David-Patterson/ [Accessed 15 September 2012].

Thompson, B. W. (1992) ' 'A way outa no way': eating problems among African-American, Latina, and white women', *Gender Soc*, **6**, 546–61.

Turner, R. J., and Avison, W. (2003) 'Status variations in stress exposure: implications for the interpretation of research on race, socioeconomic status, and gender', *J Health Soc Beh*, **44**, 488–505.

Yancey, A. K., Leslie, J. and Abel, E. K. (2006) 'Obesity at the Crossroads: Feminist and Public Health Perspectives', *Signs* **31**, 425–43.

Zikalala, F. (1994) 'Body size and culture', *Agenda*, **23**, 19–22.

14

Intimate apparel: designing intimate apparel to fit different body shapes

K. SHIN, The Hong Kong Polytechnic University, China

DOI: 10.1533/9781782422150.2.273

Abstract: Finding good fitting intimate apparel, in particular the bra, has been an issue for consumers and the industry for many years. It is apparent that achieving a good fit in bra design requires more than a repeat of current fitting and alteration procedures. Therefore, this chapter reviews a wide range of research work including bra-fitting criteria, bra sizing, breast volume, bra-fit evaluation, smart materials and production technologies. With consumer bra-fit education, standardisation of a bra sizing system, implementation of 3D body scanning technology, and the utilisation of smart materials and technology, the currently complex process of achieving a good fitting bra certainly needs to be simplified.

Key words: bra fit, breast volume, sizing, fit evaluation, fit education.

14.1 Introduction

Clothing fit is one of the most influential factors impacting on a consumer's buying decision (Chen *et al.*, 2011) although the notion of fit has never been clearly understood, largely due to its complexity (Yu, 2004). Finding good fitting intimate apparel, in particular the bra, has been an issue for consumers and the industry for many years and it is repeatedly reported that more than 70% of women are wearing the wrong size bra (Boyes, 1996; Greenbaum *et al.*, 2003; Lipton, 1996; Young *et al.*, 1994; Wood *et al.*, 2008). The fit of the bra is particularly important because not only does it influence underwear sales but also it can affect the wearer's health and well-being. An ill-fitting bra can cause discomfort, fatigue or pain (Chen *et al.*, 2010) so designing a bra which fits well and supports the breasts should be a priority for the industry.

14.1.1 Breast health and bra

Women's breast health has become a significant global health concern due to the widespread recognition and incidence of breast cancer, with about 1.5 million people being diagnosed with breast cancer in 2012 (worldwidebreastcancer.com, 2012). Consequently, a few researchers have tried to establish a link between bra wearing practices and breast health. Singer and Grismaijer (1995) concluded, after interviewing 4730 women (the cancer group of 2056 and the standard group of 2674), that women who wear bras for 24 hours per day are 125 times more likely to develop breast cancer than women who do not wear bras at all. In 2000, a study involving more than 100 participants found a link between wearing a bra and breast pain and the development of cysts. The incidence of pain fell by 7% after the women stopped wearing bras for three months (BBC, 2000). Lee *et al.* (2005) found that wearing a bra for over 12 hours per day, and wearing a bra whilst sleeping, could also increase breast cancer risk. Johnson and Barnes (2008) suggested that the bra provides an ideal environment for breast cancer cell growth by restricting the flow of oxygen, nutrients, and toxin-removing lymphatic fluids to the tissues, while also raising breast temperature. Although there has been a sceptical response from the medical community due to the lack of significant scientific evidence, the idea of the bra as a major risk factor associated with breast diseases will probably grow.

14.1.2 Bra-fit elements

Garment fit depends on five elements: grain, set, line, balance and ease (Brown and Rice, 2000). These five elements are normally used for outer-wear fit evaluation where the laws of gravity impact upon the whole garment. The bra is often required to defy gravity whilst creating an aesthetically pleasing appearance (which can be called 'good-fit'). As a result, the 'tension' exerted by negative ease (the bra band is generally 10–15 cm shorter than the actual ribcage circumference) becomes one of the most important fit elements for the bra. Achieving the optimum tension resulting in a good fit is relatively difficult, because of the different material properties used for modern bra design (Shin, 2007). This difficulty is reflected in the average bra pattern development time, which is anything from two to six months, and in the average training time for a professional bra patternmaker to be capable of producing a good-fit bra, which is anything up to 10 years (Hardaker and Fozzard, 1997). However, it is also recognized that training the consumer is equally as important as training the professional patternmaker. For example, McGhee and Steele (2010) and White and Scurr (2012) emphasized the importance of consumer education in assessing bra fit by using professional bra-fitting criteria rather than relying on the traditional bra sizing methods widely used in the market.

14.1.3 Professional bra-fitting criteria

Bra-fitting results depend on the provision of balanced support evenly exerted by the following bra components; the centrepiece, band, cup, shoulder

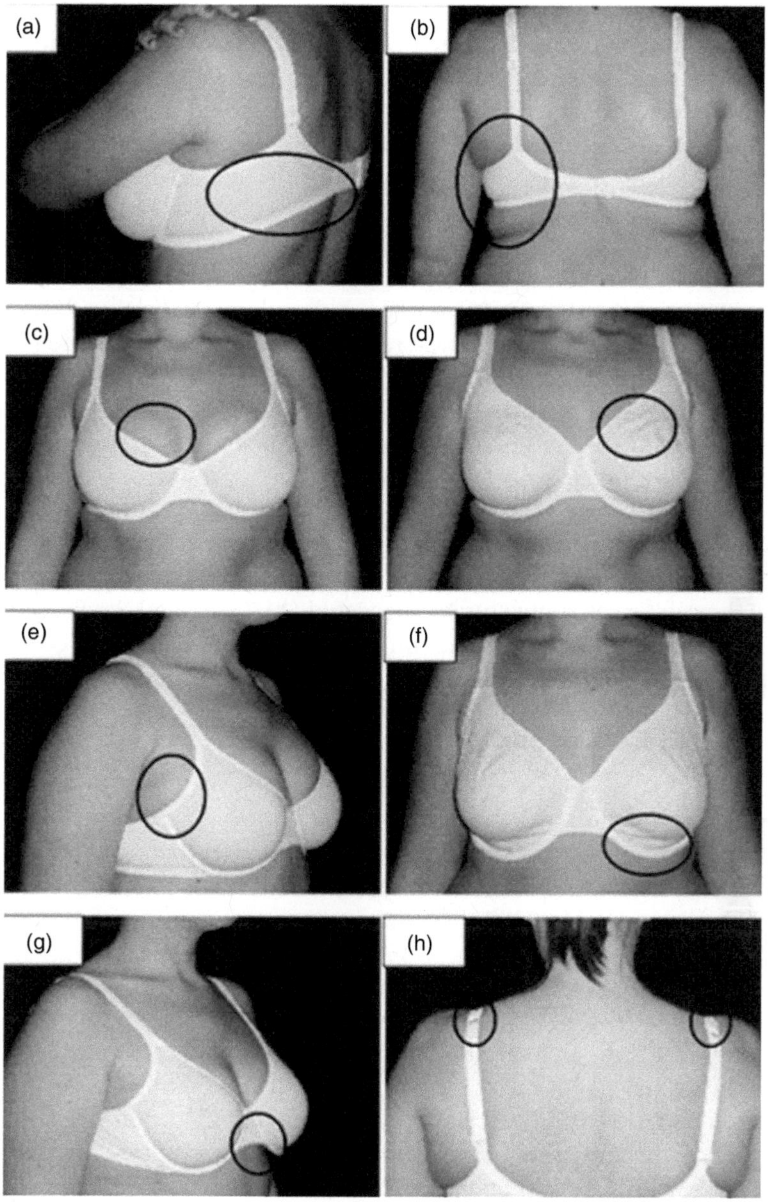

14.1 Examples of ill-fitting bras: (a) band not level; (b) band too tight; (c) vup too small; (d) vup too baggy; (e) underwire resting on breast tissue; (f) underwire resting too low; (g) front of bra pulling away from the chest; (h) shoulder straps digging in. (*Source*: White and Scurr, 2012.)

strap and underwire. A rule of thumb for bra fitting is that every component should exert a just about 'right' or sufficient amount of tension and no more. This can be elaborated as follows: (1) the centre piece of the bra should just press against the sternum, (2) the bra band should not be too tight or too loose, (3) the cup should not be too small or too big, (4) the shoulder straps should not be too tight or too loose, (5) the underwire should follow the breast root instead of pressing against the chest wall or breasts (McGhee and Steele, 2010). Figure 14.1 shows some examples of ill-fitting bras (White and Scurr, 2012). If the band is not sitting horizontally to the front and rises up (a), the band is too loose. If there is a bulge along the bra band (b), the band is too tight. If there is a double-breasts effect along the neckline (c), the cup is too small. If there are wrinkles on the cup (d), the cup is too big. If the underwire is sitting on the breast (e), the cup is too small. If the underwire is sitting outside of the inflammatory fold (breast root) (f), the cup is too big. If the centre piece (g) is lifted away from the sternum, either the cup is too small or the centre piece is too wide for the sternum. If the shoulder straps are digging in (h), the shoulder straps are too tight.

Although it is ideal to educate consumers on how to find a good-fit bra by themselves as some studies suggested (McGhee and Steele, 2010; White and Scurr, 2012), the fundamental issue which cannot be ignored is the need for a new sizing system for industry and individual consumers which can replace the traditional bra sizing system.

14.1.4 Bra sizing system and fit

Although the traditional bra sizing system covers a considerably wide size range from 28AA to 56FF, it is often criticised for its lack of clarity (Pechter, 1998; Zheng *et al.*, 2007), inadequate approximations of breast volume (Nethero, 2008; Zheng *et al.*, 2007) and the lack of allowance for the degree of firmness or consistency of the breast tissue (Nethero, 2008). This is partly because the current traditional bra sizing system is too simple to determine a bra size for the full variety of breast sizes, shapes and densities.

The traditional bra sizing system currently used in the industry was introduced in the 1930s in the US (Farrell-Beck and Gau, 2002). Using this system, bra size consists of two components; numeric band size and alphabetical cup size, which are labelled in either the imperial system or metric system or both. For instance, 34B is equivalent to 75B in the metric system. The band size for the imperial system needs to be calculated by adding 4–5 inches to the underbust girth measurement in order to make an even number, whilst the metric system uses the underbust girth measurement and rounds it up or down to the nearest five or ten (e.g., 75 cm or 80 cm) as the band size. Cup size is determined by the difference between the full bust girth and the band size. Because body measurements are rounded up or down in order to allocate the nearest band size, finding a correct bra size using the traditional bra sizing system is destined to fail.

This is supported by a study conducted by Chen *et al.* (2011). They calculated 103 participants' bra sizes, with three bra sizing calculation methods, in order to match with the actual bra size worn by their participants. The three calculation methods used for their study included the Victoria's Secret method (in other words, the traditional bra sizing system), the Washington State method (1986), and the Wright method (2002). Their results showed a zero percent (n=0) match between the calculated ideal bra size and the bra size the participants were actually wearing using the Victoria's Secret method (traditional bra sizing system) whilst a 3.9 percent (n=4) match was found with the Washington State method and a 9.7 percent (n=10) match with the Wright method. McGhee and Steele (2010) and White and Scurr (2012) concluded that the traditional bra sizing system underestimated the cup size (smaller cup) and overestimated the band size (larger band) for the majority of the participants in their studies.

14.2 Measuring breast volume

A breast measurement protocol was developed in the 1950s as a result of the development of surgical breast augmentation techniques. Interest in this area of body measurement increased in the 1950s, with the development and use of surgical techniques for breast enhancement and reduction. Plastic surgeons often use the bra as a guide to conduct an augmentation mammaplasty (Bates, 1995; Young *et al.*, 1994). However, because these surgeons are also focusing on actual breast volume differences (measured in cubic centimetres), they have increasingly tended to create and use their own breast measurement protocols in order to ensure the correct size change in volume, rather than using the two gross linear measurements of underbust and full bust girth. Subsequently, many breast volume assessment studies have been done including thermoplastic cast (Edsander-Nord *et al.*, 1996), mammography (Katariya *et al.*, 1974), MRI (Fowler *et al.*, 1990; Mineyev *et al.*, 1995), stereo-photogrammetry (Loughry *et al.*, 1987) and three-dimensional (3D) laser scanning (Kovacs *et al.*, 2006). Some of these studies have also emphasized the need to standardise the female breast measurement protocol.

14.2.1 Breast measurement protocols in a Western context

Linear body measurements have also been used to achieve increments or reductions in breast volume. After analysing 20 females whose breasts were evaluated as 'aesthetically' perfect, Penn (1955) identified an aesthetically perfect triangle which connects the sternal notch, right nipple and left nipple with each line at a length of 8.5" (21.6 cm), which could be used for breast reduction surgery. Penn's measurement values were widely used, but it was later found that this 'universal aesthetic triangle' could not be applied to many body shapes as there was little or no correlation between breast dimensions and the participant's height and weight (Westreich, 1997).

Smith *et al.* (1986) suggested a 'normal' value for breast volume and anthropomorphic measurements after studying a randomly selected sample of 55 women. Their linear measurement points are shown in Fig. 14.2. The normal values for breast volume and body measurements are shown in Table 14.1.

However, the validity of this study was also criticised as many of the participants showed significant volumetric asymmetry, and moderate degrees of ptosis. Westreich (1997) found a significant correlation between breast volume and four linear measurements – 'sternal notch-to-nipple', 'nipple-to-nipple', 'clavicle-to-nipple', and 'sternal notch-to-inframammary fold' – after studying 50 women with the so-called aesthetically perfect breast. He created a breast volume calculation formula (Fig. 14.3) for clinical application which makes use of only two of these measurements: 1) sternal notch-to-nipple distance (M-Ni) and 2) nipple-to-nipple distance (N-Ni).

Pechter (1998) developed a new breast measurement protocol using a mammary hemi-circumference as a key parameter in order to determine the cup size after examining the body measurements of 100 women. Participants' bra size was used as a measure of validity for this new method. With this

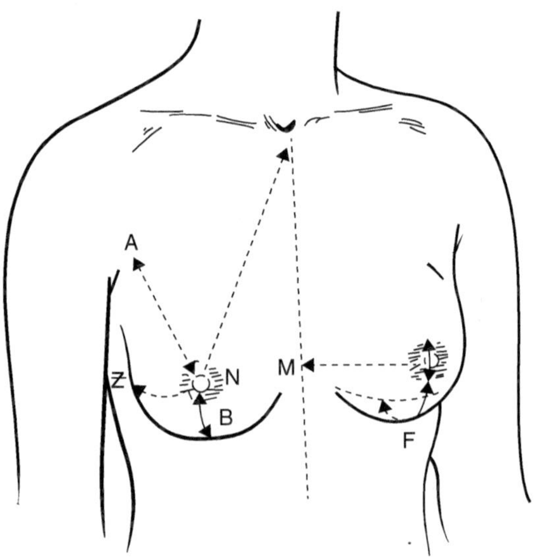

14.2 Linear measurement points. (*Source*: Smith *et al.*, 1986.)

Table 14.1 The normal values for breast volume and body measurements

Measurement	Right	Left	T	p
Z–N	9.53 + 2.30 cm	9.43 + 2.54 cm	0.84	0.402
A–N	12.70 + 2.74 cm	13.18 + 3.02 cm	–4.38	<0.001
N–M	11.14 + 1.44 cm	11.83 + 1.46 cm	–6.57	<0.001
F–N	7.44 + 1.86 cm	7.45 + 1.98 cm	–0.13	0.899
F–B	0.786 + 1.36 cm	0.922 + 1.36 cm	–2.90	0.005
Volume	273.46 + 172 cc	291.69 + 168 cc	–2.44	0.018 cc

Source: Smith *et al.* 1986.

$$\text{Volume} = (\text{M–Ni})^{1.103} \times (\text{N–Ni})^{0.811}$$

or

$$\log(\text{volume}) = [1.103 \times \log(\text{M–Ni})] + [0.811 \times \log(\text{N–Ni})]$$

14.3 Equation for breast volume. (*Source*: Westreich, 1997.)

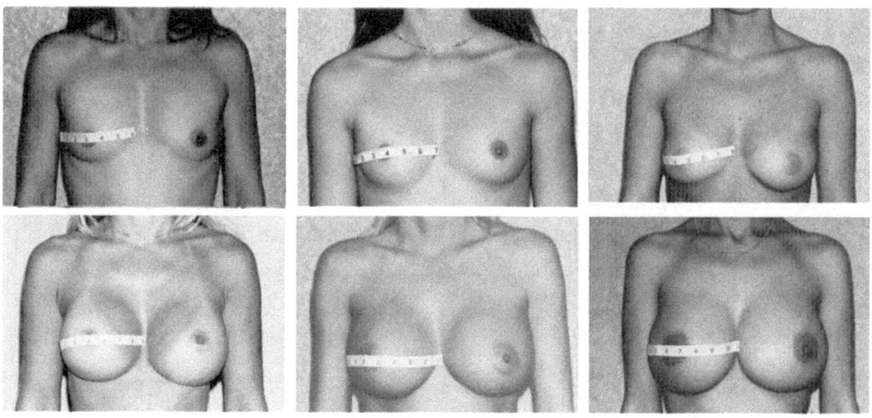

14.4 Pechter's breast cup sizing method. (*Source*: Pechter, 1998.)

method, a mammary hemi-circumference of six inches corresponds to the AA cup size, seven inches to A cup, eight inches to B cup, and so on (see Fig. 14.4). A breast implant of 300cc for each breast is required to achieve an increment of one cup size, or a one inch increment on the mammary hemi-circumference.

However, Pechter's new bra sizing method failed to define the bra band size (Kanhai and Hage, 1999), although he later identified a workable band size range as 34 to 38 (Pechter, 1999). Unfortunately, although this method is

(a) (b)

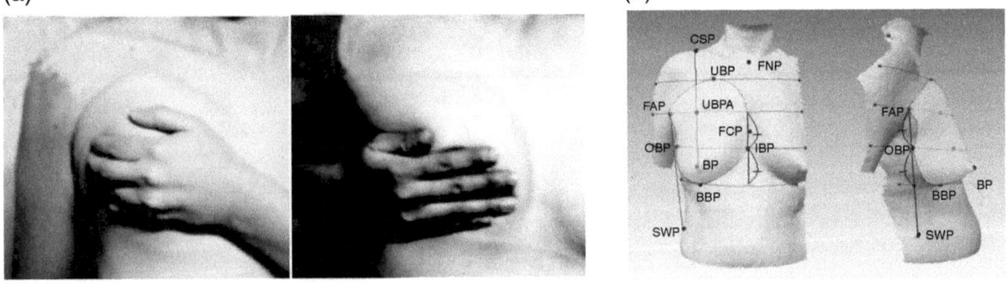

Method of defining the breast boundary — The breast boundary

14.5 Breast boundary defining method (a) and defined breast boundary (b). (*Source*: Lee *et al.*, 2004.)

simple and easy to use, it requires drastic changes in industry practice which would not necessarily result in higher sales.

14.2.2 Breast measurement protocols in an Asian context

Lee *et al.* (2004) proposed another protocol for 3D anthropometric breast parameters in order to achieve a reliable breast boundary and breast volume. This study involved measurements gathered from the naked 3D scanned body data of 37 middle-aged Korean women, whose bra size is 36A (80A), and they suggested a breast boundary defining method by pushing the breast up and inward (Fig. 14.5).

Using this protocol, having defined the breast boundary, an accurate measure of breast volume is determined by using the 3D images of the breast separated from the chest wall. In addition, the bottom breast volume and the bottom breast curve can be used to design a more comfortable and form-fitting bra. Wang and Zhang (2007) took this idea one step further and developed a breast model library based on 3D body scan data by using free form deformation technology (Fig. 14.6).

One particular advantage of these approaches is that the suggested body model parameters (Table 14.2) can be compared with the bra-fitting result. However, a broader range of sizes are required as this particular body model can only fit certain bra sizes; namely US 36AA, UK 36A or 80A, which only represents a small proportion of the global female population.

Zheng *et al.* (2007) suggested a new bra sizing protocol particularly for the Chinese female after analysing a sample the scanned 3D body data of 456 females. The key parameters used for this new sizing system are 'underbust girth' and 'breast depth width ratio'. Figure 14.7 shows the breast depth and breast width used in the calculation, and Table 14.3 shows the breast depth to width ratio which Zheng *et al.* (2007) claim can help determine bra cup size for Chinese women.

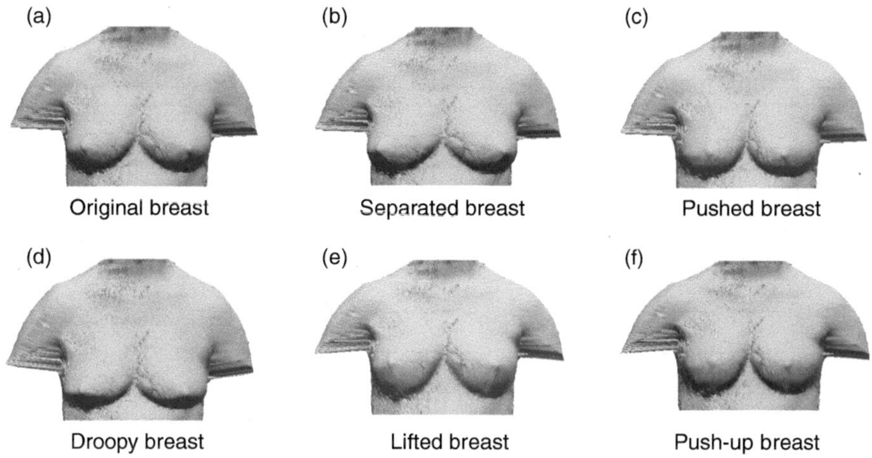

14.6 (a–f) Virtual breast shapes. (*Source*: Wang and Zhang, 2007.)

Table 14.2 Parameters of body model

Body dimensions (cm)	Bust	Underbust	Bust point distance	Bust point height	Underbust point height
(a) Original	92.1	78.6	23.5	108.5	102.7
(b) Separate	92.1	78.6	28.8	108.5	102.7
(c) Pushed	92.1	78.6	17.0	108.5	102.7
(d) Droopy	92.1	78.6	23.2	105.6	102.7
(e) Lifted	92.1	78.6	23.2	113.3	102.7
(f) Push-up	92.1	78.6	17.0	113.3	102.7

Source: Wang and Zhang, 2007.

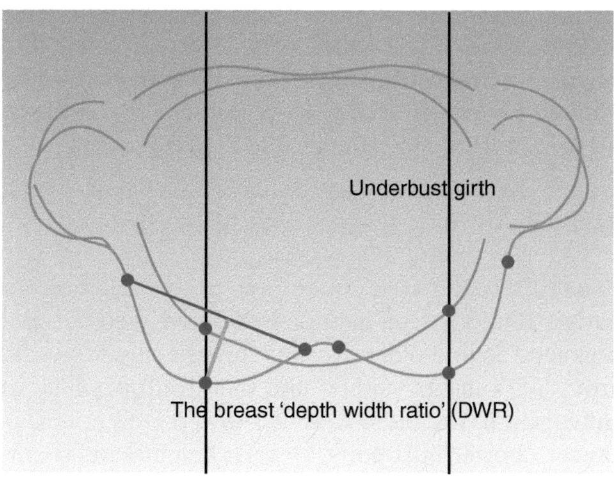

14.7 Two key control measurements for the new bra sizing system. (*Source*: Zheng *et al.*, 2007.)

Table 14.3 New bra sizing system for Chinese women

Cup size	AA	A	B	C	D	E	F	G
Breast depth/width ratio	0.25	0.30	0.34	0.38	0.42	0.46	0.50	0.55

Source: Zheng *et al.*, 2007.

Zheng *et al.* (2007) found that the bra size accommodation rate, or the degree of perfect fit, increased to 98.72% for this sample using their new sizing system, whilst the traditional bra sizing accommodation rate was 96.99%. In addition to the new bra sizing system, Zheng *et al.* (2007) also identified 103 body measurements which they suspected were relevant to achieving measurements which would accurately delineate the breast shape. They concluded that overall body build was positively correlated to underbust girth, which, in turn, was positively correlated with the participant's weight. However, this contradicts some previous studies (Katch *et al.*, 1980; Penn, 1955) which identified no correlation between height, weight and breast volume. Zheng *et al.*'s method complements the traditional bra sizing system despite the core idea being to move away from this approach. In practice, there is so far no evidence of any significant implementation of any of these new approaches led by the bra industry.

14.3 Designing intimate apparel to fit different body shapes

Designing a good fit bra starts from an understanding of the female body. Human beings are the only mammal where females of the species develop permanently enlarged breasts before the first pregnancy. Female breasts go through constant changes throughout the day as well as through their life cycle. Breasts' shape, density and volume are affected by weight changes, hormonal changes (e.g., menstrual cycle and menopause), reproduction (e.g., childbearing and nursing), and aging. These changes are expected to impact on changes in bra-fit requirements and fit satisfaction.

14.3.1 Age difference in breast configuration

Ashdown and Na (2008) compared upper torso postural changes caused by aging by comparing 3D body scan data of 40 women aged between 19 and 35 and 40 women aged 55 and older. Results showed significant differences between the group of younger women and the group of older women in their breast configuration. The older women showed greater bust point and bust point distances, shorter bust point to waist heights, and smaller upper anterior thoracic angles (a lower, fuller bust, and bust points were farther apart). These age-related physical changes often lead to older women reporting frustration over the lack of clothing choices which are in line with their changed fit requirements (Howarton and Lee, 2010). Risius *et al.* (2012) identified five key criteria which influence bra purchasing amongst older women – (1) aesthetics, (2) comfort, (3) practicalities for bra purchasing, (4) breast support and (5) psychological aspects – after interviewing 13 females aged between 45 and 65. Under the 'support' criteria, three requirements are for the bra to uplift, to give general support and to overcome breast sag.

14.3.2 Body shape and bra fit

Chen *et al.* (2010) analysed 3D scanned data and fit perception of 103 female college students in order to find out whether body-shape characteristics, including shoulder slope, bust prominence, back curvature and acromion placement (see Fig. 14.8) were significantly associated with bra-fit problems, including poor bra support and bra-motion restriction. Each body-shape characteristic was divided into three subcategories: (1) shoulder slope into 'square', 'average', and 'sloped' categories, (2) bust prominence into 'large', 'average', and 'small', (3) back curvature into 'round', 'average', and 'flat-back' categories, (4) acromion into 'backward', 'average', and 'forward' categories (see Fig. 14.8).

In their study, Chen *et al.* (2010) found the backward-acromion participants with large breasts were least satisfied with bra support, whilst the backward-acromion participants with average breasts were mostly satisfied with bra support. In addition, the result showed the round-back participants with the backward-acromion were most satisfied with their bra support. Also, a majority of the flat-back curvature participants did not have any bra-fit problem with upper-back motion restriction regardless of the types of shoulder slope, whilst the average-back participants with square shoulders indicated bra-fit problems with upper-back motion restriction. The backward acromion participants with square shoulders were most satisfied

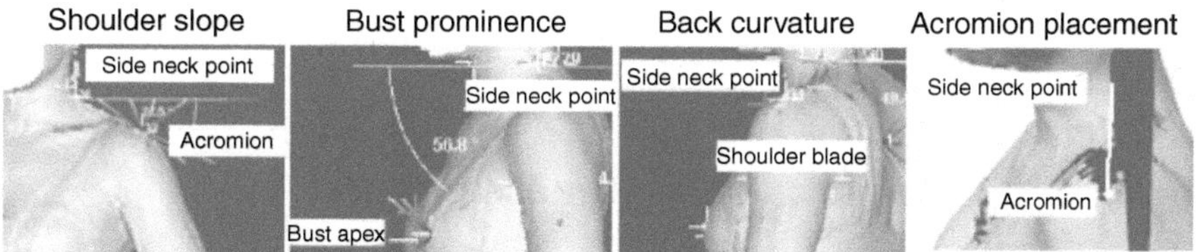

14.8 Angles of shoulder slope, bust prominence, back curvature and acromion placement. (*Source*: Chen *et al.*, 2010.)

with their bras, with no upper-back motion restriction, whilst the forward acromion participants with square shoulders were least satisfied with bra-motion restriction.

14.3.3 Ethnic difference and bra fit

Understanding of differences in body shape between Asian and Caucasian body configuration seems to be slowly emerging. After studying the scanned body data of 180 women (90 Asian and 90 Caucasian) aged between 19 and 63, Shin (2009) found significant differences in key body measurements between Asian and Caucasian women in terms of the parameters commonly used in bra design, patternmaking, and manufacture. One significant body measurement is the sternum (a key measurement relevant to gore sizing), where Asian women generally displayed a wider sternum width (more than two-tenths of an inch) and breasts which are closer together in terms of bust point to bust point (nipple to nipple) measurement.

The implications for bra sizing are that, without the ability to adjust gore size to accommodate the differing sternum widths, Asian women will find their bra does not fit well in the crucial central area (i.e. between the cups or gore) and consequently the breast will be forced against the upper inner side of the cup, leading to potential discomfort. The bust point to bust point differences (on average more than two-tenths of an inch) will mean that Asian women are likely to experience less support at the outer sides of each cup than their Caucasian counterparts. This is because their breast configuration is such that the bulk of each breast is more centrally located, with the nipples pointing in a more parallel plane. The summary bra fitting picture which emerges for Asian women is of a garment which is too loose at the lower outer edges and too tight at the upper inner edges. For Caucasian women, significant support for the outer edges of the breast is required. The success in push-up bras demonstrates the appropriateness of the design – pushing outer edges of breasts together and inwards to achieve the fashionable cleavage. The wider bust point to bust point measurements for the Caucasian participants perhaps partly explains the popularity of this type of bra amongst Caucasian consumers.

14.4 Smart materials in intimate apparel

Smart materials have one or more properties that can be dramatically changed in terms of viscosity, volume, or conductivity following certain condition changes in magnetic stress, moisture, temperature, and pH. Intelligent gels, ceramics, alloys, and polymers are emerging as smart materials with potential applications in the underwear market.

14.4.1 Shape memory foam

Amongst these smart materials, shape memory polymers and shape memory alloys are already being used for the intimate apparel industry in order to improve the fit and comfort of the bra. They are thermo-responsive materials because the deformation and recovery of their shapes occur through temperature changes.

Memory foam made of polyurethane with additional chemical additives not only compresses, but also moulds to the shape of the wearer (Memory foam reviews, 2008). Memory foam with a higher density responds to body heat and moulds itself to the shape of a warm body within a few minutes, whereas memory foam with a lower density responds to pressure and hence moulds itself to the shape of body. It is firm in cool temperatures and soft in warm temperatures. The 'Natural Accents Memory Foam bra' was introduced by Playtex early in the new millennium and followed by the 'Persuasion Memory Foam demi bra' by DKNY Intimate, 'Memory Foam bra' by M&S and 'Smart Memory bra' by Lisca, and most of these bras react to the wearer's body shape and temperature creating tailor made fit and comfort.

14.4.2 Shape memory wire

Shape memory alloys possess two extraordinary characteristics: shape memory and super-elasticity. Shape memory refers to the ability of the bra to be deformed at a low temperature and then return to its prior shape upon heating above a certain target temperature. The characteristics of super-elasticity refer to the ability to recover from large amounts of strain upon unloading after substantial deformation (Wiles and Liaw, 2000). Alloys that show super-elasticity undergo a thermo-elastic martensitic transformation, which is also a requirement for the shape memory effect. Shape memory wire made of 50% nickel and 50% titanium (Ni-Ti) possesses extraordinary characteristics such as super-elasticity and shape memory.

Applications of shape memory wires include orthodontic wires for dental braces, surgical needles, eyeglass frames, flexible headbands on headphones, and underwires for the bra (Stoeckel and Yu, 1991; Wiles and Liaw, 2000). A Ni-Ti wire can fully recover from strain values of up to 8%. When it is cooled, it becomes pliable and can be formed into almost any required shape and then when it is warmed, it remembers the programmed shape and returns to it. The super-elasticity of the underwires makes the bra comfortable to wear (Stoeckel and Yu, 1991).

14.4.3 Smart materials

The use of increasingly innovative smart materials has helped companies compete against fierce competition, and business trends have turned towards the ethical trade, involving eco-friendly production with sustainable materials. As a result, the intimate apparel sector has begun adopting eco-friendly and so-called smart materials along with the new and advanced technologies.

Common eco-friendly alternative fibres used in underwear include bamboo, Ingeo™, Lyocell/Tencel®, Modal, organic cotton, Seacell® and soy. Bamboo is a grass fibre harvested from the wild, it grows easily without the use of pesticides and fertilisers. Bamboo fibre is thermal regulating, anti-fungal, anti-static and UV resilient. It is a natural anti-microbial fibre due to its agent 'bamboo kun' which prevents bacteria from growing. However, the growing popularity of bamboo fibres has created other environmental issues such as preserving or protecting wild habitat, as well as air and water pollution related to the chemical processing of bamboo. Consequently, Broudy (2005) recommends further investigation into how 'green' bamboo really is in terms of environmental protection.

14.5 Technologies used in intimate apparel

Production technologies used in intimate apparel were significantly influenced by the fashionable and sought after 'sleek look' of the 1990s. In order to achieve a 'natural' contour for fashionable and body conscious consumers, cutting edge seamless technologies such as moulding, laminating, welding and circular knitting were utilized in the intimate apparel production, in particular in bra production (e.g., the one piece bra).

14.5.1 Seamless technology for fit and comfort

Moulded polyurethane foam with different built-in cookies for the bra cup clearly make it easier for women to change the appearance of breast volume without more drastic measures (e.g., plastic surgery). Lamination not only reduces production time and cost, but also creates a smooth surface by eliminating sewing threads and seams in the same way as moulding. Ultrasonic welding is fast, clean, and economical as there is no sewing thread involved (Branson, 2008; Shi and Little, 2000; Vincourek, 1987) and it is commonly used for replacement of bar tack in bra production. The growing popularity of seamless whole garment technology is inevitable because it offers a reduction in production costs, lower yarn consumption, flexible production, improved 'just-in-time' deliveries, higher quality, and added value (African Textiles, 2000), whilst the consumer enjoys generally greater comfort and a better fit with natural drapes and silhouettes. Kopell (2005) identified well-being and comfort as two long-term winning benefits of seamless apparel. Thus, Choi and Powell (2005) claim that seamless knitting will be one of the largest next generation knitting technologies. Bonding technology together with moulding technology and circular knitting technology made the one-piece seamless bra, which is known as the 'one-piece bra' or 'T-shirt bra', possible.

The concept of seamless technology with minimal, or no sewing at all, has attracted the underwear industry, which has been quick to adopt these advanced technologies into their production lines to help them meet consumer demand for a comfortable second skin effect without the use of unsightly seams. Despite the excitement surrounding these developments, there have been a number of challenges to overcome for bra manufacturers in particular. For example, the balance of achieving appropriate levels of tension amongst the supporting devices, and the integration of each component of the garment, were initially inferior to those of a conventional cut and sewn bra. Even today some of these challenges remain, and, for example, it is known that the wires sandwiched in a one-piece moulded bra are incapable of supporting large breasts in the same way as wires firmly held inside a cut and sewn bra using channelling tapes.

14.5.2 3D body scanning technology for design and fit evaluation

The 3D body scanning technology which started as partial-body scanning in the 1970s (Lovesey, 1974) was developed into whole-body scanning in the 1990s (Paquette, 1996). By the early 2000s, with improvements in technology and the emergence of different types of scanning technology, these 3D systems became sufficiently well-developed to have practical use in pattern-making (Mckinnon and Istook, 2002). As an anthropometric data collection tool, 3D body scanning has great advantages as it produces accurate sets of complex body measurement data in seconds (Istook and Hwang, 2001; Mckinnon and Istook, 2002; Paquette, 1996; Staples *et al.*, 1994). Also, it makes it easier to develop a sizing system from the collected anthropometric data (Beazley, 1997).

Recently, the industry has taken more of an interest in 3D body scanning for other purposes than mere anthropometric data collection, whilst the intimate apparel industry firmly focuses on the goal of creating the 'ideal' breast shape through the utilisation of the now rapidly developing new design and production technologies.

In particular, 3D body scanning technology has been used for the bra fit evaluation of different body shapes and breast volume studies (Lee *et al.*, 2004; Wang and Zhang, 2007; Zheng *et al.*, 2007), and bra fit evaluations (Chen *et al.*, 2011).

14.6 Conclusion

This chapter reviews the literature related to bra fit and the factors currently influencing the study and practice of bra fit. The chapter illustrates the exceptionally wide range of research work, including bra-fitting criteria, bra sizing, breast volume, bra-fit evaluation, smart materials and production technologies which are currently influencing this area of study. It is apparent from this review that achieving a good fit in bra design requires more than a repeat of current fitting and alteration procedures. With consumer bra-fit education, standardisation of a bra sizing system, implementation of 3D body scanning technology (from anthropometric data collection to fit evaluation), and the utilisation of smart materials and technology, the current complex process of achieving a good fitting bra certainly needs to be improved to simplify the often frustrating task of achieving a good fit. However, experience from some of the developments identified in this chapter suggest that it is the commercial imperative that will catch the eye of the industry, to the extent that good ideas will not necessarily translate into products without adding value to a company's bottom line.

14.7 References

African Textiles (2000). A revolution in seamless underwear and outerwear manufacture. *African Textiles*, 2000 (2/3), pp. 22–3.

Ashdown, S. P., and Na, H. (2008). Comparison of 3-D body scan data to quantify upper-body postural variation in older and younger women. *Clothing and Textiles and Research Journal*, **26** (4), pp. 292–307.

Bates, R. L. (1995). The efficacy of breast augmentation: Breast size increase, patient satisfaction, and psychological effects. *Plastic and Reconstructive Surgery*, **96** (5), p. 1237.

BBC News, (2000). *Claims of bra link to cancer dismissed.* Available from: http://news.bbc.co.uk/2/hi/health/998348.stm [Accessed: 20 Sept 2012].

Beazley, A. (1997). Size and fit: Procedures in undertaking a survey of body measurements. *Journal of Fashion Marketing and Management*, **2** (1), pp. 55–85.

Boyes, K. (1996). Buying the perfect bra. *Good Housekeeping*, International Magazine Corporation, New York, August 1996, p. 50.

Branson, (2008). *Ultrasonic sewing*. Available from: http://www.branson-plasticsjoin.com/textile_systems.asp#sewing [Accessed 28 March 2008].

Broudy, B. (2005). *All natural.* www.geartrends.com. Available from: http://geartrends.com/geartrends/upload/winter_ch4.2.pdf [Accessed 1 April 2008].

Brown, P. and Rice, J., (2000). *Ready-to-wear Apparel Analysis.* 3rd ed. New Jersey: Prentice Hall.

Chen, C., LaBat, K., and Bye, E. (2010). Physical characteristics related to bra fit. *Ergonomics*, **53**(4), pp. 514–24.

Chen, C., LaBat, K., and Bye, E. (2011). Breast prominence related to bra fit problems. *International Journal of Consumer Studies*, **35**, pp. 695–701.

Choi, W. and Powell, N. B. (2005). Three dimensional seamless garment knitting on v-bed flat knitting machines. *Journal of Textile and Apparel, Technology and Management*, **4**(3), North Carolina State University. Available from: http://www.tx.ncsu.edu/jtatm/volume4issue3/vo4_issue3_abstracts.htm [Accessed 31 March 2008].

Edsander-Nord, A., Wickman, M. and Jurell, G. (1996). Measurement of breast volume with thermoplastic casts. *Scandinavian Journal of Plastic and Reconstructive Surgery and Hand Surgery*, **30**(2), pp. 129–32.

Farrell-Beck, J., and Gau, C. (2002). *Uplift: The bra in America*, Philadelphia: The University of Pennsylvania Press.

Fowler, P. A., Casey, C. E., Cameron, G. G., Foster, M. A. and Knight, C. H. (1990). Cyclic changes in composition and volume of the breast during the menstrual cycle, measured by magnetic resonance imaging. *British Journal of Obstetrics and Gynaecology*, **97**, pp. 595–602.

Greenbaum, A. R., Heslop, T., Morris, J., and Dunn, K. W. (2003), An investigation of the suitability of bra fit in women referred for reduction mammaplasty. *British Journal of Plastic Surgery*, **56**(3), pp. 230–6.

Hardaker, C. H. M., and Fozzard, G. J. W. (1997). The bra design process: a study of professional practice, *International Journal of Clothing Science and Technology*, **9**(4), pp. 311–25.

Howarton, R., and Lee, B. (2010). Market analysis of fit preferences of female boomers. *Journal of Fashion Marketing and Management*, **14**(2), pp. 219–29.

Istook, C. and Hwang, S. (2001). 3D body scanning systems with application in the apparel industry. *Journal of Fashion Marketing and Management*, **5**(2), pp. 120–32.

Johnson, B. and Barnes, K. (2008). *The secret of health: Breast wisdom.* New York: Morgan James Publishing.

Kanhai, R. C. J. and Hage, J. J. (1999). Bra sup size depends on bands size. *Plastic and Reconstructive Surgery*, **104**(1), p. 300.

Katariya, R., Forrest, A. and Gravelle, I. (1974). Breast volume in cancer of the breast. *British Journal of Cancer*, **29**, pp. 270–3.

Katch, V. L., Campaigne, B., Freedson, P., Sady, S., Katch, F. I. and Behnke, A. R. (1980). Contribution of breast volume and weight to body fat distribution in females. *American Journal of Physical Anthropology*, **53**, pp. 93–100.

Kopell, M. (2005). What next for seamless bodysize clothing? *Knitting International*, **112**(1326), pp. 23–5.

Kovacs, L., Yassouridis, A. and Zimmermann, A. (2006). New aspects of breast volume measurement using 3D surface imaging. *Annals of Plastic Surgery*, p. 57.

Lee, H. Y., Hong, K. and Kim, E. A. (2004). Measurement protocol of women's nude breasts using a 3D scanning technique. *Applied Ergonomics*, **35**, pp. 353–9.

Lee, M. M., Chang, I. Y. U., Horng, C. F., Chang, J. S., Cheng, S. H., and Huang, A. (2005). Breast cancer and dietary factors in Taiwanese women. *Cancer Causes and Control*, **16**, pp. 929–37.

Lipton, B. (1996). Are you wearing the wrong size bra? *Ladies Home Journal*, Meredith Corporation, Des Moines, March 1996, p. 446.

Loughry, C. W., Sheffer, D. B., Price, T. E., Lackney, M. J., Bartfai, R. G., and Morek, W. M. (1987). Breast volume measurement of 248 women suing biostereometric analysis. *Plastic and Reconstructive Surgery*, **80**(4), pp. 553–8.

Lovesey, E. (1974). The development of a three-dimensional anthropometric measuring technique. *Applied Ergonomics*, pp. 36–41.

McGhee, D. E. and Steele, J. R. (2010). Optimising breast support in female patients through correct bra fit: a cross-sectional study. *Journal of Science and Medicine in Sport*, **13**(6), pp. 568–72.

Mckinnon, L. and Istook, C. L. (2002). Body scanning: The effects of subject respiration and foot positioning on the data integrity of scanned measurements. *Journal of Fashion Marketing and Management*, **6**(2), pp. 103–21.

Memory foam reviews (2008). *Memory foam reviews*. Available from: http://memoryfoamreviews.com/ [Accessed 10 April 2008].

Mineyev, M., Kramer, D. and Kaufman, L. (1995). Measurement of breast implant volume with magnetic resonance imaging. *Annals of Plastic Surgery*, **34**, pp. 348–51.

Nethero, S. (2008). *U.S. Patent No. US2008/0027744 A1.* Washington, DC: U.S. Patent and Trademark Office.

Paquette, S. (1996). 3D scanning in apparel design and human engineering. *IEEE Computer Graphics and Applications*, September 1996, pp. 11–15.

Pechter, E. A. (1998). A new method for determining bra size and predicting postaugmentation breast size. *Plastic and Reconstructive Surgery*, **102**(4), pp. 1259–65.

Pechter, E. A. (1999). Bra cup size depends on bands size. *Plastic and Reconstructive Surgery*, **104** (1), p.300.

Penn, J. (1955). Breast reduction. *British Journal of Plastic Surgery*, **7**, pp. 357–71.

Risius, D., Thelwell, R., Wagstaff, C., and Scurr, J. (2012). Influential factors of bra purchasing in older women. *Journal of Fashion Marketing and Management*, **16**(3), pp. 366–80.

Shi, W. and Little, T. (2000). Mechanism of ultrasonic joining of textile materials. *International Journal of Clothing Science and Technology*, **12**(5), pp. 331–50.

Shin, K. (2007). Patternmaking for the underwired bra: New directions. *The Journal of the Textile Institute*, **98**(4), pp. 301–18.

Shin, K. (2009). *The origins and evolution of the bra.* PhD thesis, University of Northumbria.

Singer, S. R. and Grismaijer, S. (1995) *Dressed to kill: The link between breast cancer and the bra*. Pahoa: ISCD Press.

Smith, D. J., Palin, W. E., Katch, V. L. and Bennett, J. E. (1986). Breast volume and anthropomorphic measurements. *Plastic and Reconstructive Surgery*, **78**(3), pp. 331–5.

Staples, N., Pargas, R. and Davis, S. (1994). Body scanning in the future. *Apparel Industry Magazine*, **55**(10), pp. 48–55.

Stoeckel, D. and Yu, W. (1991). Super-elastic Ni-Ti wire. *Wire Journal International*, (March), pp. 45–50.

Vincourek, V. (1987). On to ultrasonic sewing. *Textile Asia*, **18**(10), pp. 57–61.

Wang, J. and Zhang, W. (2007). An approach to predicting bra cup dart quantity in the 3D virtual environment. *International Journal of Clothing Science and Technology*, **19**(5), pp. 361–73.

Washington State University and Cooperative Extension Service (1986). *Actionwear: Clothing for fitness: Choosing a sport bra*. Washington State University Extension Service, Pullman, Washington.

Westreich, M. (1997). Anthropomorphic breast measurement: Protocol and results in 50 women with aesthetically perfect breasts and clinical application. *Plastic and Reconstructive Surgery*, **100**(2), pp. 468–79.

White, J. and Scurr, J. (2012). Evaluation of professional bra fitting criteria for bra selection and fitting in the UK. *Ergonomics*, **55**(6), pp. 704–11.

Wiles, K. E. and Liaw, P. K. (2000). The fatigue behaviour of shape-memory alloys. *Journal of the Minerals, Metals, and Materials Society*, **52**(10), pp. 45–51.

Worldwidebreastcancer.com (2012). Breast cancer statistics worldwide. Available from: http://www.worldwidebreastcancer.com/learn/breast-cancer-statistics-worldwide/ [Accessed 10 October 2012].

Wood, K., Cameron, M., and Fitzgerald, K. (2008). Breast size, bra fit and thoracic pain in young women: a correlational study. *Chiropractic & Osteopathy*, **16**(1). Available from: doi:10.1186/1746-1340-16-1 [Accessed 15 September 2012].

Wright, M.C.M. (2002). Graphical analysis of bra size calculation procedures. *International Journal of Clothing Science & Technology*, **14**, pp. 41–4.

Young, V. L., Nemecek, J. R. and Nemecek, D. A. (1994). The efficacy of breast augmentation: Breast size increase, patient satisfaction, and psychological effects. *Plastic and Reconstructive Surgery*, **94**(7), pp. 958–69.

Yu, W. (2004). Subjective assessment of clothing fit. In: Fan, J., Yu, W., and Hunter, L. (eds.), *Clothing appearance and fit: Science and technology*. Cambridge: Woodhead Publishing Limited, pp. 31–42.

Zheng, R., Yu, W. and Fan, J. (2007). Development of a new Chinese bra sizing system based on breast anthropometric measurements. *Industrial Ergonomics*, **37**, pp. 679–705.

15
Headwear: designing headwear to fit the size and shape of Western and Asian populations

R. BALL, The Hong Kong Polytechnic University, China

DOI: 10.1533/9781782422150.2.292

Abstract: Designing headwear today requires knowledge of head and facial anatomy, anthropometrics, ethnic variation, fit and comfort, design technology and manufacturing. The psychological and physical aspects of creating a fashionable piece of headgear are discussed. The different types of head shape data tools are compared and a review of current design resources are presented. The impact of 3D scanning and CAD software in the design of headgear is explored. Future trends in headgear design including 'Asian Fit' products and online retailing introduces new opportunities for the industry.

Key words: design, anthropometry, headform, shape, 3D, China, headgear.

15.1 Introduction

Throughout human history, people have worn head coverings to serve as protection from weather and physical danger. The types of products that we wear on our heads give people important information about who we are and what we do. Historically, the apparel industry has been concerned with hats for personal adornment or to establish rank or position. Royal crowns are probably the best recognized of all symbolic headgear, a recognition that extends to the honorary crowning of beauty queens in competitions. Headgear since the time of the Romans has announced other symbolic ranks, including military titles, judicial office, religious affiliation, or tribal citizenship – including citizenship in voluntary 'modern tribes' like the Shriners, or the Red Sox Nation.

The head is the center of our identity and home to our senses. Vision, hearing, smell and taste are all located in the head. These delicate organs are exceptionally vulnerable, and so head product producers must pay diligent attention to their comfort and protection. The face also functions as our physical identity; it is how we remember people and in turn how we are remembered ('I never forget a face') and consequently we devote much time, money and effort to enhancing its attractiveness. Any successful headwear product must understand and enhance our self perception and vanity.

Head size and shape is determined by gender, race, and age. Designing for each of these groups requires an understanding of their specialized requirements. Any successful piece of headwear carefully resolves three design challenges; the first challenge is to understand the sizing and head shape of the target user group, the second challenge is creating headwear that is comfortable and stylish, which is more elusive and difficult to measure, and the third challenge involves manufacturing and production; can it be made? Are the materials comfortable and hygienic? Is the cost acceptable?

These ingredients must be seamlessly integrated with each other, and to achieve this they must all be considered at every stage. All three criteria must be satisfied to achieve success. If any of the three remain unresolved, the product will fail. There is never a point where a designer can decide to think about one of the ingredients 'later'. Fashion, fit, and production must all be integrated right from the start.

Contemporary headwear, used in both work and leisure activities, continues to respond to challenges inspired by the ebb and flow of fashion trends, new materials and technology and by the development of new sports and leisure activities. In today's always connected era of digital technology, an entire range of new possibilities is revealed, as a variety of headphones and headsets join the repertoire of fashionable things to be worn upon the head (see Fig. 15.1).

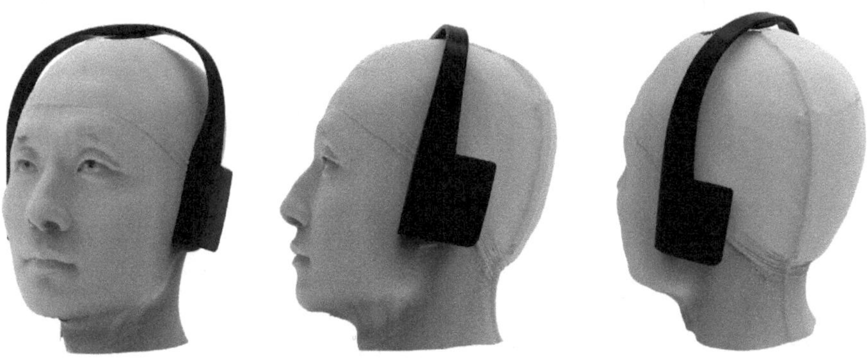

15.1 Scan & Print. Designing headgear with Digital Human Models, CAD software and 3D Printing.

15.2 Types of headwear

Headgear protects us from the driving rain and the blazing sun, it lets us change our 'look' instantly and hides us when we do not want to be seen. We all wear multiple types of headgear at some point during our lives, mainly by choice but occasionally through regulation; wearing a wedding veil is a personal choice but wearing a motorcycle helmet is legal requirement.

15.2.1 Hats

Hats are the most common apparel item worn on the head. They have evolved in many different shapes as defined by their use and materials, from the wide brimmed Stetson of the American west to the tiny fez of the Middle East. These iconic shapes evoke a strong emotional response reflecting societal values and defining moments in history. 'Sitting so prominently on the head and in such close proximity to the face, the hat can mesh with its wearer and his or her identity in an extraordinary way: each is accentuated by the other. It is this iconic vividness that perhaps gives the hat its prominence in our memories of events and personalities. Hats can act as a frame for some very human moments in history' (Design Museum, 2011).

In the history of Western fashion, the passion for hats peaked in the late 1890s, when any respectable man or woman would have required multiple hats for different business and social occasions. When busy hatters applied industrial processes to the art of hat making, the first mass-produced consumer headgear was born. Many of the inventions of the nineteenth century hat maker can still be seen in use today. Hats continued to be an essential part of business and society until the post war period, when hairstyles like the bouffant and beehive spelled the end of women wearing hats (Stone, 2008). The 1960s and 1970s were a celebration of hair for both genders and hats were reserved for special events like the Royal Ascot Race or fashion week in Paris. Milliners battled to create ever more outrageous bespoke designs. Wearing a hat became a performance intended to amaze and delight the paparazzi.

Hats have enjoyed a renaissance in recent years as a must have fashion accessory, with new materials and designs appearing on the market regularly. The demand may be due in part to the raised awareness of the dangers of exposure to UV rays. Technology has also permitted new advances in the design of hats. The availability of high resolution scanners in combination with 3D computer aided design (CAD) software is driving the development of new methods for creating headgear. United Nude's low resolution scanning of a classic driver's cap produced an innovative new faceted design inspired by the F-117 fighter called the Stealth hat (United Nude). This reverse engineering of classic shapes using scanning and 3D software is an exciting area for new development.

15.2.2 Baseball hats/caps

The most popular hat today for both sexes is the baseball hat, produced annually in quantities of tens of millions. The New York Yankees baseball hat, with its dark navy blue color and white iconic logo, adorns heads in every country and every continent across the globe. The baseball hat benefitted from the rise of hip-hop, as rappers embraced professional sports apparel as part of their image. When the baseball hat crossed over to women's fashion it solidified its position as one of the most popular global fashion items ever produced. The majority of baseball hats produced are the low cost type adorned by logos of pizza stores and farm equipment manufacturers, and their main feature is an adjustable headband at the back. The other end of the market is the 'almost bespoke' fitted baseball cap (New Era Cap Ltd.). These sleek design icons are purchased from a specialty store and can cost 50 USD each. The fitted cap is manufactured in a range of sizes, so the store is required to warehouse hundreds of different hats to meet normal demand. Collectors will pay hundreds of dollars for a treasured hat in mint condition protected in its hard clamshell package.

15.2.3 Protective helmets

Some of the oldest forms of headgear are armored helmets, intended to protect soldiers from rocks, arrows and spears. The modern military continues to be a driving force in the development of protective headgear. The technologically sophisticated ballistic helmets of today respond to ever-greater challenges of armed conflict, applying advances in composite materials, optics and electronics. The use of protective helmets for non-military applications is a relatively recent innovation. Around the turn of the twentieth century, football players adopted a form of leather 'head harness' to protect against abrasions and 'cauliflower ear' (Giffin, 1992). Protective helmets used in team sports are in crisis today, as the design of helmets has not kept pace with the advances in sports sciences, nutrition, and year round training. The bigger, faster, stronger athletes of today are at a greater risk of career ending concussions than ever before.

Hardhats for work applications appeared as early as 1919, manufactured out of canvas and glue, to give miners protection from falling rocks and, more importantly, as a place to mount a light for hands free pick axing. Protective helmets for motorcycles were created soon after the first motorcycle appeared in the 1900s, although they were initially not much more than a leather hat intended to keep the head warm and the hair tidy. Today, a wide variety of protective helmets are available for many different sports and working conditions. The demand for specialized headgear continues to grow, driven by athletes, scientists and the military seeking to extend the limits of human performance.

15.3 Head physiology

The head can be divided into two main areas: the upper skull area and the lower facial area. For our purposes of describing apparel for consumers and the impact of shape differences we will concentrate on the upper skull area. At its very top, the head contains the skull, brain, and skin and very few muscles. This is the area of the body where our underlying skeletal structure lies closest to the surface. The skin sits directly on top of the hard skull. Headgear, like shoes, has always required a level of precision fit that is not necessarily needed for other parts of the body. Much of the human body is soft. Padded by muscle and fatty tissues, the middle sections of the body can be successfully squished into different shapes by means of corsets or elastic bands, or sculptured directly into fashionable contours by means of rigorous diet and exercise. None of this is possible for the head. Rigid and bony, the head is immutably resistant to the forces of fashion in a physical sense, at the same time as it represents the pinnacle of fashion in being the central focus of the human sense of self.

15.3.1 Skull growth

In newborn babies, the bones of the top and the sides of the skull are not yet joined together, but are separated by membranes. The flexible skull of the baby permits it to compress in order to pass through the constricted human birth canal (Roberts, 1999). Only 63% of its mature size at birth, the skull reaches 88% after two years. By the age of ten, the skull will have grown to 95% of its final size, and by age 18 the skull will reach more than 98% of its final size (Nellhaus, 1968). As the baby matures, the skull bones begin to join together. Generally before the age of two, the last of the open fontanels or 'soft spots' will have closed. The development of the rigid interpenetrating 'sutures' connecting the plates of the skull takes far longer. In old age, the sutures become entirely rigid.

15.3.2 Frankfurt plane

So how do we define what is face and what is skull? Anthropology provides the answer with the Frankfurt plane (Fig. 15.2). The Frankfurt plane (named for the location of the conference that defined it) is an imaginary surface that runs through the middle of the skull, touching on four standard points. These points are the bottom point of each eye orbit (infraorbitale) and the center of the ear hole (tragion). As a base for standard reference, the Frankfurt plane permits the methodical development of different kinds of anatomical data. For example, it allows height to be measured accurately, eliminating variations that can result when the head is held at different angles. Above all, it offers the considerable advantage that it can always be located for every human head, even though individual head sizes, shapes

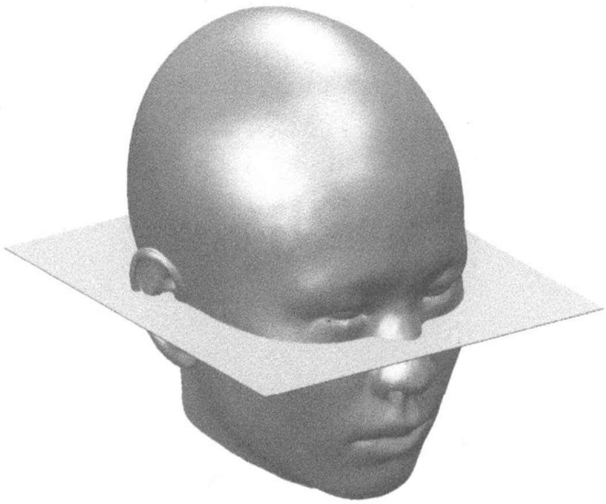

15.2 Frankfurt Plane. The Frankfurt plane serves as a datum line for head measurements.

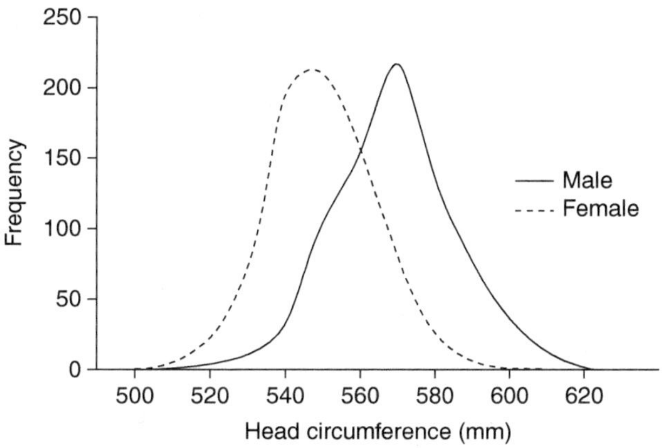

15.3 Adult Chinese Head size distribution by gender. Circumference measurements from the SizeChina database show a significant difference in head size between males and females.

and overall geometry vary widely. It is a convenient place to separate the head from the face. It is an industry rule of thumb that the 'head' begins approximately one inch (50 mm) above the Frankfurt plane.

15.3.3 Size range distribution

Statistical anthropometric research has shown that the distribution of different sizes in a population generally falls into a 'normal' or 'bell curve' shape, meaning that the vast majority of people are reasonably similar to each other in terms of any given measurement (Fig. 15.3). Very few individuals fall at the extremes of being exceptionally tall, short, thin, or fat. Most people are average. In practice, however, products worn on the head are exceptionally sensitive to pressure, so that even small variations in size or shape can create a product that feels uncomfortable. Many factors influence physical comfort – elasticity of materials, heat buildup, tactile perceptions of softness – and it is for these reasons that headwear manufacturers offer their products in different sizes.

15.4 Sizing systems for headgear

Companies use a variety of systems to size their offering of hats, helmets and wigs to their users. Hat sizes vary between countries and are expressed either in hat maker's increments (e.g. 7–5/8) or in metric diameter (61 cm) or inches (24"). Driven by the demands of retailers seeking to limit the inventory they need to warehouse, most manufacturers embrace the quintet system of extra small/small/medium/large/extra large, but unfortunately each manufacturer establishes their own sizing increments and there is no standardization. Lack of standardization is compounded by the difficulty of getting accurate internal measurements from a piece of headgear once it has been manufactured. Sizing remains a challenging issue.

15.4.1 Ethnic shape variations

Univariate head shape variation between ethnic groups is well known historically, but numbers do not provide details of shape. High-Resolution 3D scan technology provides the best answer to capturing shape. Today's 3D scanning technologies allow for the collection of high resolution on head shape. Two recent studies, the CAESAR survey and the SizeChina project, have for the first time provided large datasets to compare head shape differences (Ball, 2010). The results are significant to the design of headwear. The western or Caucasian head is a long narrow egg shape, whereas the Chinese or Asian head shape is almost round like an orange (Fig. 15.4). Flexible headwear such as cloth hats are able to conform to these shape differences rather easily. However, if the headwear contains molded components or stiffer materials then fit problems become severe. In molded products like a bicycle helmet, completely different product models are required. High resolution 3D data on head shape for many other ethnic groups has not yet been recorded.

15.4.2 Comfort and fit in headgear

In its broadest sense, an item of headwear can be understood to 'fit' when a user who is trying it on feels that it is the correct size and shape, physically

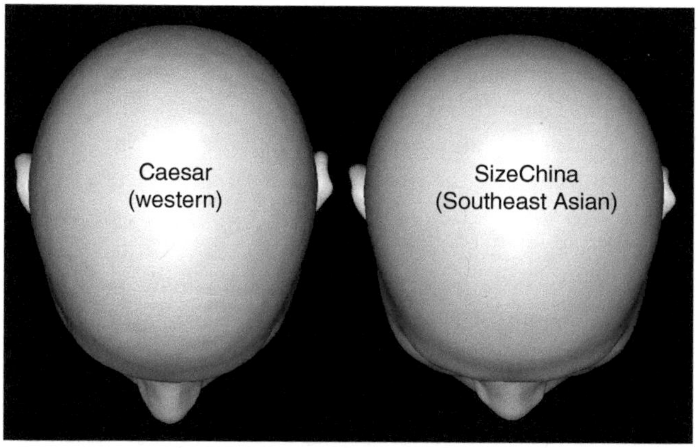

15.4 Ethnic Head Shape Variation. The result of the CAESAR (Western) and the SizeChina 3D surveys show a significant headshape difference.

comfortable, emotionally satisfying, and stylistically appropriate. None of those issues is straightforward. The criteria for physical comfort, for example, vary widely for different products. Products that touch the body may be 'loose fit', like clothing, 'firm fit', as in hiking boots, or 'precision fit', as in military aviation helmets, where even a few millimeters of misalignment can throw off delicate helmet-mounted displays. A single product may require different fit in different places, as when a pair of sunglasses must rest lightly on the bridge of the nose, but grasp the temples of the head tightly in order to stay put. All of these types of fit require a thorough understanding of the size and shape of the underlying body. This is particularly important when inelastic parts of the body, like the head, are in contact with inelastic products, like safety helmets. Even small variations of size and shape are crucial under those conditions.

In terms of the end user, acceptance of headgear products essentially relies on what marketing and sales people call the 'mirror test', where a customer visits a store and uses a mirror as the primary tool to make a purchase decision. In this complex process, the users first try on the headgear to assess the interior shape. Is it the correct size and shape? They then evaluate comfort aspects. Are the interior pads soft enough? Do the materials irritate the skin? Does it feel hot? Finally they judge the emotional aspects of the color, materials and form of the product as it appears on them, and assess it in terms of personal vanity, and for the social messages communicated to its intended audience of other people. Does it look good on me? Will my friends laugh at me? The product will also be mentally compared to competitive products for factors such as 'value' and 'style'.

15.5 Head anthropometrics and design

What design measurements do designers need to use? The key dimensions of head length, width and circumference are easy to collect with a tape measure and calipers. The face represents a much bigger challenge. Facial anthropometrics must include landmarking the process of locating the structure of the facial bones underneath the soft tissues (Fig. 15.5). Products designed for the head must be supported by the underlying hard bony skull. Soft cartilage areas like the nose or ear can only support very little pressure. The face contains complex areas of geometry such as the nose bridge or the channel that joins the ear to the head, which are very challenging to measure with simple hand tools. These complex areas are best captured using 3D scanning technology.

Hair is important to our physical appearance, and much time and effort is invested to present it at its most flattering configuration. Be it luxurious bouffant or economy class crew cut, the properties of hair – length, thickness and texture – affect the fit and performance of head mounted products.

15.5.1 Sizing information for headwear design

How does a designer find information for the task of designing headwear? Most anthropometric information is not well suited to design use. Numerical dimensions, spreadsheets and unprocessed 3D scans cannot be immediately integrated into the established methodology of design, which centers on hand sketching, model-making, digitizing, CAD, 3D printing, manufacturing and animation. Designers are often unwilling or unable to spend large amounts of time reviewing specialized anthropometric data in order to extract from it the information relevant to their practice. Professional design consulting

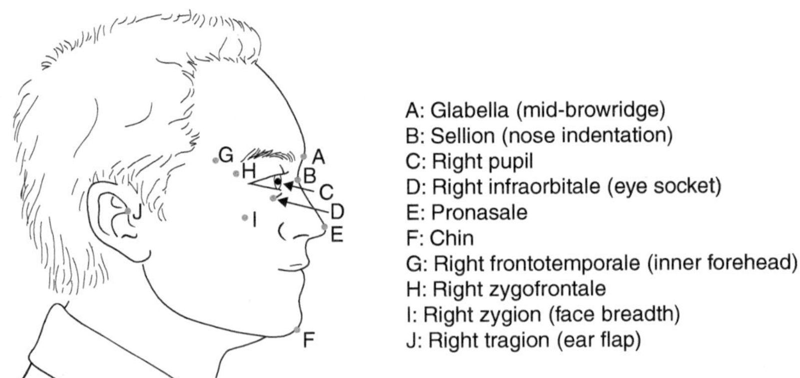

15.5 Facial landmarks. Each product needs to establish the critical landmarks for fit.

Table 15.1 Table of head dimensions. Circ/width/length for 5%, 50% and 95% for adult Chinese males and females

		Head circumference (mm)	Head width (mm)	Head length (mm)
Male	5%-tile	539	147	177
	50%-tile	565	159	188
	95%-tile	593	171	200
Female	5%-tile	522	141	169
	50%-tile	545	151	180
	95%-tile	570	161	191

firms are under ever increasing demands to design better and faster. The research phase of a design project where basic anthropometric data is gathered can be a short as a few weeks, or even a few days in rush projects. As a result, the products created by designers rarely benefit from anthropometric insight.

15.5.2 Charts and tables

'Univariate' or 'one-dimensional' anthropometric data consist of straightforward lists of numbers arranged into charts and spreadsheets. These numbers summarize research findings about the dimensions of human bodies (see Table 15.1). For example, a chart might explain that a '95%-ile male' has a head circumference of 61 cm. This means that 95% of all men surveyed were below 61 cm in head circumference. Univariate data are compact, precise, and relatively easy to understand (Gordon and Bradtmiller, 1989).

One of the disadvantages of univariate charts is that they fail to provide access to the original data behind their summaries. They present statistical means or averages of research findings, and no information about the raw data that contributed to the summary. They also convey little of the nuance of 3D shape. A person may have any kind of three-dimensional head shape, which affects the design of headwear to fit their needs. Univariate data on head circumference or face length give only limited clues about that shape, because a person with a 95%-ile face length will not necessarily have a 95%-ile head circumference. Percentiles have an important place in the design process, but they are easily misused. Designers will often assume that if they make a product that fits 95% of the population then everyone is covered, when in practice the people on the far end of the range, the small size, will find the product far too big. Often, designers are unaware that this practice is fundamentally flawed. In addition, the

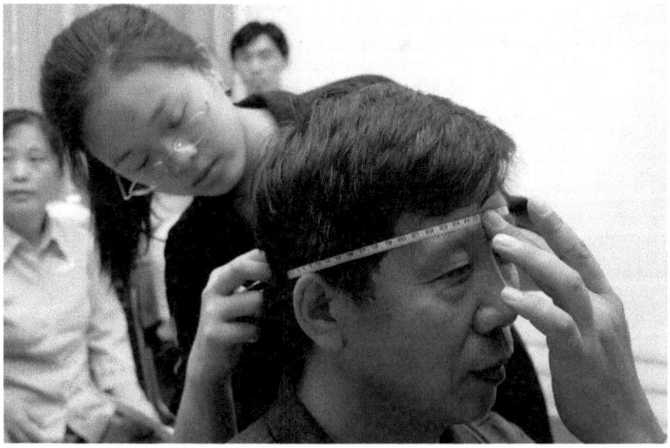

15.6 Taking head circumference measurement. Flexible tailors tape is placed 50 mm above the Frankfurt Plane and measurement recorded manually.

numerical form of univariate data is often unappealing to designers. They must study and 'translate' dry charts into their visual equivalents before the information can be assimilated into their visual and spatial approach to problem-solving.

15.5.3 Sizing surveys

While charts and tables provide a starting point to establish sizing, designers must take care to confirm their target group is well served by the sizing range. In order to confirm sizing it is recommended to take a personal survey of the target customer group. If, for example, you are designing sunhats for children it is a good idea to arrange a visit to a school and spend a few hours taking circumference measurements before you begin designing, and return later for fit testing of prototypes, carefully documenting the collected circumference measurements in a spreadsheet. A sample size of at least 20 subjects is recommended to get a representative cross section. Over time these surveys can provide a valuable design resource. Skipping the fit testing process sets the stage for fit problems once the product has been introduced into the market place.

15.5.4 Taking a head measurement

While the Frankfurt plane is important when scientifically measuring subjects, a simpler method is needed in everyday settings. Because sizing systems are based on circumference measurements, a tailor's flexible cloth tape measure provides a simple and easy to use tool. Position the tape measure just above the eyebrows and pull the tape measure tight to compress the hair. Take three separate measurements to confirm accuracy, and record the average measurement (see Fig. 15.6).

15.6 Head models

Designing headgear is a physical task requiring multiple fit checkpoints during the development of a headgear product. Physical head models are difficult to find and often frighteningly inaccurate care must be taken at the very beginning to choose the correct 'Headform'. Head models are increasingly available in 3D digital CAD formats.

15.6.1 Styrofoam wig stands

When designers need a head model to use for creating headwear, many purchase an inexpensive Styrofoam head model from a stores fixture shop. These heads are intended for use as display stands for hats, wigs and helmets, and are not accurate in size or shape. They are not suitable for use as a design tool and will lead to many mistakes and false assumptions.

15.6.2 Hat makers blocks

A hat maker's block consists of a two piece mold usually made from wood. The exterior half forms the outside of a specific hat style, with an interior half or plug giving the hat its size and shape. These tools have remained basically unchanged for hundreds of years and vary widely from set to set. They are a production tool not suitable for design and development purposes.

15.6.3 Headforms

What should a designer use, then, when an accurate physical model is required for designing headwear? There are tools called Headforms that are commercially available and accurately represent the shape of a human head. There are many different types of Headforms commercially available. Some Headforms represent just the upper part of the head, and others are facially featured. The Headforms come in a variety of materials and hardnesses. There are testing Headforms, which are used to physically test head mounted products, and reference Headforms, which are used to visually check a helmet fit and head coverage

15.6.4 Reference headforms

Reference Headforms that represent the median shape of a specific population are suitable for use in the design of headwear. The EN960 European Headforms are the most widely accepted reference Headforms used in most helmet standards. They come in a set of thirteen sizes as measured

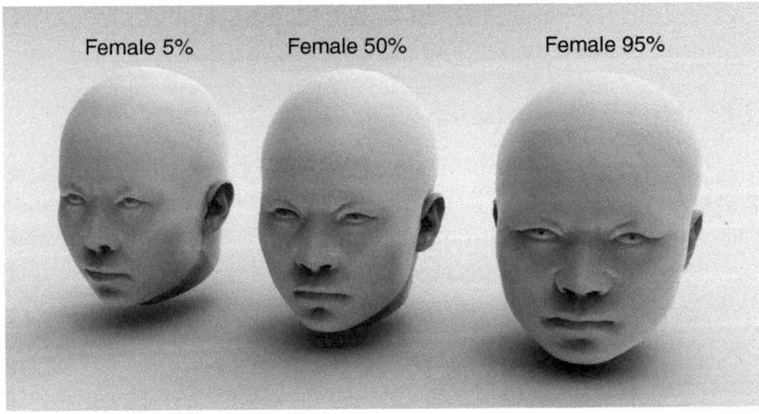

15.7 Facially Featured Headforms. These Adult Chinese facially featured Headforms are available as digital files or hard models.

by circumference, starting from 50 cm circumference. They are commercially available, and are suitable for casting, measuring as a base to develop an interior shape that is technically accurate (CADEX, 2006). The first set of ten Chinese reference Headforms developed from the SizeChina 3D anthropometric survey are now commercially available (SizeChina, 2012). This new set of Headforms is intended for use in developing 'Asian fit' head products. A new set of facially featured reference Headforms based on gender and percentiles has been recently introduced for the Adult Chinese population (see Fig. 15.7).

15.6.5 3D datasets

Anthropometric data standards are currently undergoing a quiet revolution with the emergence of digital 3D scanning as a measurement tool. More accurate and more consistent than manual measurements, 3D scanning also captures far more information. For the first time, the full spatial shape of the human head can be recorded. 3D data are well adapted to use by designers. Because the information exists as computer data, it can be readily displayed in the form of a pictorial image, making it intuitively easy to understand the forms being described. It can also be manipulated within the computer's virtual environment, offering compatibility with the 3D design programs used by designers (see Fig. 15.8).

There are drawbacks to 3D scan files. As a new product, they are not yet fully integrated with design software. Head scans also have data holes and gaps in areas such as the nose, the ears and underneath the chin. These areas require extra work to fill and repair holes, to bring them up to a design ready

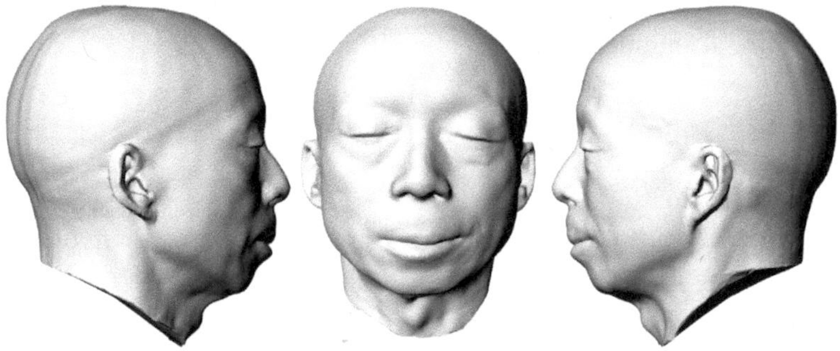

15.8 High-resolution 3D scans of Chinese Adults. These scans typically contain 300 000 datapoints. This high-resolution data is required for product that require exact fit such as earphones, eyeframes and breathing apparatus.

level. In addition, they are expensive, which is a problem for students or small design studios.

15.6.6 Related products

Digital 3D models of the human body are commercially available at a number of popular websites. Their primary use lies in entertainment, gaming and architectural rendering applications, where compact file size is of primary importance. Included among the 3D models available from these sources are a few models of heads, but they are only visual representations. They lack the dimensional accuracy needed for headwear design applications.

15.7 Future trends

The headgear manufacturers have historically focused on western market but Asia's booming economies are driving demand for new 'Asia Fit' products. Headgear brands are searching for new ways to sell their fit sensitive products online.

15.7.1 Ethnic fit

The head shape differences between east and west are significant. Western headgear does not fit Asians, so manufacturers are creating 'Asian fit' designs to enter the massive new South East Asian markets. India will surpass China as the most populous nation in 2025, and poses new challenges in understanding head size and shape. The Latino population will become the single largest ethnic group in the US in early 2020, and research is needed to determine if significant head shape differences exist.

15.7.2 Online shopping

Many traditional fit-sensitive products such as shoes, gloves, and hats were thought to be unsuitable for online shopping due to their demanding fit requirements. However, the success of online footwear sales will increasingly drive the online headwear sales. Headwear manufacturers and online retailers are challenged to create tools for determining size and shape on line.

15.7.3 Digital head models

The first generation of Digital Head Models (DHM) of Chinese and Caucasian populations have recently been made available commercially. These DHMs offers 'plug and play 'compatibility for CAD software and 3D printing, which will drive innovation in the design of headwear.

15.8 Sources of further information and advice

There are few resource books that deal directly with the head and face; usually the head data is included as an appendix in a larger full body survey. Access to this data is further hampered by many of the publications being out of print. Here are a few that the interested reader might look into:

Bradtmiller B., Friess M., A head-and-face anthropometric survey of US respirator users. Anthrotech technical report for NIOSH/NPPTL. 2004.Croney J., Anthropometrics for designers. London: Batsford, 1971.

Du L., Zhuang Z., Guan H., Xing J., Tang X., Wang L., Wang Z., Wang H., Liu Y., Su W., Benson S., Gallagher S., Viscusi D. and Chen W., Head-and-Face anthropometric survey of Chinese workers. Annals of Occupational Hygiene 2008;**52**(8):773-82.

Farkas L.G., Anthropometry of the head and face. New York, USA: Raven Press, 1994.

15.9 References

Ball, R. (2010), 'A comparison between Chinese and Caucasian head shapes', *Applied Ergonomics*, vol. **41**, no. 6, pp. 832–9.

CADEX (2006), *Catalogue – EN 960 Headforms*. Available at: http://www.cadexinc.com/.

Design Museum (2011), *Fifty hats that changed the world*, Conran Octopus Ltd., UK.

Giffin, C.S. (1992), 'Wrestler's Ear: Pathophysiology and Treatment', *Annals of Plastic Surgery*, vol. **28**, no. 2.
Gordon, C. and Bradtmiller, B. (1989), *1988 Anthropometric Survey of US Army Personnel: Methods and Summary Statistics*, US Army Natick Research, Development and Engineering Center, Natick, MA, USA.
Nellhaus, G. (1968), 'Head Circumference from Birth to Eighteen Years', *Pediatrics*, vol. **41**, no. 1, pp. 106.
New Era Cap Ltd., *Fitted cap history* [Homepage of ncw era cap.ltd.]. Available at: http://www.neweracap.com/en_US/Explore/New_Era_Heritage.aspx [Feb 2013].
Roberts, J.M. (1999), 'Diagram C' in *The Illustrated History of the World*, Oxford University Press, New York, pp. 155.
SizeChina (2012), *Adult Chinese reference head forms* [Homepage of SizeChina]. Available at: http://www.sizechina.com/products.php [Feb, 2013].
Stone, E. (2008), *The Dynamics of Fashion,* 3rd edn, Fairchild Books, NY.
United Nude, *Stealth Hat* [Homepage of united nude]. Available at: http://www.unitednude.com/news/2011/stealth-hat-17 [Feb. 2013].

Index

active wear, 250–1
adulescent, 79
affordable well-fitting clothing, 215
age distribution, 247
ageing, 156, 207
AlvaInsight, 9, 10
Alvanon, 7, 9, 49, 157, 199
AlvaScan, 10
American Society for Testing and Materials (ASTM), 159–60
anthropometry, 190
apparel design
  Asian and Caucasian for two different ethnic groups, 169–87
    examples for specific styles for different ethnic populations, 183
    future trends, 183–5
    sizing requirements for different ethnic populations, 178–82
    sources of further information and advice, 186–7
    understanding different body shapes and case of North American college students, 175–8
  intimate apparel to fit different body shapes, 273–88
    measuring breast volume, 277–82
    smart materials, 284–6
    technologies, 286–8
  overweight and obese consumers shape and sizing that fits this specific market, 153–66
    future trends, 166–7
    sizing requirements, 159–63
    specific styles, 164–6
    understanding different body shapes, 154–9
  plus size Black and Latino women and implications of body shape and size, 256–70
    causality and reality, 261–6
    clothing designers who hire Black and Latina models, 266–8
    future trends, 269–70
    models in fashion world, 257–61
  sizing systems, apparel design and, 17–31
    apparel industry methods of creating ready-made garments, 22–4
    development of sizing systems, 20–2
    emerging mass customisation model for clothing sizing, 27–30
    future trends, 30–2
    relationship of apparel products to the body, 18–20
    standard sizing and body shape variation, 24–7
apparel market
  segmenting male consumers by body shape and other factors, 221–34
    female French population, 224–8
    male French population, 228–33
    techniques to cluster body shapes, 221–3
apparel product data exchange standard (APDES), 184
apparel size
  body shape and its influence on, 3–16
    applying the data, 10–12
    background to sizing, 4–7
    body shape research, 7–10

apparel size (*cont.*)
consumer experience, 13–14
future trends, 14–16
percentage of women by key body shapes in the 2003 Size USA Project, 8
reality of retail, 12–13
Archetype Solutions, 15
arthritis, 136
Asian
Caucasian designing apparel for two different ethnic groups, 169–87
examples for specific styles for different ethnic populations, 183
future trends, 183–5
sizing requirements for different ethnic populations, 178–82
sources of further information and advice, 186–7
understanding different body shapes and case of North American college students, 175–8
designing headwear to fit the size and shape of Western populations, 292–306
future trends, 305–6
head anthropometrics and design, 300–3
head models, 303–5
head physiology, 295–8
sizing system for headgear, 298–9
types of headwear, 294–5
ASTM D6240–98, 180
ASTM D6458–99, 180
ASTM international, 186
athletic type, 228–30

bamboo, 286
baseball hats, 294–5
Black models, 266–8
Black women
plus size, implications of body shape and size for apparel design, 256–70
causality and reality, 261–6
clothing designers who hire Black and Latina models, 266–8
future trends, 269–70
models in fashion world, 257–61
Body Benchmark Study, 65, 76
body fat, 154–5
body image, 175–6
body mapping, 148
body mass index (BMI), 58, 74, 156, 175, 224, 257
body volume index development and, 60–8
body volume index vs, 58–76
BVI value assessment, 71–3
how prepared the world is for measuring obesity technically, 75
implications for the apparel industry, 75
realism of adopting BVI, 73–4
research on the use of BVI, 68–70
significance of cultural and ethical variations, 74
body measurement, 197
chart, 210
drop -12 cm for 180 cm male, 211
standards, 179
body morphotypes, 210
body scan
data, 160
survey, 158
technology, 30
body shape, 206–7
female French population, 224–8
correct location of measurements used in the study, 225
distribution of body shapes, 226
French vs American study, 227
representative individuals for each cluster, 226
sizing a size 40 garment, 227
influence on apparel size and consumer choices, 3–16
applying the data, 10–12
background to sizing, 4–7
body shape research, 7–10
consumer experience, 13–14
future trends, 14–16
reality of retail, 12–13
male French population, 228–33
cluster 1 the athletic type, 228–30
cluster 2 the normal type, 230–1
cluster 3 the portly type, 231

cluster 5 the potbellied type, 232
cluster 6 the seriously potbellied type, 233
cluster 4 the squat type, 231–2
male population clusters, 229
overweight and obese consumers sizing to design apparel that fits specific market, 153–66
future trends, 166–7
sizing requirements, 159–63
specific styles, 164–6
plus size Black and Latino women and implications of size for apparel design, 256–70
causality and reality, 261–6
clothing designers who hire Black and Latina models, 266–8
future trends, 269–70
models in fashion world, 257–61
segmenting male consumers in apparel market and other factors, 221–34
techniques to cluster body shapes, 221–3
understanding different body shapes, 154–9
fat distribution on females, 155
front view of Alvanon forms, 157
side view of Alvanon forms, 158
straight, moderate and curvy, 159
understanding pregnant women's sizing and style preferences, 235–53
case study in pregnant women in Hong Kong, 246–53
identifying target markets, 236–8
key stages in pregnancy, 238–43
maternity wear brief history, 244–5
meeting the needs of target market, body measurement and fit satisfaction, 245–6
pregnancy and apparel choice, 243
Body Shape Analysis, 7
Body Shape Assessment Scale (BSAS), 155
Body Shape Denim, 11
body shapes, 175–8
bra fit, 283–4
angles of shoulder slope, bust prominence, back curvature and acromion placement, 283
designing intimate apparel, 273–88
measuring breast volume, 277–82
smart materials, 284–6
technologies, 286–8
North American college students from Asian and Caucasian ethnic groups, 175–8
BMI and WHR vs different body types, gender and ethnicities, 177
body type characterisation by human body measurements, 176
Caucasian vs Asian students weight, 178
body size
causality and reality among Black and Latino women, 261–6
brief sociology of plus size women in US, 261–2
Latinas and plus size industry, 266
what plus size women in fashion world really saying, 262–5
plus size Black and Latino women and implications of body shape for apparel design, 256–70
clothing designers who hire Black and Latina models, 266–8
future trends, 269–70
models in fashion world, 257–61
body type, 175–6
Body Volume Index (BVI), 59, 65
body mass index and the development of, 60–8
additional research and validation, 65–6
BMI as a measure of body shape, 60–1
collaborators on Shape GB-The National Childrenswear Survey, 67
3D scanners, BVI and online developments, 67
differentiation between two individuals with exactly same BMI but different weight distribution, 63

Body Volume Index (BVI) (*cont.*)
leaflet used by Select to recruit 11 000 participants for the SizeUK project, 64
leaflet used to promote The Body Benchmark Study, 66
origins of BVI, 61–2
research and development of BVI concept, 62–6
significance of 'national' and 'benchmark' as reference criteria in sizing, 68
women with BMI of 30 highlighting different body shape, 61
body mass index vs, 58–76
how prepared the world is for measuring obesity technically, 75
implications of BVI for the apparel industry, 74–5
realism of adopting BVI, 73–4
significance of cultural and ethical variations, 74
research on the use of, 68–70
Body Benchmark Validation Study, 69–70
financial implications of obesity and the costs to the health sector, 70
research on BVI prior to the Body Benchmark Study, 68–9
value assessment, 71–3
benefits of BVI in measuring body shape, 72–3
how BVI works, 71–2
initial results from BVI compared to manual methods, 70–1
Bodymetrics, 15
bonding technology, 287
bone size, 190–1
bone structure, 190–1
bra, 273–4
fit, 276–7
body shape, 283–4
elements, 274
ethnic difference, 284
bra sizing system, 276–7
breast boundary defining method, 280
breast health, 273–4
breast measurement protocols, 277
Asian context, 280–2
breast boundary defining method and defined breast boundary, 280
new bra sizing system for Chinese women, 282
parameters of body model, 281
two key control measurements for bra sizing system, 281
virtual breast shapes, 281
Western context, 277–80
equation for breast volume, 279
linear measurements points, 278
normal values for breast volume and body measurements, 279
Pechter's breast cup sizing method, 279
breast volume calculation formula, 278
breast volume measurement, 277–82
protocols in Asian context, 280–2
protocols in Western context, 277–80
British Standard Institution (BSI), 208
bust-to-hip ratio (BHR), 224

CAESAR (Civilian American and European Surface Anthropometry Resource) survey, 37, 298
Cámara Nacional de la Industriadel Vestido (CANAIVE), 38
Canadian case study, 204–5
Canadian General Standards Board (CGSB), 204–5
caps, 294–5
Caucasian
Asian designing apparel for two different ethnic groups, 169–87
examples for specific styles for different ethnic populations, 183
future trends, 183–5
sizing requirements for different ethnic populations, 178–82
sources of further information and advice, 186–7
understanding different body shapes and case of North

American college students, 175–8
celebrevity, 87
Centre for Disease Control (CDC), 166, 192–3
Charming Shoppes, 8
children
classification of body shapes, 113–18
body types classification for age 7–12, 116
classification rules based on gender and age groups, 115–18
decision tree for female lower body (age7–12), 117
decision tree for female upper body (age 7–12), 113, 114
profile of lower body types: female (age7–12), 116
profile of upper body types: female (age7–12), 114
regression tree validation, lower body: female (age7–12), 116
regression tree validation, upper body: female (age7–12), 113
small size for upper body: female (age7–12), 118
clustering body shapes and sizes, 108–12
age vs cluster groups for lower body: female (age7–12), 111
age vs cluster groups for upper body: female (age7–12), 110
body characteristics of females (age7–12), 112
cluster groups for lower body: female (age7–12), 110
cluster groups for upper body: female (age7–12), 109
clustering for lower body for female sample size (age 7–12), 110–11
clustering for upper body for female sample size (age 7–12), 108–10
scatter plot of cluster groups according to height and bust for females (age7–12), 109
scatter plot of cluster groups according to height and hip for females (age7–12), 111
summary of cluster sampling technique for sample group (age7–12), 112
different body shapes, 97–9
differences in height growth in girls and boys, 98
physical growth and development, 97–9
stages of growth and development from childhood to adulthood, 98
future trends, 123–4
key dimensions identifications, 104–6
female samples (age 7–12), 106–8
principal component analysis extraction for females (age7–12), 104
principal component analysis technique to identify key dimensions, 104
principal component analysis with Varimax rotation for females (age7–12), 107
results of components analysis, 104–6
scree plot of female sample population (age7–12), 105
sizing requirements, 99–103
anthropometric analysis findings, 103–4
applications of particular body shapes and sizes for sizing system development, 103–4
body shapes and size analysis, 99–101
continuous data, 102–3
descriptive statistics of selected body dimensions for sizing system, 102
list of body dimensions according to ISO8559/1989, 100
sample population distribution according to four variable body shapes and samples, 101

children (*cont.*)
stages of sizing development for sample population of children, 100
sizing system development and designation, 118–23
distribution graph of bust girth vs height for female upper body (age7–12), 119
distribution graph of height vs hip girth for female lower body (age7–12), 120
female size designation tables (age7–12), 123
female sizing system for lower body garments (age 7–12), 119–20
female sizing system for upper body garments (age7–12), 119
large size for lower body: female (age7–12), 122
large size for upper body: female (age7–12), 121
medium size for lower body: female (age7–12), 122
medium size for upper body: female (age7–12), 121
size validation according to gender and age, 120, 123
small size for lower body: female (age7–12), 121
understanding sizing, body shapes and apparel requirements for, 95–124
circular knitting technology, 287
classification and regression tree (CRT) predictive modelling technique, 113
cluster body shapes
techniques, 221–3
methodology to identify body shape clusters, 223
co-fitting, 216
consumer choices
body shape and its influence on, 3–16
applying the data, 10–12
background to sizing, 4–7
body shape research, 7–10
consumer experience, 13–14
future trends, 14–16
reality of retail, 12–13
psychological and sociological factors, 78–90
need to attract attention, 86–7
need to be authentic, 81–2
need to be ethical, 88
need to constantly change appearances, 82–3
need to expose the body, 83–4
need to look attractive, 84–6
need to look marginal and rebellious, 80–1
need to look sporty, 87–8
need to look young, 79
sizing systems, apparel design and, 17–31
apparel industry methods of creating ready-made garments, 22–4
development of sizing systems, 20–2
emerging mass customisation model for clothing sizing, 27–30
future trends, 30–2
relationship of apparel products to the body, 18–20
standard sizing and body shape variation, 24–7
contour
weight distribution, 197–8
back silhouette, 197
leg stance, 198
seat stance, 198
stomach silhouette, 197
thigh stance, 198
cortisol, 262
crotch height, 193
Curve ID, 11, 30
Curves and Chaos, 266
Cyberware technology, 38, 41

3D body scanning technology, 37, 185, 287–8
3D computer aided design (CAD) software, 294
3D datasets, 304–5
3D photonic scanning, 184

David's Bridal, 8
density variation, 191–2
Department of Commerce Office of Textiles and Apparel, 159–60
Design for Ageing Well, 134, 144
development, 97
digital anthropometric data, 184
digital head models, 306
disposable fashion, 4, 133

e-sizing, 217
e-tailoring, 216
  sizing requirements, 215–16
ectomorph, 196
EN 13402, 180–1
endomorph, 196
ethnic fit, 305
ethnic groups
  Asian and Caucasian designing apparel, 169–87
    body type classifications by various researchers, 173
    examples for specific styles for different ethnic populations, 183
    future trends, 183–5
    selected examples of body type variations, 172
    sizing requirements for different ethnic populations, 178–82
    sources of further information and advice, 186–7
    understanding different body shapes and case of North American college students, 175–8
ethnic shape variations, 298
  result of CAESAR and SizeChina 3D survey shows significant headshape difference, 299
European Union, 180–1
extimacy, 86, 89

fashion market, 179
fashion newspapers, 187
Fashion Police, 13
fat mass index, 192
fat tissue, 163
Fatshion Chic, 266–7
Fifty and Still Sexy, 85
figure variation, 207
first trimester pregnancy, 239–40
fit, 4, 190
fit evaluation
  designing intimate apparel to different body shapes, 273–88
    measuring breast volume, 277–82
    smart materials, 284–6
    technologies, 286–8
  headwear design of size and shape of Western and Asian populations, 292–306
    future trends, 305–6
    head anthropometrics and design, 300–3
    head models, 303–5
    head physiology, 295–8
    sizing system for headgear, 298–9
    types of headwear, 294–5
Fit Logic, 13, 14
Fitthinking, 217
'flat' pattern cutting, 139
footwear size conversion, 214
'form' pattern cutting, 139
Frankfurt plane, 296–7
free form deformation technology, 280
French Sizing Campaign, 222

GABIFRESH, 263
Gap, 12
garment fit, 251
  percentage of satisfaction of our respondents with clothing size and fit, 251
garment layering system, 138
Get My Size, 14
gloves, 214
Gore-Tex, 139
grade rules, 6, 23
green behaviour, 88
growth, 97

H-type figure, 206
hair, 300
hat makers blocks, 303

hats, 214
head anthropometrics, 300–3
  charts and tables, 301–2
    table of head circumference, 301
  sizing information for headwear design, 300–1
    facial landmarks, 300
  sizing surveys, 302
  taking head measurement, 302–3
    head circumference measurement, 302
head models, 303–5
  3D datasets, 304–5
    high resolution 3D scans of Chinese adults, 305
  hat makers blocks, 303
  headforms, 303
  reference headforms, 303–4
    facially featured headforms, 304
  related products, 305
  styrofoam wig stands, 303
head physiology, 295–8
  Frankfurt plane, 296–7
    schematic diagram, 297
  size range distribution, 297–8
    adult Chinese Head size distribution by gender, 297
  skull growth, 296
head shape
  headwear design to fit the size of Western and Asian populations, 292–306
    future trends, 305–6
    head anthropometrics and design, 300–3
    head models, 303–5
    head physiology, 295–8
    sizing system for headgear, 298–9
    types of headwear, 294–5
head size
  headwear design to fit the shape of Western and Asian populations, 292–306
    future trends, 305–6
    head anthropometrics and design, 300–3
    head models, 303–5
    head physiology, 295–8
    sizing system for headgear, 298–9
    types of headwear, 294–5
headforms, 303
headgear
  sizing system, 298–9
    comfort and fit in headgear, 298–9
    ethnic shape variations, 298
headwear
  fit evaluation of size and shape of Western and Asian populations, 292–306
    future trends, 305–6
    head anthropometrics and design, 300–3
    head models, 303–5
    head physiology, 295–8
    protective helmet for snowboarding, 293
    sizing system for headgear, 298–9
  types, 294–5
    baseball hats and caps, 294–5
    hats, 294
    protective helmets, 295
height, 191
hierarchical clustering methods, 222
high resolution 3D scan technology, 298
Hohenstein Institutes, 38
human body mass, 191
Human Solutions, 15, 38, 41, 54
hyperconsumption, 82
hypersexual fashion, 81

infants *see* children
Ingeo, 286
Intellifit Corporation, 8
International Organisation for Standardisation (ISO), 186, 208
intimate apparel
  apparel design to fit different body shapes, 273–88, 282–4
    age difference in breast configuration, 282–3
    body shape and bra fit, 283–4
    ethnic difference and bra fit, 284
    measuring breast volume, 277–82
  bra-fit elements, 274

bra sizing system and fit, 276–7
breast health and bra, 273–4
professional bra-fitting criteria, 274–6
examples of ill-fitting bras, 275
smart materials, 284–6
shape memory foam, 285
shape memory wire, 285–6
technologies, 286–8
3D body scanning technology for design and fit evaluation, 287–8
seamless technology for fit and comfort, 286–7
invisible force, 89
ISO 2068:2005, 185
ISO 3635, 179–80, 180–1
ISO 8559/1989, 99
ISO TC133, 99, 119
technical committee, 179–80

JC Penney, 12, 22
juvenile delinquent, 80

Kinect technology, 30
Kohl's, 12
Kurt Salmon Associates, 13

Lands' End, 12
laser based scanners, 41
Latino women
plus size, implications of body shape and size for apparel design, 256–70
causality and reality, 261–6
clothing designers who hire Black and Latina models, 266–8
future trends, 269–70
models in fashion world, 257–61, 266–8
lean muscle tissue, 163
Levi Strauss, 9
Levi's, 12, 30
Lyocell, 286

Made4Me.com, 14
magnetic resonance imaging (MRI), 277
male apparel
examples of sizing systems, 211–14
key issues for comparison, 190–200
appropriate size chart, 199
bone size and structure, 190–1
contour and weight distribution, 197–8
fat mass index, 192
fit, 190
height, 191
overweight, 192–3
proportional body areas, 193–5
proportional body areas and personality types, 195–7
size identification, 198–9
size inflation, 199–200
standard sizing problems, 199
weight and density variation, 191–2
sizing requirements, 189–218
body measurement chart, 210
different body shapes, 200–7
future trends, 214–18
size adaptation and sticky information, 210–11
sizing system, 207–10
male body shapes, 200–7
Canadian case study, 204–5
changes among young males, 202–4
physical appearance, 202
problems with manufacturing standard sizes, 203–4
virtuality vs reality, 203
different approaches, 201–2
four male figures, 206
research plan, 205–7
body shape, 206–7
body shape and ageing, 207
posture and figure variation, 207
theory of body shapes, 200–1
male consumers
apparel market by body shape and other factors, 221–34
cluster of males in French population, 234
female French population, 224–8
male French population, 228–33
techniques to cluster body shapes, 221–3
mammary hemi-circumference, 278–9
mammography, 277
Marks & Spencer, 11, 61

mass customisation, 27, 214–15
maternity wear, 244–5
  criteria, 251–2
    importance of various criteria for purchase, 252
  market, 243
maturation, 97
Me-Ality, 15, 186
Memory Foam bra, 285
mesomorph, 195
Metail, 15
Microsoft Kinect, 42
mitts, 214
'modified hourglass' shape, 10
moulded polyurethane foam, 286–7
moulding technology, 287
muscle tissue, 191
My Best Fit, 14
My Shape, 14
My True Fit, 14
My Virtual Model, 14

National Centre for Health Statistics, 186
National Child Measurement Programme (NCMP), 66
National Health and Nutrition Examination Survey (NHANES), 41
National Institute of Standards and Technology (NIST), 184, 186
national sizing survey
  applications for apparel, 49–55
    cross-section extracted from 3D body data, 54
    3D body scans illustrating difference in body shape, 55
    3D visualisation, 53–5
    development and refinement of garment sizing strategies, 50–2
    new product development, 52–3
    population distribution by seat for SizeUSA women, 51
    population distribution by waist for SizeUSA women, 51
    sizing standards and guidelines, 49–50
  data analysis, 45–9
    analysis of body dimension and shape characteristics, 47–9
    body measurement extraction, 45–7
    3d body scan data, 48
    measurement extraction parameters, 46
  data collection, 43–5
    demographic information collection, 44–5
    participant experience, 44
    registration and engagement of participants, 43–4
  other applications, 55–6
  survey design, 38–43
    data privacy, 42–3
    goals, 38–40
    population sampling strategy, 40–1
    technology, 41–2
  techniques, data analysis and apparel product development, 35–57
    initiatives, 36–8
    value of sizing surveys, 36
Natural Accents Memory Foam bra, 285
Net Space database, 185
Next, 61
Nike, 12
normal type, 230–1
NPD Group, 9
nylon, 139

O-type figure, 207
obese consumers
  overweight consumers shape and sizing to design apparel that fits specific market, 153–67
    future trends, 166–7
    sizing requirements, 159–63
    specific styles, 164–6
    understanding different body shapes, 154–9
obesity, 257
older consumers
  aesthetic needs, 144–7
    design detail, 147
    fabrication/handle/drape, 146–7
    fit, 145
    greater colour choice, 145

less transient styling/fewer styles, 146
positive image of ageing, 144–5
style lines, 145–6
future trends, 149–50
needs of the body, 147–9
cutting for movement, 148
moisture management/thermal regulation, 148
multi-functionality, 148–9
psychological 'feel good factor,' 147–8
senses, 149
size and shape, 148
sizing requirements, 138–44
body scanning and pattern block development, 140–1
co-design, 141
co-design prototype development, 142–3
context, 138–9
cutting for movement, 141–2
need for effective garment design specification, 143–4
textile innovation driving more tailored sportswear, 139–40
understanding sizing, body shapes and requirements of apparel, 132–50
understanding their different body shapes, 134–8
'feel good factor,' 137–8
movement/agility, 136
physical changes related to ageing, 135–6
sensory considerations, 136
vital signs, 137
one-dimensional anthropometric data *see* univariate anthropometric data
one size fits all, 18
online shopping, 305–6
osteopenia, 135
osteoporosis, 135
'out-of-bed' look, 89
overweight, 192–3
obese consumers shape and sizing to design apparel that fits specific market, 153–66
future trends, 166–7
specific styles, 164–6
understanding different body shapes, 154–9
sizing requirements, 159–63
measurements for size 18W in inches, 161

Pechter's new bra sizing method, 279–80
personality types, 195–7
Persuasion Memory Foam demi bra, 285
philosophy of black, 88
physical appearance, 202
Plus Model Magazine, 263
plus size
blogosphere, 263
clothing designers, 266–8
overweight and obese consumers
body shape to design apparel that fits specific market, 153–66
future trends, 166–7
sizing requirements, 159–63
specific styles, 164–6
understanding different body shapes, 154–9
plus size models
fashion world, 257–61
effects of fame, 260–1
fashion industry as an art world, 258–9
plus size women vs plus size models, 259–60
images of models, 259
polyester, 139
portly type, 231
posture, 207
potbellied type, 232
pregnancy, 238
activities, 243
apparel choice, 243
duration, 239
first trimester, 239–40
body profile illustration, 240
key stages, 238–43
second trimester, 240
body profile illustration, 241
third trimester, 240–3
abdominal extension during pregnancy, 242
body profile illustration, 241

pregnant women
case study in Hong Kong, 246–53
active wear and sports wear, 250–1
age range distribution in percentage for our sample, 248
criteria for maternity wear, 251–2
garment fit and satisfaction, 251
matrimonial status distribution of our sample, 247
nationality distribution of our sample, 247
revenue range distribution in percentage for our sample, 248
need for new garments, 249–50
percentage of respondents with specific clothing need by trimester, 250
physical changes, 248–9
perceived physical changes for different body parts per trimester, 249
serving as a model to be measured, 252–3
percentage of respondents who need to discuss with husband, doctor, friends or family, 253
understanding shape, sizing and style preferences, 235–53
identifying target markets, 236–8
key stages in pregnancy, 238–43
maternity wear brief history, 244–5
meeting the needs of target market, body measurement and fit satisfaction, 245–6
pregnancy and apparel choice, 243
principal component analysis (PCA), 104, 222
professional bra-fitting criteria, 274–6
Project Runway, 13
proportional body areas, 193–5, 195–7
chest and arms silhouette, 195
crotch length, 194
neck and shoulders, 194
personality types, 195–7
three basic body shapes from Sheldon's Somatotyping Theory, 196
torso proportions, 193
protective helmets, 295
psycho-sociological factors
influencing consumers' choice, 78–90
need to attract attention, 86–7
need to be authentic, 81–2
need to be ethical, 88
need to constantly change appearances, 82–3
need to expose the body, 83–4
need to look attractive, 84–6
need to look marginal and rebellious, 80–1
need to look sporty, 87–8
need to look young, 79

reference headforms, 303–4
retailer's size chart, 160–1
'Right Fit' jeans, 11

Scanatars, 54
Seacell, 286
seamless technology, 286–7
Sears catalogue, 22
second trimester pregnancy, 240
Select Research, 37, 49, 59
seriously potbellied type, 233
Shape GB, 37, 49, 59, 68
shape memory alloys, 285
shape memory foam, 285
shape memory wire, 285–6
Sheldon's somatotyping theory, 195–7
size chart, 199
size identification, 198–9
size inflation, 199–200
size range distribution, 297–8
SizeChina project, 298
SizeGERMANY, 38
SizeKorea, 186
SizeMexico, 44
SizeThailand, 186
SizeUK, 37, 68, 186
SizeUSA, 37, 41, 45, 55, 187
sizing
understanding pregnant women's body shape and style preferences, 235–53
case study in pregnant women in Hong Kong, 246–53

identifying target markets, 236–8
key stages in pregnancy, 238–43
maternity wear brief history, 244–5
meeting the needs of target market, body measurement and fit satisfaction, 245–6
pregnancy and apparel choice, 243
sizing requirements
different ethnic populations, 178–82
key body measurements areas for clothing sizing systems, 181
selected examples for international clothing size conversions, 182
selected examples of international sizing standards, 180
technology integration with international sizing systems in global environment, 185
male apparel, 189–218, 207–11
different body shapes, 200–7
examples of sizing systems, 211–14
future trends, 214–18
key issues for comparison, 190–200
sizing standards, 179
sizing systems, 207–10, 211–14
apparel design, customer choice and, 17–31
detailed arc measurements, depths, widths needed for effective body shape categorisation, 32
future trends, 30–2
apparel industry methods of creating ready-made garments, 22–4
body size identification of men's figure type, 208
development, 20–2
advertisement in November 1915 McCall's Magazine, 21
emerging mass customisation model for clothing sizing, 27–30
communication of sizing and fit with mass customised clothing, 29–30
conceptual model adjusted to address issues of mass-customised system, 28
mass customisation and sizing, 27–8
mass customised production, 29
mass customised sizing and fit, 29
global categorisation of four male figure types, 209
hats, gloves and mitts, 214
sizes, 214
men's belts, 212–13
belt sizes, 213
men's shirt, 212
sports shirts, T-shirts and dress shirts, 212
men's shoes, 214
men's suits, 212
specialty sizes, suits and pants, 212
relationship of apparel products to the body, 18–20
clothing is provided using many different sizing strategies, 19
standard sizing and body shape variation, 24–7
conceptual framework for research on sizing and fit, 26
SKORCH Magazine, 263
skull growth, 296
slops, 20
smart materials, 286
Smart Memory bra, 285
somatotypes theory, 236–7
sports wear, 250–1
squat type, 231–2
standard sizing problems, 199
Stealth hat, 294
stereo-photogrammetry, 277
stick-thin models, 258
straight-size models, 258
street fashion, 87
Styku, 15
style preferences
understanding pregnant women's body shape and sizing, 235–53
case study in pregnant women in Hong Kong, 246–53
identifying target markets, 236–8
key stages in pregnancy, 238–43
maternity wear brief history, 244–5
meeting the needs of target market, body measurement and fit satisfaction, 245–6
pregnancy and apparel choice, 243

styrofoam wig stands, 303
synthetic miracle fibres, 139

Tactel, 139
Target, 12
target market, 236–8
  body measurement and fit satisfaction, 245–6
    business wire. Alvanon releases most extensive Chinese body measurement study, 245
  different body figures, 237
  endomorphy, mesomorphy and ectomorphy body shapes, 236
[TC]^2, 15, 53
Tencell, 286
Textile/Clothing Technology Corporation [TC]^2, 37
The Cutter Up, 20
theory of anthropometric structures, 200–1
theory of body shapes, 200–1
thermoplastic cast, 277
third trimester pregnancy, 240–3
three dimensional (3D) laser scanning, 277
travelling tailors, 215–16

ultraviolet (UV) rays, 294
United Nude's low resolution scanning, 294
univariate anthropometric data, 301
universal aesthetic triangle, 278
Upcloud, 15
US Department of Health and Human Services' Office of Minority Health, 262

V-type figure, 206
vanity sizing, 7, 24, 134, 148
virtual fit, 31
virtual reality (VR), 203

waist-hip ratio (WHR), 176, 224
Wal-Mart, 12
websites, 186–7
weight, 191–2
  analysis, 176
  density variation, 191–2
    frame size based on height and weight, 192
  distribution, 197–8
Western populations
  designing headwear to fit the size and shape of Asian populations, 292–306
    future trends, 305–6
    head anthropometrics and design, 300–3
    head models, 303–5
    head physiology, 295–8
    sizing system for headgear, 298–9
    types of headwear, 294–5
white light scanning technologies, 41

X-type figure, 207
XOJane, 264–5

Y-type figure, 206